Professional Services Agreements

Second edition

ice
Institution of Civil Engineers
publishing

Professional Services Agreements

Second edition

A guide for construction professionals

Rachel Barnes
Beale and Company Solicitors LLP, UK

Published by ICE Publishing, 40 Marsh Wall, London E14 9TP.

Full details of ICE Publishing sales representatives and distributors can be found at: www.icevirtuallibrary.com/info/printbooksales

First published 2000 (Leslie Edwards and Rachel Barnes)
Second edition 2012

Also available from ICE Publishing

NEC3: A User's Guide. J. Broome. ISBN 978-0-7277-4109-7
Dispute Boards: Procedures and Practice.
G. Owen and B. Totterdill. ISBN 978-0-7277-3508-9
ICE Dispute Resolution Procedures 2011.
Institution of Civil Engineers. ISBN 978-0-7277-5763-0

www.icevirtuallibrary.com

A catalogue record for this book is available from the British Library

ISBN 978-0-7277-4158-5

ICE Publishing is a division of Thomas Telford Ltd, a wholly-owned subsidiary of the Institution of Civil Engineers (ICE).

Associate Commissioning Editor: Jennifer Barratt
Production Editor: Imran Mirza
Market Development Executive: Catherine de Gatacre

Typeset by Academic + Technical, Bristol
Printed and bound by CPI Group (UK) Ltd, Croydon, CR0 4YY

Acknowledgement

The author would like to acknowledge the contributions of Leslie Edwards to the first edition. It was his idea to produce a guide for construction professionals confronted with non-standard professional services agreements and he produced the initial draft of the first edition, providing valuable practical examples and input from the consultant's perspective. She would also like to thank Kris Hannah again for uncomplainingly typing all the revisions (and revisions to revisions) and the new material for this second edition, and for her help in presentation and achieving consistency. Also, the author is grateful to the partners of Beale and Company for their encouragement and support and in particular to Christopher Cullen, a consultant to the firm, who read the whole manuscript and provided valuable comments.

Contents

Foreword to the first edition

The RIBA first published a professional services agreement for its members in 1872. The conditions ran to some 16 pages – and the pages were little bigger than those of a pocket diary. More remarkable still, it was only 100 years later that a substantial revision was printed. In the last 30 years, a series of standard forms of agreement has been published by both the RIBA and ACE – each edition having a shorter shelf life, and tending to be longer than the last.

Moreover, nowadays, the construction professional who wins a new job, or is hoping to win one, is often presented with a bespoke form of contract. The contract documentation may (sometimes) be short and to the point, or it may be delivered in lever arch files. Beware any professional who does not read through this documentation from beginning to end. Unusual, indeed unconscionable, liabilities may be lurking within the small print.

This book should be on the desk of all those reviewing professional services agreements before they are signed, and those who refer to them once they are signed. There has been a need for a book of this kind for some time, and Leslie Edwards and Rachel Barnes have taken on a daunting task. As is evident from the wealth of helpful detail and examples, they have enormous experience of their subject.

Bidding for a job is not only about assessing the fee to undertake the services. It is also important to assess the risk of everything not going to plan. Only in that way can the construction professional in the twenty-first century remain competitive. This book will be an invaluable tool in making that assessment.

Frances A. Paterson
Former Director of Legal Affairs at the Association of Consulting Engineers
May 2000

Foreword to the second edition

Entering into a contract for professional services, particularly in the construction and engineering sector, can seem simple but for the unwary or inexperienced professional it can be a minefield. If there is no cause for complaint and the professional is fully paid up, in one sense how the contract was formed and what its terms were does not matter. However, it is vital and it is professional for professionals to get their contracts right so that if something does go wrong, and if it can often do so, there can be protection and clarity. In the Technology and Construction Court, we see often the results of unfortunately drafted agreements between professionals and their clients.

This is the second edition of Rachel Barnes' book on professional services contracts and one can understand why she and her publishers, after the success of the first edition, see a need for a new edition. There has been a significant series of developments since the first edition. As the economy takes time to recover, construction clients will be seeking to secure tight agreements with their professionals, be they architects, engineers or surveyors. Counter-intuitively, economic hard times lead to more litigation and when the better times arrive that leads usually to a significant increase in development projects in which all construction professional disciplines share.

There is a great need for a book like this particularly in the construction and engineering field, which contains a step by step explanation of how contracts are formed, what the scope of the obligations and rights are, how liabilities can arise and the ways in which disputes can be resolved and risks can be managed. Additionally, there are good and above all practical explanations of determining who the client is (which can be complicated in large developments with the use of agents and shell or single purpose vehicles), intellectual property rights, insurances, guarantees, collateral warranties and potential liabilities to third parties and the CDM Regulations.

For both the inexperienced and the experienced construction professionals, this book will in its second edition continue to provide an essential service to protect them against the many pitfalls and to point them in the right direction when things go wrong. I have absolutely no hesitation in recommending this book as one which all such professionals should have in their offices.

The Honourable Mr Justice Akenhead
Judge in Charge – Technology and Construction Court
Royal Courts of Justice

Preface to the first edition

The idea for this book arose when it was realised that there was no readily available material for professionals on how to identify and manage risks in professional services agreements. It was also realised that such information that was available often did not permeate to the levels in a professional organisation that should be concerned with such issues.

This book therefore attempts to describe comprehensively, and relatively simply, the more usual risks inherent in professional services agreements and how such risks might be managed. Also included are collateral warranties, assignments and novations, each of which could incorporate risks similar to those arising from professional services agreements. Of necessity the book also describes the cover provided by typical professional indemnity insurances.

It is hoped that the book will provide a practical, easily understood source of reference for those at any level in a professional organisation who need to understand and deal with such matters.

The book's basic aims are therefore to

- make all appropriate levels of a professional organisation aware of the potential risks inherent in professional services agreements, possible methods of managing those risks and the scope and limitations of professional indemnity insurance
- provide sufficient background information to enable a professional to realise when it needs advice from insurers, brokers or lawyers, and to understand that advice.

As a result, the book will assist the professional to make the correct commercial decisions on whether or not to accept the risks inherent in those agreements.

It is also hoped that this book will demonstrate the need to establish, review regularly and, if necessary, amend management procedures to achieve the foregoing aims.

Leslie Edwards and Rachel Barnes
September 2000

Preface to the second edition

In this second edition, I have brought the law up to date and included some new cases, such as the Co-operative Retail Services case and the Technotrade case, where these help illustrate the matters being discussed. I have also commented on the amendments to the payment provisions incorporated in the Local Democracy, Economic Development and Construction Act 2009. Most of the material in the book has been expanded and some new provisions in non-standard appointments are also considered. There is more on intellectual property and a paper on Building Information Modelling is attached as an appendix.

There is now a separate chapter on novation to a design and build contractor because of the particular problems this raises and because it is being used much more frequently than in 2000. Confidentiality also has its own chapter, as does the client's obligations.

The chapter on the CDM Regulations has been extensively re-written with the objective of giving designers an overview of the 2007 Regulations as well as a detailed description of their own duties. The chapter on collateral warranties now refers to the CIC standard forms 2003, and those are also reproduced in the appendices.

Finally, there is a new chapter on the NEC's Professional Services Contract – PSC3 – providing a full commentary on that form of contract and to go alongside the comments in the main text on the standard clauses in the ACE Agreement 1: 2009, the RIBA Standard Agreement 2010 and the RICS Standard Form 2011.

Rachel Barnes
2012

About the author

Rachel Barnes is a consultant in Beale and Company Solicitors LLP and joined the firm in 1973.

For the first ten years or so as a practising solicitor Rachel was mainly concerned with construction litigation. This involved defending consulting engineers in high court cases and in arbitrations in the UK and overseas. From 1985, Rachel has worked on the non-contentious side of the firm's practice, giving advice to construction professionals and others on professional appointments, collateral warranties and third party rights for all types of projects. She also advises on building contracts, novation agreements, bonds and tenders, architectural competitions and joint ventures.

Rachel also lectures, and topics have included the CDM Regulations and the Construction Act payment provisions. She assisted the Association for Consultancy and Engineering in drafting the ACE conditions of engagement 1995 and later editions and the agreements published in 2009. She is a member of the RIBA specialist practice panel and of the Construction Industry Council liability panel.

She is a regular contributor to *Building* magazine, and with Clive Marsden of Alan Baxter & Associates LLP contributed the chapter on Engineers' Appointments in the *ICE Manual of Construction Law*.

Abbreviations

ACA	Association of Consultant Architects
ACE	Association for Consultancy and Engineering
ACoP	Approved Code of Practice
ADR	Alternative dispute resolution
ARB	Architects' Registration Board
AIA	American Institute of Architects
BIM	Building information modelling
BPF	British Property Federation
BSI	British Standards Institution
CIAT	Chartered Institute of Architectural Technologists
CIBSE	Chartered Institute of Building Services Engineers
CIC	Construction Industry Council
CIMAR	Construction Industry Model Adjudication Rules
CLLS	City of London Law Society
CML	The Council of Mortgage Lenders
EEA	European Economic Area
ECC	Engineering Construction Contract
FIDIC	International Federation of Consulting Engineers
HSE/HSC	Health and Safety Executive/Commission
ICE	Institution of Civil Engineers
JCT	Joint Contracts Tribunal Ltd
MF	Model Forms
NEC	New Engineering Contract
NHBC	National House-Building Council
PSC3	Professional Services Contract
RIBA	Royal Institute of British Architects
RICS	Royal Institution of Chartered Surveyors
TeCSA	Technology and Construction Solicitors Association

Professional Services Agreements
ISBN 978-0-7277-4158-5

http://dx.doi.org/10.1680/psa.41585.001

Chapter 1
Introduction

1.1. General

Effective internal risk management is crucial to the continuing existence of all organisations offering professional services in the construction industry. These include consulting engineers, architects, quantity surveyors, surveyors and project managers.

As with all commercial concerns, it is to be hoped that professionals already have procedures (and review them periodically) to ensure the following are achieved in an effective manner.

- The receipt of invitations to bid for work.
- The satisfactory negotiation of fees that achieve a reasonable proportion of successful assignments and that also provide an opportunity for making an acceptable profit.
- The undertaking of the assignments won and achieving those profits.
- Obtaining payment for work done and within a reasonable time.

Professional organisations currently in business are clearly getting much of the foregoing elements right. For those that feel they need advice, there is a reasonable number of reference books and similar documents available. Such matters are therefore not considered in detail in the text.

Those professionals who provide construction litigation and litigation support services, however, will realise that a significant part of their income derives from the fact that other professionals accept overly onerous terms in their agreement with clients to provide professional services.

The text of this book is therefore focused on the risks inherent in professional services agreements and associated issues and the liabilities these could give rise to which would adversely affect a professional's opportunity for maximising profit.

1.2. Liability

Liability for professional organisations can arise in a number of ways and these are explained in the text. Liability to the client will generally arise as a result of a breach of a term of the contract with the client. A professional may also owe a concurrent duty in tort to the client to use reasonable skill and care in carrying out services – see Section 4.1. Other liabilities to clients and third parties can also arise in tort. Statutes can impose duties on individuals and organisations with civil or criminal liability if there is a breach. The occasions on which this can happen are explained in the text, which also includes a general description of the sort of damages that could be payable.

Sometimes liabilities are recognised and accepted as commercial risks. Sometimes liabilities are not recognised. It is essential that the exposure to possible liability is understood and that proper management procedures are in place for handling it.

Professionals will often be obliged by a client to maintain professional indemnity insurance so that the client is reassured that claims should be covered by insurance. This is sound commercial practice anyway. It is essential, however, for both professionals and clients to realise that the insurance policy that is in place when any claim comes in will be the relevant policy, and that it is possible that at that time there could be different restrictions on cover, or possibly no cover at all, for certain claims. This topic, too, is dealt with in the text.

Even if the professional has professional indemnity insurance, claims could be made which are outside professional indemnity insurance cover, or within the cover but below an excess or above the ceiling of the indemnity provided. This will result in direct uninsured costs. Even if a claim is currently inside insurance cover, increases in premiums and/or the excess and/or other conditions or exclusions could be imposed on future insurance cover if high-value or frequent claims have been made.

1.3. Standard conditions of engagement

Standard conditions of engagement for professionals are written in terms discussed with, and which are generally acceptable to, the professionals and professional indemnity insurers for the professional sector concerned. The division of risks between clients and professionals within the standard conditions has therefore been considered, and the risks carried by the professional will generally be covered by professional indemnity insurance.

In the text, extracts from the following standard conditions are included for comparison and ease of reference.

- ACE (Association for Consultancy and Engineering) Agreement 1: Design 2009 Edition Second Revision ('ACE Agreement 1').
- RIBA (Royal Institute of British Architects) Standard Conditions of Appointment for an Architect 2010 as amended by Amendment 1 – 1 October 2011 ('RIBA 2010').
- RICS (Royal Institution of Chartered Surveyors) Standard Form of Consultant's Appointment November 2011 ('RICS 2011').
- The CIC (Construction Industry Council) Consultants' Contract Conditions second edition October 2011 ('CIC').

Another standard form which professionals are asked to sign is that produced by the Institution of Civil Engineers as part of their New Engineering Contracts ('the NEC'). The form for the professional is the Professional Services Contract and as it is in its third edition (June 2005 with amendments June 2006 and September 2011) it is known as NEC3 Professional Services Contract or PSC3, and it is as 'PSC3' that it will be referred to in the text. It is a contract that is in a very different form from those of the professional institutions and so, although some of its provisions are discussed when the other standard forms are discussed, it is considered in some detail in Chapter 9.

While standard conditions of engagement are not always entirely suitable for a particular appointment, and some might be stronger than others in some areas, particular problems arise in relation to non-standard conditions of engagement and collateral warranties. Provisions may not be comprehensive, and can be omitted altogether. The wording may be onerous and can impose severe or unacceptable risks on professionals. If the risk is uninsured, any successful claim would have to be met by the professional from other sources, for example, cash flow, loans, sale of capital assets, etc. The same applies to excesses for each insured claim. This could have a catastrophic impact on profitability or even the continued existence of an organisation.

The following chapters describe the matters that should be included in most professional services agreements, the consequences of their non-inclusion, or of the incorporation of more onerous wording. Ways are also suggested as to how those risks might be managed.

1.4. The commercial nature of the management of agreements

A decision on how to tackle incomplete or onerous wordings in professional services agreements should always be a commercial one. It can depend on many factors. For example, a professional might consider that a relatively low-priced job in an urban area could cause a large potential liability if it goes wrong, whereas a high-priced but technically simple job in a rural area might be straightforward with minimum risk. Some clients may be one-off; others may have a continuing workload for which the professional would like to be considered. Past experience may indicate that some clients are amenable to discussion or have agreed changes to their agreements in the past; other clients may be inflexible. The professional might consider that the work is an absolute necessity to maintain minimum staff workload. The professional might have several other risky contracts in hand and not want to increase exposure to risk. The professional may have a particular expertise not readily available elsewhere and hence be in a stronger negotiating position.

It is therefore most important to recognise that a decision to accept any particular risk is for the professional, as only the professional will have all the technical and commercial information necessary and have had a lawyer's and/or insurer's advice (where appropriate) to make such a decision.

Professional Services Agreements
ISBN 978-0-7277-4158-5

http://dx.doi.org/10.1680/psa.41585.005

Chapter 2
The client

2.1. Identification of the client

It is essential to identify the client so that the professional knows to whom contractual obligations are owed. The extent of the obligations undertaken and the level of risk can also be affected because of the nature of a particular client. For example, the professional may not wish to be engaged directly by a known 'difficult' client, or a client of doubtful or unknown liquidity or of dubious 'pedigree'. On other occasions, the professional may be prepared to be engaged by a particular client as long as some other, reputable, organisation is responsible for fees. Elementary problems concerning identity of the client can arise. Some of these problems will be considered in this chapter.

2.2. Agent acting for the client

A professional can believe it is appointed by one person as a client, whereas in fact that person is acting as an agent for another person or organisation but does not clearly state that fact, perhaps because the agent genuinely considers that it is 'trade and custom' knowledge in the sector concerned. Although the professional may be entitled to enforce obligations such as payment against an agent (for example, if the identity of the principal is not known), difficulties can arise if, for example, the client refuses to pay and the agent is impecunious, or if the client argues that things done by its agent are outside its authority and therefore do not bind the client. It is prudent, therefore – if it is thought that a person is acting as an agent for a client (perhaps because invoices are to be sent to a third party) – to clarify the position and the extent of the agency as soon as possible. Where an agent is formally appointed on behalf of a number of parties, the appointment itself will usually set this out and confirm the extent of the agent's authority, for example, to act in all matters on behalf of its principals in relation to that project. The position can be less clear where there is no formal appointment document and, in these circumstances, the ultimate client, not the agent, should be asked to clarify the position. Always address invoices to the ultimate client unless specifically authorised by the client to address them to the agent. Where a professional is employed for expert work in relation to arbitration or litigation, the instructing solicitor will usually make clear the identity of the client, that he is acting on behalf of the client, and that it is the client who is responsible for the professional's fees. However, if the solicitor does not do so, as a result of Law Society rules rather than the principles of agency, the solicitor must reimburse the professional's fees, whether or not money is received by that solicitor from the ultimate client. More rarely, a solicitor may act for a client in engaging professionals and letting the main construction contract. In those cases, the usual agency principles apply.

A professional cannot enforce obligations such as payment against an agent if the agency and the identity of the client are known. He has to go against the client unless the contract states the agent will be liable or the agent acts outside his authority. The professional can sue the agent if he did not know he was an agent and thought he was the principal.

The situation in which a professional might be an agent for a client, and a further discussion of the principles of agency, are set out in Chapter 13.

2.3. Client acting on behalf of, or for the benefit of, third parties/Panatown/Contracts (Rights of Third Parties) Act 1999

A client can often commission a project on behalf of other parties, such as another company in the same group, or for a particular purchaser or tenant, and can also enter into complicated trust arrangements with funders or others. None of these will be formal agency arrangements of the type described in Section 2.2.

Professionals and contractors have been alerted, by cases such as *St Martins Corporation Limited* v. *Sir Robert McAlpine & Sons Limited (1994)* and *Alfred McAlpine Co.* v. *Panatown (2000)*, that substantial damages could be recovered from them either by clients (who have suffered no loss) for and on behalf of certain owners/occupiers or directly by an owner/occupier who was not the client and who had no contractual relationship with the professional or contractor. Such damages will normally only be recoverable where there is a close relationship between the client and owner/occupier and the professional or contractor knows that the development was being built for the benefit of that owner/occupier.

The reason why, in such situations, the client has suffered no loss (and therefore has no grounds on which to sue the professional) is usually that he no longer has an interest in the property that is the subject of the claim (for example, because he has sold it) or (as in the St Martins case) has passed his interest to another party prior to any breach of contract occurring without assigning (effectively) his rights under his contract with the professional.

The question of whether or not the 'no loss' argument can apply where the rights to sue the other party to a contract have been assigned is considered in Section 19.2.

The right of the client to recover on behalf of an owner/occupier, which is an exception to the rule of privity, will not exist where the professional or contractor (as the case may be) has given a direct collateral warranty to the owner/occupier unless the collateral warranty provides otherwise. It does not matter that the damages recoverable under that collateral warranty could be different or less generous from those that would have been recoverable if the client had sued on behalf of the owner/occupier.

The right to recovery could arise in circumstances where the client could be acting, or could decide subsequently to act, on behalf of, or for the benefit of, others. This should be considered for each appointment and, if it appears to be the case, legal advice should be sought as to whether or not any limitation should be put on the client's right to recover damages on behalf of those third parties.

One way to limit the client's recovery of damages to those the client alone has suffered would be to add the following clause to the appointment document:

> The Client confirms that it is entering into this agreement wholly on its own behalf and not on behalf of, or for the benefit of, any other party and agrees that in the event of any claim arising out of, or in connection with, this agreement it shall be entitled to recover from [the professional] only the losses, if any, it has itself suffered.

The Contracts (Rights of Third Parties) Act 1999 confers rights on a third party to enforce any term of a contract if the contract expressly provides that it may do so, or if the term purports to confer a benefit on the third party unless it appears on a proper construction of the contract that the parties did not intend the term to be enforceable by the third party. The Act applies to most sorts of contracts and therefore applies to professional appointments, collateral warranties, building contracts, sub-contracts and novation agreements.

The third party has to be expressly identified but can be referred to as a member of a class or answering a particular description. Thus, the Act is wide enough to catch not only named parties but also all parties referred to generically, such as 'contractors, consultants, tenants, purchasers', etc.

Some terms which come within the Act will be fairly easy to identify and the professional will then be able to assess the risks involved, for example, 'The [professional] will provide [the named contractor] with [certain specified drawings] at a certain time and [the named contractor] shall be entitled to enforce this term'. The professional will know that if he provides information late, the contractor (as well as the client) will be able to sue direct for damages.

Other terms which may come within the Act may not be so clear and/or the consequences of any breach may be difficult to assess, for example, a term such as 'The contractor and all members of the professional team may in their own right enforce any term of this contract which is for their benefit or is necessary for the proper performance by them of their obligations'. This will give rights to the contractor and other members of the professional team to sue the professional direct, but it may not be clear which terms in the appointment are for their benefit and what the consequences of any breach of those terms may be. It will become more difficult still if the third parties who are given the right to enforce a term are referred to as 'tenants', 'purchasers', etc. because any tenant or purchaser will then be able to sue.

The third party has the rights to the same remedies as the client, for example, damages, injunctions and specific performance.

Various limitations on the third parties' rights are permissible; for example, it would be possible to limit a tenant's right to the cost of repairs. Otherwise the same limitations apply to the third party as they do to the client. All the defences and rights of set-off which would have been available if the client had brought the claim would also be available to a claim brought by a third party.

Once a third party has been given a right, it cannot be altered or extinguished unless that third party consents. It is, however, possible to provide that this part of the Act will not apply. Clients may exclude in their contracts the necessity of obtaining third party consents because they will not want the practical difficulty of obtaining a tenant's or sub-contractor's consent, for example.

One of the objectives of the Act was to remove the need for collateral warranties – see Chapter 21. Third party rights schedules in place of collateral warranties are used by some clients – see Section 21.13 – but, in the main, collateral warranties continue to be used for establishing express obligations between professionals and funders, purchasers and tenants.

Excluding the provisions of the Act would remove any uncertainties as to whether obligations in the appointment, such as an obligation to give information to the contractor for passing to a sub-contractor, gave that sub-contractor a right to enforce the term directly. Most professional services agreements, collateral warranties, standard building contracts and novation agreements exclude the provisions in the Act expressly and all such contracts ought to be checked to see that the Act is excluded. Typical wording to exclude the Act in a professional services agreement is as follows:

> Nothing in this Agreement confers or purports to confer on any third party any benefit or any right to enforce any term of this Agreement pursuant to the Contracts (Rights of Third Parties) Act 1999.

2.4. Receivers as clients

Construction contracts often contain provisions relating to the receivership or liquidation of a party. Such provisions may, for example, result in automatic termination of the employment of an insolvent contractor or give an employer the right to terminate the insolvent contractor's employment. Certain sub-contracts may cater for such events, but this is rarely the case in professional appointments, for example, where professionals are engaged by a design and build contractor. Banks that fund developments often take security over the development property and require the right to appoint receivers to complete the project in the event of the insolvency of the developer or contractor.

A receiver must be an individual and a qualified insolvency practitioner.

The appointment of a receiver will often have no immediate effect on the professional's appointment. It will be a matter for the receiver to decide whether the company should continue with existing contracts, including any appointments.

Care must be taken by professionals when contracting with receivers, administrative receivers or liquidators. This can be by way of an entirely new appointment to complete a project that was under way at the time of insolvency, or it can arise out of an existing appointment or collateral warranty.

Any attempt by an insolvency practitioner to avoid personal liability should be resisted. If a professional is appointed by a receiver, or has an appointment taken over by a receiver in circumstances where personal liability has not been excluded, the professional should either be paid in advance or invoice on a regular basis and have the ability to terminate or suspend the services immediately if invoices remain unpaid after a short period.

Usually a receiver has a right, but not an obligation, to take over a professional appointment, or to enter into a fresh contract on identical or similar terms. Under the latter arrangement, the receiver may well have no obligation to discharge any fees due and arising under the original appointment. The professional should ensure that, on insolvency of its employer, the professional does not lose copyright or give a licence to an insolvency practitioner without either outstanding fees being paid or a fresh contract being entered into.

2.5. Joint clients

Occasionally a professional may be asked to act for joint clients, that is two or more clients. It is important to establish, preferably in the appointment itself, the responsibilities of each of the clients. In relation to payment of fees, the joint clients should be made jointly and severally liable for all fees and additional payments, so that if one does not pay, the other(s) are wholly responsible. If, as sometimes occurs, the joint clients wish to pay only a proportionate part of all monies due, provision should be included for the professional to terminate if one of the clients does not pay its share. A similar right to terminate should be included if one of the joint clients is wholly responsible for the payment of fees and is in default of those obligations.

Each of the client's obligations should be considered carefully. Who is responsible for providing information or making decisions? Does the decision of one bind the other? What happens if there is a conflict of instructions or a delay in instructions being given or decisions being made? These matters should be addressed in the professional's appointment and a mechanism included for resolving them or for termination if this is not possible. One solution might be for the joint clients to appoint a project manager who has complete authority to act on their behalf. The professional could then proceed on the basis of the project manager's instructions and if one client did not agree with what the other client had done or not done in relation to the project that dispute would be restricted to the clients and would not involve or concern the professional. This would not, however, resolve any problems concerning the payment of fees.

Where a professional is engaged by a joint venture made up of two or more parties or partners, the potential problems for a professional are the same as where there are joint clients. If, however, a special joint venture company has been formed, that company will usually be the client and the same assessment of the risks of accepting an appointment from such a client needs to be made as for any other company client. It is perhaps more likely that such a company will be a 'shell company' – see Section 2.8. In that case, particular consideration will need to be given to the joint venture company's ability to pay, remembering that a shell company may cease to have any significant assets or may even cease to exist when its purpose, the construction of a project, is completed, but there may remain outstanding fees still to be paid. Guarantees should be sought from one or more of the joint venturers.

2.6. Design and build contractor clients

Professionals are often engaged by a contractor employed in a design and build capacity. The main difficulties in such appointments can relate to: (i) the design obligation taken on by the contractor which the contractor will seek to pass on to professionals, and (ii) the question of inspection or supervision during construction. Unless expressly stated otherwise in the contractor's contract, a contractor will have undertaken an obligation to ensure that the construction will be fit for its intended purpose. This is a different and more onerous obligation from that imposed on designers under the common law, which is to use reasonable skill and care in carrying out their design (including specifications and drawings) and any inspections – see Sections 4.1 and 4.2. However, commonly used standard conditions for such contracts, such as JCT (Joint Contracts Tribunal Ltd) and ICE, state that a contractor's design liability shall be the same as a professional's. Others, such as ACA (Association of Consultant Architects) and ACA/BPF (British Property Federation), stipulate a fitness for purpose obligation.

While a professional will normally undertake periodic inspection for a client, no inspection or supervision will normally be undertaken for a design and build contractor. Construction and supervision of the works are the contractor's responsibility and there are dangers inherent in the professional undertaking an obligation of this nature – see Section 5.8.

2.7. The Crown, Government, local authorities and other statutory body clients

The Crown can enter into contracts with non-government bodies and individuals. The Attorney-General will enforce obligations under those contracts on the Crown's behalf and may be sued under the same rules of court as other individuals. The Crown does, however, retain certain privileges, for example, no injunction or order for specific discovery may be granted, although the court can make declarations concerning the parties' rights. The Crown may be required to provide documents under the rules of discovery or disclosure, but it may claim public interest immunity to justify withholding some or all of those documents.

The 'ultra vires' doctrine does not apply to contracts entered into by the Crown as the Crown has the power to enter into contracts without any statutory authority. The professional therefore does not have to check whether the Crown has power to enter into any form of contract. The powers of government ministers to enter into contracts may, however, be limited by statute or statutory instrument and this would need to be checked.

The main concern for a professional contracting with a public authority will be whether the relevant authority is permitted to enter into the proposed contract, as the powers of all authorities are defined by statute. In particular, the powers of local authorities derive from the Local Government Act 1972, although later statutes have regulated the way in which local authorities can enter into contracts, for example, by providing for PFI arrangements or the certification of contracts. Normal professional appointments should not create difficulties, but if the appointment covers more than the usual professional services these powers should be checked.

2.8. 'Shell company' clients

For some projects a new company will be set up. However, even though the parent or holding company may be substantial, it cannot be assumed that the parent company will ensure the new company fulfils its obligations. Indeed, the new company may well have been set up to take on the project so as to shield the parent from liability. Problems can also arise when it comes to payment of fees. The new company may be formed with only, say, £100 share capital and have no other assets at all – the funding for the project coming from third parties. Further, the professional should not assume that the company entering into the appointment necessarily owns the site. Even if it does, it is likely that the site will be charged to funders. In these circumstances, there would be no automatic recourse to the assets of any parent or holding company to meet claims, including unpaid professional fees, against the new company. It is always worthwhile making enquiries as to the financial standing of the new company, such as one might with any new client. Certain public information may be easily obtainable from credit references searches, company searches or land registry searches to ascertain the ownership of the land or site. These can be done at little cost, although the information gathered may not always be up to date.

Where the shell company is located overseas, particularly where it is in an offshore tax haven, it may not be possible to obtain any financial information in this way.

If there is concern about any shell company client's ability to pay, the schedule for the payment of professional fees should reflect this so as to reduce the credit risk. For example, time for payment should be kept as short as possible, time for payment should be stated to be of the essence and a specific right to terminate for non-payment included. A better alternative would be to obtain a third party guarantee, perhaps from the substantial parent company so as to underwrite the new company's obligations. Unless this is done, a claim for recovery of fees against the parent needs to rely on technical arguments based on the principles of agency: that is, that the shell company was merely the agent of the parent. If the shell company goes into insolvent liquidation, there might then be a route to the directors of the company on the basis of 'wrongful trading' or if personal assurances were given about payment. Such cases are rarely straightforward and, in any event, any recovery from the directors would simply be part of the liquidation and be divided among all the company's creditors. It is better to avoid these difficulties. No client putting forward an insubstantial company as a contracting party should object to giving a parent company guarantee.

2.9. Another professional as a client

Sometimes one professional will employ a sub-consultant to perform some of the professional's services. It is important to remember that, even though the sub-consultant's appointment may be 'back to back' with the professional's, the professional will have obligations to the sub-consultant, particularly in relation to payment, the provision of information and taking decisions. The appointment document between the professional and sub-consultant should therefore deal with these specifically – see also Chapter 12. In the past, sub-consultancy appointments provided that any payment to the sub-consultant would depend on payment by the ultimate client ('pay when paid'). Such clauses are prohibited by the Housing Grants, Construction and Regeneration Act 1996 ('the Construction Act 1996') where the professional services agreement is a 'construction contract' as defined in the Construction Act 1996, except where insolvency is concerned – see Section 25.3 for more comments on 'construction contracts'. Professional services agreements will usually be 'construction contracts'.

2.10. Use of correct name

Once a client is identified, the correct name should be used in any professional services agreement, including the correct form of 'Limited', 'Ltd', 'plc', 'LLP' '& Partners', etc. Otherwise the appointment might refer to a non-existent organisation, a different one within a group, or even a different organisation entirely from the one intended.

Professional Services Agreements
ISBN 978-0-7277-4158-5

http://dx.doi.org/10.1680/psa.41585.013

Chapter 3
Forming the appointment

3.1. General – forming a contract

In order for there to be a contract (or appointment) between a professional and a client, there has in law to be an offer to do something by one party (for example, to carry out certain services) and an acceptance of that offer by the other party. This turns the offer to carry out the services, for example, into a contractual promise. There also has to be consideration for the promise – generally an obligation to pay for the services – and an intention to create legal relations. Other formalities may be required in any particular case if that is the intention of the parties, for example that the appointment should be in writing and signed as a deed.

Contract formation is sometimes thought to be a question of offer and acceptance: has an unconditional offer been unconditionally accepted? If an offer is made and the other party does not accept it but proposes some other terms, this is a counter-offer which is then open for acceptance or rejection. If accepted there is a contract on the terms of the counter-offer. If further proposals are made in reply, this is a counter-counter-offer which is open for acceptance or rejection. This describes the process of negotiations. Acceptance can be indicated informally and there can be a stipulation in the offer itself as to what is to constitute acceptance, for example 'Acceptance shall be indicated by signing and returning a copy of this letter'. In a situation of any complexity, however, where the existence or terms of a contract are being disputed, this approach of determining whether there has been an unconditional offer and acceptance is unlikely to be helpful. It is more likely that a contract will have come into existence, if at all, as a result of work being carried out. This can be called 'deemed acceptance' or 'acceptance by conduct'; that is, if a party in response to an offer goes ahead and starts the services, in the case of a professional, or the client pays in accordance with the terms of the professional's offer, a contract may well have been formed. Professionals should be very careful therefore about commencing work if the form of contract has not been agreed – as illustrated in the case described below.

The basic test for determining whether there is a contract is whether the parties intended to create legal relations. This has to be judged objectively. What the parties say they intended is usually inadmissible. The parties must be assumed to have acted as reasonable business people and what they said and did is to be assessed in this light and in the light of what, as reasonable business people, they would be assumed to have thought and known. As a result, inferences as to what they intended may have to be drawn from their behaviour. In *RTS Flexible Systems* v. *Molkerie Alois Müller Gmbh*

& Co KG (UK Production) (2010) the court inferred that the parties had waived the requirements, in the MF/1 terms, that the contract had to be executed before it came into effect, although there was no evidence before the courts that the parties had in fact considered the point.

This case involved the supply, installation and commissioning of an automated system of packaging yoghurt pots. The work was commenced pursuant to a letter of intent. Following the expiry of the period covered by the letter of intent, the work continued, while the parties negotiated over the detailed terms of the contract. It was envisaged that the contract would be on the MF/1 conditions, clause 48 of which provides that a contract will only come into effect on the contract being executed. No contract was ever executed, although most of the terms were agreed in correspondence.

The judge at first instance held that a contract had been reached between the parties on limited terms, which did not include the MF/1 conditions. He ruled that, in circumstances where the work had been completed, it was unrealistic not to conclude that a contract had been formed. The Court of Appeal concluded that no contract had been formed. The Supreme Court decided that a contract was formed on the MF/1 conditions (as amended by agreement).

As was said in the Supreme Court:

> The different decisions in the courts below and the arguments in this court demonstrate the perils of beginning work without agreeing the precise basis upon which it is to be done. The moral of the story is to agree first and to start work later.

The judge at first instance commented that the case was another example of the perils of proceeding with work under a letter of intent. It might sometimes be thought that to start work with a letter of intent is better than having nothing at all. The object of a letter of intent is indeed to enable work to start without the parties being committed (beyond a certain point) unless and until the terms of a full contract are agreed. In practice, however, once work is started, it may be difficult to stop.

The problem of whether there is a contract, and if so on what terms, in circumstances where the work has been carried out, will be more likely to arise as an issue in adjudications now that the amendments to the Construction Act 1996 have come into force. The Construction Act 1996 stipulated that, for a contract to be subject to the Act, it had to be in writing or evidenced in writing. The purpose was to limit difficult issues over contract formation. The amendments to the Act (by the Local Democracy, Economic Development and Construction Act 2009) remove this requirement, and will thus make it possible for all the issues debated in this case to be referred to an adjudicator for a decision.

A potentially misleading approach to contract formation is to consider whether the parties have agreed on all essential terms. It would be wrong to conclude that a contract was formed once the parties had agreed all the terms that a judge (or arbitrator or adjudicator) considered essential. What is relevant is what terms the parties intended to agree before considering themselves contractually bound. There may, of course, be terms that seem so important that one would not expect the parties to intend to be contractually bound without their being agreed, but everything must be judged in the light of the admissible evidence. There may also be situations where, if an obviously crucial term has not been agreed, the contract would be void for uncertainty.

The objective approach permits a robust approach to the evidence. The old-fashioned literalist approach of dictionaries and grammars and syntactical analysis is out of favour. What someone actually wrote may be considered less important than what they, as the hypothetical reasonable business people, must be assumed to have meant. The Supreme Court was clearly indicating, in *RTS Flexible Systems* v. *Molkerie Alois Müller Gmbh & Co KG (UK Production) (2010)*, that, where work has been carried out, the likelihood is that a contract was formed somewhere along the way. A fair amount of confidence on the part of adjudicators or anyone else may be needed when called upon to decide how and when this may have occurred. This will often need to move beyond a literal analysis of the correspondence and other relevant documents, but must not get carried away: their conclusions should still be based upon, and not inconsistent with, the admissible evidence.

A very simple form of contract – where a professional offers to carry out certain services for an agreed fee, which is accepted by the client – can be entered into, but it is more often the case that contracts contain lengthy provisions detailing the rights and obligations of the parties and extend to matters well beyond the services to be performed and the fee to be paid.

A professional services agreement should define in writing all the terms and conditions under which work is to be carried out, the services to be undertaken and the obligations of each party to the other. The appointment will often be on the basis of standard conditions such as those published by the ACE, RIBA, RICS and NEC, or the client may wish to use its own non-standard conditions of engagement. Such an appointment will then constitute the contract between the professional and the client.

It is clearly advisable to have an agreed written appointment before work commences. In practice this is not always possible, no matter what a quality assurance or management system might require, particularly if an early start is necessary. Where this happens and commercial risks have been assessed and accepted, the best procedure is that a letter should be sent to the client as soon as possible stating the professional's terms and conditions and other matters agreed, requesting the client's confirmation and, if appropriate, stating that work has commenced.

3.2. An appointment comprising several documents/ incomplete drafts

Where initial proposals have been made concerning an appointment and agreement reached in correspondence, perhaps by reference to other documents, the correspondence and the documents (if any) will comprise 'the appointment'. If a formal document is then produced which is to comprise 'the appointment', it is important that either all the matters previously agreed are incorporated or that the correspondence and documents are specifically referred to and agreed as forming part of the appointment. Frequently an appointment document will contain a provision which states that the document contains all the agreed terms and/or supersedes all previous agreements and arrangements between the parties. Such a provision should not be accepted if, in fact, the appointment document does not include, or incorporate by reference, everything that has been agreed.

It is a common feature of many proposed appointments that reference is made by the client to other documents (such as the brief, the proposed contract between the contractor and the client, or forms of collateral warranties) but these documents are not provided with the draft appointment. Such documents must be provided before the appointment is agreed and signed as they might contain unacceptably onerous conditions and obligations. For example, the appointment might say that the professional must 'comply with all the obligations relating to the Services as stated in the Brief' but the brief has not been provided or has not been finalised. It is essential to see it (and in its final form) so that the obligations can be assessed. Are they, for instance, practicable? Are they within the professional's discipline? Do they have an effect on the fee or on how the professional is proposing to carry out the work? If the brief has not been finalised, how does the appointment deal with changes to the required services and/or the fee?

Another example is that the professional must 'carry out the Services having due regard to the obligations of the Client contained in . . . [various other contracts]'. What is 'due regard'? At the very least it would require the professional to look at the client's obligations in those documents to see whether the things the client has to do affect what the professional has to do, how he does it and when he has to do it. All of which would mean that the professional needs to be given these documents. If there could be a significant effect, the client should be warned and the matter discussed and any changes agreed before signature. The very task of looking at such things as development agreements and draft leases could be a time-consuming and expensive one for the professional.

Many draft appointments initially omit fundamental details, such as the basic services to be provided, possible additional services, and the basis of the fee calculation and reimbursement. Even if the professional is content with all the other conditions, it must be made clear to the client that any agreement to the appointment depends on agreeing those details. If the services details are absent, the professional should supply a draft. This has two advantages.

(*a*) The professional will be able to set out in detail the services to be provided (which are not always what the client understands will be provided).
(*b*) The professional's own wording can be used or the services adopted from an appropriate standard conditions of engagement.

Draft appointments should also be checked to see that they contain the provisions that are essential to the professional, for example time limits for payment of fees, revisions of fees in appropriate circumstances, copyright clauses, information needed from the client, termination provisions, etc. These should be put forward by the professional at an early stage of the negotiations and, if not accepted, will alert the professional to potential problems.

3.3. Tenders

Professionals can be invited to tender for certain work and often the tender documents will include the form of appointment or the conditions that will apply if the tender is accepted by the client.

An invitation to tender does not normally constitute an offer of an appointment which can be accepted by the successful tenderer, but constitutes ‘an invitation to treat’. It is therefore not necessary for such an invitation to state that the client does not undertake to accept the lowest or any tender. It is the professional’s tender that is ‘the offer’ as far as contract law is concerned and acceptance of that tender by the client then completes the contract.

In some circumstances an invitation to tender can constitute an offer, for example where a client agrees to accept the lowest tender.

EC and UK legislation has laid down specific procurement rules and requirements concerning the tenders for particular sorts of work (such as works contracts over a certain value let by a public body or utility) but these are beyond the scope of this book.

3.4. Letters of intent

It can often take a considerable period of time for contracts to be negotiated and executed, particularly where the details of the contractual obligations are extremely important or complex. Alternatively, a building contractor may need to sub-contract elements of its contractual obligations, such as groundworks, and the tender price will depend on the cost of the sub-contractor’s work. The contractor will not want to enter into contracts with the sub-contractors until its own tender has been accepted. Sometimes a client will not want to enter into any long-term contracts until funding has been obtained. Similar issues can occur in relation to appointments between a client and a professional.

In an attempt to resolve these problems, the use of a letter of intent has become widespread, particularly with contractors, but it can also be used for professionals. By the letter of intent the parties indicate their intention to enter into a formal contract and may agree that in the interim the work can begin.

The contractual status of these letters depends on their precise terms and on the principles for forming a contract discussed in Section 3.1. They may simply state an intention to enter into a contract at a later date without a specific request to undertake work or to enter into a sub-contract. On the other hand, they often create an obligation to pay for any work performed in reliance on them. Thus letters of intent have often been regarded as short-term contracts, existing only until the main contracts are entered into. The letter is regarded as an offer, which is accepted by the contractor or professional either expressly or through its conduct in performing the work or services.

Although a party issuing a letter of intent will often draft it in a way that tries to avoid creating rights and obligations on either party, the courts will strive to reject this argument where the parties have spent considerable time and effort in reliance on the letter. If no basis of payment is stated in the letter, the contracting parties will be entitled to claim payment for work done on a *quantum meruit* basis – see Section 3.9. If there is a basis of payment proposed in the letter but it is unacceptable, it must be resolved to the satisfaction of both parties and a new letter of intent agreed before any work commences.

It will be apparent that letters of intent can have the same effect as contracts, although the courts have described their status in different ways. The letters can, however, only act as contracts where it is clear that the parties intended to be bound by their terms and that they intended them to have legal status.

Thus, where a professional is asked to start work under a letter of intent, the basis of payment must be acceptable and the work to be undertaken sufficiently particularised. Sometimes whole appointments can be carried out under letters of intent and this may be acceptable. If it is not, it would be prudent to have a time limit within the letter for agreeing the formal contract and provisions for what should happen, for example in relation to payment and the scope of the services to be undertaken, if no formal agreement is reached within that time limit.

Letters of intent can be very complex and contain almost as many obligations as the final appointment documents – including, for example, the obligation to give licences, collateral warranties, etc. In this event, each obligation should be considered as if it were contained in the final appointment document.

Generally, letters of intent are designed to be short-term arrangements until the final contract or appointment has been entered into or, less often, a project or commission aborted. It is important to ensure that when the final contract or appointment is entered into, it supersedes the letter of intent and that the letter of intent is of no further effect. Normally, the final contract or appointment will be retrospective, governing the rights and obligations of the parties from inception of the services. The professional should ensure that any payment not received under the letter of intent is payable under the final appointment.

The perils of entering into or working on a letter of intent are well illustrated by the case of RTS Flexible Systems referred to in Section 3.1.

3.5. Incomplete contracts

The courts cannot enforce incomplete contracts. If therefore there are terms without which the contract cannot be enforced, the contract is incomplete and void for uncertainty, as has been said. It is for the parties to decide whether they wish to be bound and if so by what terms, whether important or unimportant. It is the parties who are, as the judge put it, 'masters of their contractual fate'. The more important the term is, the less likely it is that the parties will have left it for future decision or 'to be agreed', but there is no legal obstacle which stands in the way of the parties agreeing to be bound now while deferring important matters to be agreed later. This happens, for example, when parties enter into 'heads of agreement'.

3.6. Implied terms

The courts are prepared to imply terms into a written contract where a contract is otherwise incomplete. The justification for incorporating the terms into the contract is that the parties intended that the terms should have been included from the outset.

In assessing the intention of the parties, the courts will be looking at two indicators of intention against the background of the negotiation of the contract. The first indicator is the 'business efficacy' test. This test involves looking at the contract objectively and considering whether the contract would be workable without the term sought to be incorporated. If the missing term is essential to the contract, it is likely that the parties would have intended to include it from the outset, and the term will be incorporated. The second indicator is the 'officious bystander' test. If it would be obvious to a third party watching the negotiations taking place that a term ought to be included, and the parties appear merely to have overlooked it, perhaps because it was obvious, the term will be included.

The courts will, however, never incorporate an implied term where it contradicts a term which has been expressly included in the contract. It will also be hard for a party to argue for the inclusion of an implied term where the contract appears to have been carefully and comprehensively drafted.

Terms may also be implied into contracts based on custom or trade practice, where both parties are familiar with the relevant trade. In order for such terms to be incorporated, the trade practice must be widely known and both certain and reasonable, and it must be a well-established custom in the particular trade.

Terms may also be implied into a contract based on a previous course of dealings between the parties, particularly where the parties have contracted with each other on several occasions before on the basis of those terms.

In professional services agreements which are silent about the client's obligations other than that of paying the professional, there would usually be implied a term that the client would give the professional the information, decisions, consents, approvals and assistance he needs in order to perform the services.

3.7. 'Entire' contract

Some non-standard conditions contain 'entire' agreement or contract clauses, such as:

> Each party acknowledges that this contract together with all documents referred to in this contract represents the entire understanding and constitutes the whole

agreement between the parties relating to the subject matter of this contract and without prejudice to the generality of the foregoing excludes any warranty, condition or other undertaking implied at law or by custom.

The effect of such a clause is that the parties are limited to their rights under the express terms of the appointment and neither party can argue that there are implied terms. This could be dangerous because, as set out in Section 3.6, such terms could be necessary to give business efficacy to an appointment. For example, if there was no obligation in the appointment on the client to give any necessary information to the professional, this would usually be implied. If an 'entire contract' clause had been included in the contract, the professional would not be able to bring a claim should the client not provide information and this resulted in delays and extra cost to the professional.

Some entire agreement clauses state that the agreement shall supersede or exclude 'any prior promises, agreements, representations, undertakings or implications whether made orally or in writing relating to the subject matter of this Agreement...'. The Court of Appeal in *Axa Sun Life Services plc* v. *Campbell Martin Limited (2011)* considered such a clause and held that it was effective to exclude collateral warranties but, notwithstanding the reference to 'representations', was not effective to exclude or supersede misrepresentations as to matters that are not the subject of the terms of the agreement. It followed that the clause had no effect on misrepresentations of fact, nor would the clause exclude implied terms that are intrinsic to the agreement.

(Regarding the exclusion of collateral warranties, it should perhaps be said that such a clause would not affect a requirement in the contract to give a collateral warranty.)

The court was bound by a decision of the court in 2010 that an entire agreement clause of this kind is effective in accordance with its terms. (It should be noted, however, that the respondents to the action reserved their right to seek to argue in the Supreme Court that the decision was wrongly decided.)

The contract in which this clause appeared was subject to the requirements of the Unfair Contract Terms Act 1977 (UCTA) and therefore the clause was subject to the requirements of reasonableness, if it could be shown that any collateral warranties or representations affected the performance reasonably expected of a party. If it did in fact exclude misrepresentations (contrary to what the court found), it would be subject to the requirement of reasonableness in Section 3 of the Misrepresentation Act. The court in this case held that the clause was reasonable.

If misrepresentations are excluded by an 'entire agreement' clause, a professional would also not be able to rely on any written or oral misrepresentation made prior to it entering into the contract, unless the statement was fraudulent – see Section 3.8. Clients need expressly to exclude fraudulent misrepresentations from 'entire' agreement clauses or the clause itself may be struck down because it fails the 'reasonableness' test under the UCTA – see Chapter 13.

3.8. Misrepresentation

During the negotiation of an appointment, many statements may be made about how the project or the services will be carried out, or the basis on which the parties are to carry out their obligations. Care needs to be taken about such statements and there can be a temptation to overstate facts where the professional is under pressure to win the commission. Equally, the professional may be concerned to know, if he has agreed to enter into an appointment which, for example, contains a higher duty of care, that all other professionals are to sign up on the same basis so that there is a level playing-field, and may seek a specific assurance of this from the client. The professional might also have quoted his fee on the basis that the procurement route chosen by the client will not be design/build or that there will be no suspensions of the project beyond a certain number of months.

Some of these statements may be expressly included in the final appointment, and if they turn out to be untrue the injured party may be able to sue for breach of warranty or condition. If, however, any of the preliminary statements not included in the final appointment turn out to be untrue, the injured party can only sue the party making the statements for misrepresentation. Some oral statements can be terms of the contract, depending on the intentions of the parties, but it is assumed for the purposes of this section that oral statements are only representations.

A representation is a statement of fact, as opposed to a statement of opinion or intention; a misrepresentation is simply an untrue representation. The distinction between what is fact and what is opinion can be difficult to establish. In one case a forecast given by Esso to a purchaser of a filling station as to the likely number of gallons that would be sold in a year was held by the Court of Appeal to be a statement of fact as Esso had special knowledge and skill in the forecasting of throughput of petrol.

As to intention, a statement of intention is not a statement of fact, but a person who misrepresents his present intention does make a statement of fact because the state of his intention is a matter of fact.

The motives of the maker of the statement are irrelevant in classifying misrepresentations, but they will be relevant in assessing the level of compensation payable to the injured party.

In order to sue on the misrepresentation, the injured party must show that the statement induced it to enter a contract, and that it would not have entered the contract but for the misrepresentation. The injured party has no cause of action if it knew that the statement was untrue or if it only learned about the statement after making the contract.

Once a misrepresentation is established, it must be considered whether it was made fraudulently, negligently or innocently. A fraudulent misrepresentation is made when the person making it knows or believes it to be untrue or makes it recklessly; that is, careless as to whether it be true or false. Such a misrepresentation constitutes the tort of deceit. A negligent misrepresentation is made when the person making it is careless as to whether it is true or false. An innocent misrepresentation is made when the person making it honestly believes it to be true.

In the 2009 case of *Fitzroy Robinson* v. *Mentmore Towers*, the architects' representation was that a particular person would work on a job. This statement was true at the time it was made but became a fraudulent misrepresentation as it was not corrected when that person handed in his notice. However, the damages were probably low as it is difficult to see what loss the client had suffered as a result.

The effect of a misrepresentation is to make the contract voidable by the injured party; that is, it can choose whether to set it aside and treat it as if it had never been made (this is called rescission) or to continue with the contract. The contract is not automatically void. The choice will not be open for ever, though, and if the decision to rescind is made it must be brought to the notice of the representor. The right to rescind the contract can be lost in certain circumstances.

The right to rescind the contract is lost

(*a*) by affirmation – where the person to whom the representation has been made chooses to proceed with the contract with full knowledge of the facts
(*b*) by lapse of time – this may be treated as evidence of affirmation
(*c*) where it has become impossible to restore the parties to their pre-contract positions, for example, where the subject matter of the contract has altered in some substantial manner
(*d*) where a third party would suffer if rescission was permitted.

The level of damages will depend on the nature of the misrepresentation. An injured party can claim compensation for all losses resulting from a fraudulent misrepresentation, including those to put the claimant in the position he would have been in had the tort not been committed and aggravated damages (as compensation for injury to feelings), but not for exemplary damages or punitive damages.

Losses resulting from a negligent misrepresentation will usually be restricted by the courts to reasonably foreseeable losses. The Misrepresentation Act 1967 introduced a claim for damages for negligent misrepresentation and the measure of damages is the measure of damages for the tort of deceit, i.e. damages will be assessed as if the statement had been made fraudulently. The burden of proof then falls on the party in default to show that the misrepresentation was not made fraudulently. At common law the remedy for innocent misrepresentation is rescission of the contract with no damages being awarded as innocent misrepresentation is not a tort. However, under Section 2(2) of the Misrepresentation Act the court has a discretion to award damages in lieu of rescission.

It is not possible to exclude or limit liability for fraudulent misrepresentation but it is possible in respect of negligent or innocent misrepresentation, provided it is included as a term in a contract and subject to the constraints of the Unfair Contract Terms

Act 1977 or Unfair Terms in Consumer Contracts Regulations. The Misrepresentation Act specifically provides that any contract which contains a term which would exclude or restrict any liability or remedy available as a result of a misrepresentation is of no effect except insofar as it satisfies the 'reasonableness' test under the Unfair Contract Terms Act 1977. (See also Chapter 16 on limiting liability.)

In January 2010, in the case of *EDS* v. *BSkyB*, the judge found that EDS had made fraudulent misrepresentations as to its ability to deliver a project within a certain timetable and in particular that it had carried out a proper analysis to enable it to make this statement. The judge also found that it was as a result of these misrepresentations that BSkyB had been induced to enter into the contract with EDS. The damages that could be payable as a result have been estimated at £200 million or more. There was a limit of liability in the contract to £30 million but both parties have accepted that such a limit is not effective to limit liability for fraudulent misrepresentation.

3.9. Payment on a *quantum meruit* basis

If there is a contract but no fee has been expressly agreed, the law will impose an obligation to pay a reasonable sum on a *quantum meruit* basis provided the circumstances show clearly that the work was not to be done gratuitously. This principle may extend to services performed in anticipation that negotiations will lead to the conclusion of a contract provided the services were requested or acquiesced in by the recipient.

Where there is no contract, payment on a *quantum meruit* basis may be due on restitutionary principles if there is held to have been unjust enrichment. The case of *Stephen Donald Architects Limited* v. *King* (2003) illustrates this. In that case, the client sought to redevelop a property into a number of flats with a photographic studio, and approached the architect to provide the design services. The architect carried out some work, obtained planning permission for the development, and was paid £15 000. However, the client determined the architect's engagement before a building contract was entered into because he was concerned that he would not be able to finance the development as envisaged in the architect's drawings. The architect brought proceedings to recover its fees and sought, *inter alia*, a payment on a *quantum meruit* basis. The client contended that the true nature of the project was that the redevelopment was to be a joint effort and that either the architect would not be paid until completion of the project or the architect would receive one of the flats on completion in lieu of his fees.

The court held it was a prerequisite of a *quantum meruit* payment that a party be unjustly enriched and the intentions of the parties might be highly relevant to the question of whether the enrichment had been unjust. There was nothing unjust about being visited with the consequences of a risk that one had consciously run. On the evidence, the whole idea of the architect's activities in relation to the development was that through a joint effort, in which the client would make the property available for redevelopment and shoulder the burden of discharging any loans for the redevelopment to proceed, and the architect would undertake the necessary architectural work and assist in raising finance, the client and the architect would bring about a situation whereby the property could be redeveloped to the benefit of both. The architect therefore assumed the risk in relation to his work up to the point at which a building contract might have been concluded, that the client might choose not to proceed. That risk had eventuated, and there was nothing unjust in the client retaining the benefit of the planning permission without paying compensation additional to that already paid. Accordingly, the claim for a *quantum meruit* payment failed.

The general principle is that a successful party in a *quantum meruit* claim can recover the reasonable value of any services provided at the date on which they were provided. 'Reasonable value' may include an element to represent the profits of the provider of services, although this is not always available. This approach has been confirmed by the courts (*Sanjay Lachhami & Another* v. *Destination Canada (UK) Limited (1997)*), which held that a fair value ought to provide for a reasonable or normal profit margin over and above the costs incurred in carrying out contract works. The court also held that only costs which were reasonably and necessarily incurred should be reimbursed, which indicates that not all costs can be claimed, particularly where the supplier has not acted efficiently. The court did, however, hold that any pricing level put forward by a building contractor should be considered in assessing what the fair value for the work should be, even if applying this level would mean that the contractor would suffer a financial loss. This principle would appear to be relevant, for example, where a professional charges on an hourly rate basis. The court would consider these rates when assessing a *quantum meruit* claim, and it would be open to the court to find that there should be no profit margin as a result.

There has been some discussion of whether the party receiving the services needs to receive any benefit therefrom so as to entitle the party providing the services to a *quantum meruit* claim. The widely held view is that the receiving party must obtain at least a tangible benefit, so that a *quantum meruit* claim fails if the party undertaking services has only carried out preparatory work of no value to the receiving party.

An action for *quantum meruit* may allow recovery of a reasonable sum as additional remuneration for extra work even where the contract entitles the client to order additional work. This is discussed in relation to variations in Section 5.12.

There may also be an entitlement to a re-valuing of the original price to be paid on a *quantum meruit* basis because the subject matter of the contract was fundamentally different at the time the contract was entered into.

IN Duncan Wallace in his work *Construction contract: principles and policy in tort and contract* (1986) has said 'There is nothing new about the "scope and cardinal change theories". Since the earliest days the courts have felt no difficulty holding that, in principle, and on appropriate facts, alterations of the work in a construction project could be of a kind which would have the effect of dispensing with the original contract prices altogether . . .'.

He summarised the law as follows.

(*a*) There can be little doubt that, both in the UK and US, work can be ordered in a construction project which is 'outside the contract'. One way this can happen is where extra work is ordered which has no connection with the contract work, either in description or purpose or time . . .
(*b*) Equally, additional compensation for changes over and above the prices permitted by a tightly restricted valuation clause may in principle be recoverable if the nature or totality of the changes could be said to be 'outside the contract' . . . The price of the unvaried work will not, however, be affected.
(*c*) Changes may be so substantial that in principle *quantum meruit* can replace the contract price altogether in cases where the final work no longer bears a resemblance to the original work.

These principles would apply equally to a contract for professional services.

3.10. Speculative work

A client may request that a professional undertake work on the basis of the professional being paid only if the project proceeds. A letter of appointment is still necessary, not least to record what was actually required to be undertaken initially and to set out the arrangements agreed should the project proceed. It should include a provision whereby if the project does proceed the client will pay the professional's fees to date, the basis of fees being recorded in advance in the appointment. It may be appropriate for fee rates to be higher than normal to reflect the risks inherent in the speculative nature of the initial work. It is important to note, however, that the professional will be liable if he has been negligent in performing that work or in giving advice, even though there has been no payment.

3.11. Trading conditions, purchase orders and similar

Sometimes clients will seek to order or accept offers of professional services by means of *pro forma* purchase order forms which are not designed for the provision of professional services; for example, they may be intended principally for the engagement of contractors or the suppliers of materials. These forms may incorporate trading conditions, usually on the reverse side. Such trading conditions should not be accepted.

If a professional has submitted an initial proposal including terms of engagement and the client confirms acceptance with a purchase order (which incorporates these trading conditions), the purchase order (and trading conditions) will constitute a counter-offer. The question will then be whether the professional by commencing work has accepted the counter-offer – see Section 3.1. Generally the answer will be that he has, and thus has accepted the trading conditions. To avoid this, the client should be told that the trading conditions cannot be accepted and the professional should not start work until an appropriate form of appointment has been agreed.

3.12. Checking revised drafts

Non-standard professional services agreements will often go through several revisions. All amendments should be checked to ensure that they correctly incorporate what has been agreed.

It is also necessary to check the rest of the document. It is dangerous to assume that because amendments are highlighted or indicated in different type that these are the only parts that need to be checked. Amendments may have been made to other parts and no indication made that this has happened. A final recheck should also be carried out before signature.

3.13. Appointments signed towards the end of a commission

A check is particularly important where an appointment is not being signed until near, or even after, the completion of the services. Leaving aside the question of whether the professional might be better off without the appointment (if the appointment contains terms that are particularly onerous to the professional, it may be better if it is never agreed and signed), it is essential that the appointment accurately records what has been done. Things may have happened during the project to alter what was originally agreed. This is particularly important in relation to the services schedule – the services actually performed may well not be those originally envisaged.

It is also arguable that if an appointment is signed, either under hand or as a deed, the applicable limitation periods start from the date of signing and not the date otherwise prescribed by law, which could be earlier – see Section 3.15. This could be dealt with by adding a term to the effect that the limitation periods will be those that would have applied if the appointment had been signed when the services commenced.

Ancillary documents such as collateral warranties should be checked before signing to ensure they are in favour of the people to whom the professional has agreed to give them, and are in the form of any agreed draft. If they are being signed after the duties have been completed, any references to continuing duties, such as 'the professional will continue to exercise reasonable skill and care', should be deleted or the courts may interpret this as meaning there were further duties, such as an obligation to revisit a design in the light of later technical developments. (See also Chapter 21 on collateral warranties.)

3.14. Signing under hand or as a deed

Whether the appointment is to be signed under hand or as a deed must be checked early on. The main difference relates to limitation periods, it being six years for the former and 12 years for the latter – see Section 3.15. If it is the professional's policy not to sign appointments and warranties as a deed, this should be resolved during negotiations.

In general, contracts can be made informally, either orally or in writing (that is, 'under hand'), although some statutes prescribe that certain contractual documents must be made in writing, and furthermore they must be made by deed. (This is the same as signing a document 'under seal'.) There are no such statutory requirements applying to professional appointments, although many appointments are required by clients to be signed as a deed or under seal to take advantage of the longer limitation period.

There are no specific formalities for executing a contract in writing. There may be either a single document which both parties sign, or alternatively there may be two copies of the document, with each party signing its copy, and the copies are subsequently exchanged. A duly authorised signatory can sign on behalf of a company, limited liability partnership or a partnership, and it is usual to state that he or she is signing in this capacity. The signature of an authorised signatory does not have to be witnessed.

The formalities for executing a deed depend on whether an individual, limited liability partnership or a company is executing the deed.

A document executed by an individual as a deed must state clearly that it is intended to be a deed, and it must be signed by the individual in the presence of a witness attesting to the individual's signature. Alternatively, the individual can direct that the deed be signed by another person, but two witnesses are required in this situation. Finally, the document must be delivered as a deed, which means that the individual must state in writing his intention to be bound by it as a deed. There is no longer a requirement that an individual must affix his seal.

For a limited liability partnership executing a deed two members must sign or one member in the presence of a witness who attests his signature.

If a company is executing a deed, the document must be signed by a director and by the company secretary, or by two directors or by one director whose signature is made in the presence of a witness attesting to that director's signature.

In all these cases, the document must state that it is executed by the company or the limited liability partnership. Only if the company has not been incorporated under the relevant Companies Acts is it still necessary to affix the company's seal to the document. The document must also be delivered as a deed.

As a matter of corporate and limited liability partnership governance, it is advisable to have two people sign a deed, given the additional significance and risk associated with a deed.

If the contract is executed as a deed, the general contractual requirement for some form of consideration actually to be exchanged between the parties does not apply. A gratuitous promise made in a deed can be enforced. Thus, where no consideration or payment is to be made to a professional signing a collateral warranty – see Chapter 21 – he will be asked to sign it as a deed, or the warranty will be unenforceable by the other party. However, if some consideration is given, even if this is as little as £1, this would be sufficient consideration. If receipt of that sum is acknowledged, even if it is not paid, that too would be sufficient consideration.

3.15. Effect of signing appointment under hand or as a deed

The time during which a professional can be sued for breach of contract is six years after the breach if the appointment has been signed under hand or 12 years if the appointment has been executed 'as a deed' (or 'under seal'). Signing as a deed or under seal should, therefore, be resisted if possible. If an appointment or a collateral warranty is being signed as a deed long after the services have been started, or even after their completion, consideration should be given to reducing the number of years for bringing any claims – see Section 13.6.

It is important that all the appointments relating to a particular project are signed in the same manner. For example, where the professional is a lead consultant, all sub-consultants' appointments should be signed as deeds if the professional's appointment is so signed. If that is not so, the professional could find that he can be sued by the client but is unable to sue the sub-consultants under their appointments because the claim is time-barred.

Professional Services Agreements
ISBN 978-0-7277-4158-5

http://dx.doi.org/10.1680/psa.41585.029

Chapter 4
Obligations imposed on the professional

In this chapter, some matters which commonly occur in appointments are considered. It will be realised that in professional appointments (as well as in documents produced by the professional such as specifications) loose or careless wording, or wording that might have more than one meaning, or mean one thing to a technical person and another to a layman, can carry inherent risks, and the legal interpretation of certain words can carry different implications to everyday or even technical usage.

4.1. Duty of care in contract and in tort

The standard of care imposed on a professional at common law is to carry out its services with 'reasonable skill and care'. The test is that of the ordinary skilled and competent practitioner in the relevant profession. This is imposed regardless of whether it is specifically referred to in the appointment. It is also imposed on the provider of services by the Supply of Goods and Services Act 1982, Section 1. If a professional fails to use the requisite degree of skill and care specified in his appointment he will be in breach of contract and will have been negligent. If no express duty of care is incorporated in the appointment, the common law duty of care will be implied as a term of the contract. A failure to use reasonable skill and care can also constitute the tort of negligence – see later in this section for the general principles concerning the duty of care in tort owed to the client.

All of the standard conditions of engagement include a contractual obligation relating to the duty of care. Some of these impose a duty of care that is the equivalent of the common law duty, often with the additional duty to proceed with 'diligence', that is, within a reasonable time.

ACE Agreement 1 provides:

> F2.1 The Consultant shall exercise reasonable skill, care and diligence in the performance of the Services.

RIBA 2010 provides:

> 2.1 The Architect shall exercise reasonable skill care and diligence in accordance with the normal standards of the Architect's profession in performing the Services and discharging all obligations under this clause 2.

PSC3 duty is described thus:

> 21.2 The *Consultant's* obligation is to use the skill and care normally used by professionals providing services similar to the *services*.

This duty is discussed in Chapter 9.

Alternative wordings to 'reasonable skill and care' are often proposed in non-standard conditions of engagement, such as

- 'all skill and care'
- 'all proper skill'
- 'due professional skill'
- 'due skill and care'.

These are clearly different from 'reasonable skill and care' and could impose a higher duty. 'Proper' skill, for example, could be a higher degree of skill than 'reasonable' skill.

Appointments can impose a higher and more onerous duty of care, such as:

> The professional shall exercise the skill, care and diligence to be expected of a properly qualified [relevant discipline] experienced in carrying out work of a similar size, scope and complexity to the Project.

In such cases, the professional is more likely to be in breach of this duty which could lead to more successful claims, higher damages and ultimately increased insurance premiums.

RICS 2011 has adopted a similar higher duty of care:

> 3.1 The Consultant exercises in the performance of the Services the Requisite Standard.

The Requisite Standard is defined as:

> The standard of reasonable skill and care to be expected of an appropriately qualified professional consultant of the discipline specified in the Appendix holding itself out as having the competence and resources to perform the Services and who is experienced in providing services in connection with works of a similar size, scope, nature, complexity and value as the Project.

When deciding whether a professional has been negligent, the question will be: has the professional done or not done what no other ordinary skilled and competent practitioner in the same profession acting with ordinary competence would have done or not done in the same circumstances? In the Bolam case, which provided a definitive test for negligence, the test was stated as being:

> The test is the standard of the ordinary skilled man exercising and professing to have that special skill. A man need not possess the highest expert skill at the risk of being found negligent. It is well established law that it is sufficient if he exercises the ordinarily skill of an ordinary competent man exercising that particular art.

The tests under the higher duty wording set out above will be those set out in the relevant definition, which are different from the common law. Nothing in the common law requires a professional to demonstrate he has 'appropriate qualifications', for example, or to hold himself out as or describe himself out as having the requisite competence to perform the services.

Whether or not such a higher duty is acceptable will depend on many factors, including the nature of the project, the experience of the particular firm, any professional indemnity insurance restrictions and whether all other professionals, sub-consultants, sub-contractors, or others with design or other responsibilities that may affect the services, have undertaken an equivalent duty of care.

CIC imposes 'the Standard' on professionals. This is a higher duty and is defined as:

> The reasonable skill, care and diligence to be expected of a competent consultant of the relevant discipline undertaking the relevant role who is experienced in providing similar services in connection with projects of a similar size, scope and complexity to the Project.

However, CIC is intended for use by experienced clients and professionals undertaking major commercial property development projects and thus is different in a number of respects from those forms of appointments published by the ACE, RIBA and others. For such a contract a higher duty of care might well be acceptable.

It is important, however, that all the professionals working on the same project sign up to the same level of care, or the professional with the higher duty of care could be found to have been in breach of that higher duty when the other professionals are not in breach. An assurance from the client or his solicitors that this is the case should be sufficient unless the appointment contains an entire agreement clause – see Section 3.7.

It is difficult to understand why the RICS has imposed a duty on its members which is higher than and different from that imposed by the common law. Further, in the case of RIBA 2010 it is not clear whether their wording intends to say that all the other obligations in the agreement are to be performed with that duty of care, so that the architect will only be liable if he fails to exercise that duty of care. It is not thought this will be the case as the other obligations are drafted as stand-alone separate obligations, and such a qualification to what are otherwise strict obligations should be clearly stated.

Any provision implying a continuing obligation to exercise reasonable skill and care once the services have been completed should not be accepted without a full appreciation of what this might require – for example, an obligation to keep the design under review or warn of problems revealed by later knowledge. Be careful too of using statements that 'the professional is a specialist in . . .' The professional will then be judged as such and therefore against higher standards and not against the criteria for using reasonable skill and care.

Wherever possible, more onerous clauses than 'reasonable skill and care' should not be agreed to unless the risk in practice is minimal or other commercial considerations apply.

An appointment should contain no statement or obligation that the professional owes a duty of care to any third party. This can occur in collateral warranties or as a result of giving rights to third parties – see Chapter 21 – but there can also be separate statements in appointments, for example, that the professional owes a duty of care to another company in the same group as the client, or to a specified purchaser or tenant, etc., or even to a contractor. This could give rise to a duty in tort to that person. There is a significant difference between the obligations owed by a professional to his client under his contract and the obligations a professional may owe to a third party. The appointment governs (or should govern) the relationship between the client and the professional only and, as has been said, if the professional is in breach of his contractual duty of care, the client has a remedy which is an action for breach of contract.

It also needs to be understood that the professional will also owe his client a duty of care as a result of the law of tort and will be negligent if he fails to discharge that duty. The test is the same as for a breach of the contractual duty of care – the client would have to establish what the scope was of the duty owed to him, and that there has been a breach of that duty. He cannot bring a claim unless he has actually suffered a loss. In a professional engagement there is normally little or no distinction between the duty owed to a client in contract and that owed in tort, and the losses recoverable would be the same. The duty in tort is a Hedley Byrne-type duty – that is one 'akin to a contract' – see Section 14.4. The time limits for bringing such claims are different, however – see Section 16.6.

It is possible for the scope of the duty owed to the client in negligence to be greater than or different from that under the contract where the duty arises outside the contract. Say, for example, there was no obligation under the appointment to give advice on some aspect of waterproofing but the professional did so – the result could well be that the courts would decide that the professional had assumed a responsibility for giving advice about waterproofing and therefore would be liable to his client in negligence (although he would not be liable in contract) if that advice was not given carefully.

Similar principles can apply to third parties – a professional can during the course of a project assume a responsibility to them, if, for example, he gives advice voluntarily to a tenant about his proposed fit-out works or to a contractor about a method of construction. The scope of the duty assumed will depend on the circumstances of the case. A professional, however, does not owe duties in a vacuum. First, it is necessary to establish whether he owes a particular person a duty of care, which depends on whether the damage is foreseeable, whether there is a relationship of proximity between the parties and whether the imposition of the duty would be fair, just and reasonable. If the professional is found to owe a duty to a particular person, it will be his relationship to that person that determines the scope of his duty and thus the kind of loss for which he is liable.

The principle was emphasised by the House of Lords in the case of *South Australia Asset Management Corporation* v. *York Montague Limited (1997)*.

This was a case about valuers' negligence and what the measure of loss payable by the valuers should be. Lord Hoffman said:

> ... Before one can consider the principle on which one should calculate the damages to which a plaintiff is entitled as compensation for loss, it is necessary to decide for what kind of loss he is entitled to compensation.

He continued:

> A duty of care does not exist in the abstract. A plaintiff who sues for breach of a duty imposed by the law (whether in contract or tort or under statute) must do more than prove that the defendant has failed to comply. He must show that the duty was owed to him and that it was a duty in respect of the kind of loss which he has suffered.

He then quoted from another landmark case of the House of Lords on professional negligence, *Caparo* v. *Dickman*, which involved a claim against auditors.

> It is never sufficient to ask simply whether A owes B a duty of care. It is always necessary to determine the scope of the duty by reference to the kind of damage from which A must take care to save B harmless.

So if a third party can establish that the professional owed it a duty to take reasonable care, then unless the scope of the duty is determined to be that 'akin to a contract' or a special relationship exists as described in Section 14.4, and as a result there is a duty to prevent economic loss, the third party will be entitled to recover damages for personal injury or damage to third party property such as chattels and building contents only, and not economic losses. Economic losses are not normally recoverable in tort. Economic losses refers to the financial loss suffered by a claimant which does not flow from any damage to his own person or property. Repair or replacement loss in relation to a defective building is pure economic loss – only where the defect has caused damage to another item of property can liability be established on the basis that this was physical damage. The case of *Murphy* v. *Brentwood District Council (1991)* held that the cost of rectifying a defect in a building and the physical damage caused to the same building by that defect was economic loss – see also Section 14.3, Section 14.12 where the economic loss suffered was physical damage caused by flooding due to undersized drainage and Section 14.16 where the cost of stabilising the site was also classed as economic loss.It will be seen from this that the third parties to whom a professional may owe a duty of care and the scope of that duty are complex questions and in particular may or may not include a duty in respect of economic loss. So professionals should not accept a term in an appointment which states or agrees that the professional owes a general non-specific duty of care to third parties. A breach of such a term could result in the payment of substantial damages in addition to any damages payable to the client when such damages might not be payable at all or not in respect of economic losses. The courts would try to give some meaning to these words and in the absence of a description of any particular duty might interpret this as imposing a wide-ranging duty of care to the specified third party. In order to discharge such a duty, and thus not become liable to that third party for failure to exercise the appropriate level of skill and care, the professional would have to consider throughout the project the interests of that third party and how best as a professional to protect those interests and act accordingly. This could not only involve the professional in much additional work (for which there would be no fee unless this was provided for in the appointment) but could also put the professional in a position where the interests of the third party conflict with those of the client.

4.2. Warranties for fitness for purpose and warranties generally

'Fitness for purpose' requirements often arise in non-standard conditions of engagement and collateral warranties. They are important enough for detailed consideration.

Building contracts are governed by the Supply of Goods and Services Act 1982 (as amended), and there are also warranties which are implied into building contracts by common law. The warranties to be implied are as follows.

(*a*) The materials used in the works and the completed works themselves will be reasonably fit for the purpose for which they are required.
(*b*) The materials used will be of 'satisfactory quality' (the Sale and Supply of Goods Act 1994).
(*c*) The work will be carried out in a good and workmanlike manner.

A contractor under a design and build contract may take on a 'fitness for purpose' obligation in relation to design and seek to pass this on to professionals engaged directly to assist it, or attempt to impose a site supervisory role on the professionals that could involve the professional in considering the 'fitness for purpose' aspects of the contractor's work.

It should be noted that the ICE Design and Construct Conditions of Contract 2nd Edition 2001 and the JCT Design and Build Contract 2005 revision 2 2009 do not impose fitness for purpose in relation to design. The ACA and ACA/BPF forms do contain fitness for purpose obligations.

In the ICE form the design obligation is as follows.

8(2)(a) In carrying out all his design obligations under the Contract including those arising under sub-clause (2)(b) of this Clause (and including the selection of materials and plant to the extent that these are not specified in the Employer's Requirements) the Contractor shall exercise all reasonable skill care and diligence.

In the JCT form:

2.17.1 Insofar as its design of the Works is comprised in the Contractor's Proposals and in what the Contractor is to complete in accordance with the Employer's Requirements and these Conditions (including any further design required to be carried out by the Contractor as a result of a Change) the Contractor shall in respect of any inadequacy in such design have the like liability to the Employer, whether under statute or otherwise, as would an architect or, as the case may be, other appropriate professional designer holding himself out as competent to take on work for such design who, acting independently under a separate contract with the Employer, has supplied such design for or in connection with works to be carried out and completed by a building contractor who is not the supplier of the design.

If this obligation is undertaken by the contractor, there will be no problem for a professional in accepting a similar obligation provided 'the separate contract with the Employer' would be deemed to have included a duty of care obligation only. A warranty for fitness for purpose as to design should, however, be resisted.

In some cases these implied terms can be displaced. The Court of Appeal (*Rotherham MBC* v. *Frank Haslam (1996)*) has stated that, where a client has produced its own design, specification and arrangements for inspection and approval, a client cannot claim to be relying on a contractor's judgement. Of particular interest to designers is the statement that:

> . . . where the efficacy of a building or other object depends upon a designer it is the designer who may be expected to bear the responsibility for ensuring the suitability of the components incorporated into it . . . the designers of the building indisputably were . . . [the] architect and engineer

and that:

> the specification . . . coupled with the architect's entitlement to test, inspect and approve materials, denied the likelihood that the contractor had been relied upon.

In other words, the contract documents took away the contractor's fitness for purpose obligations in those circumstances.

The Court of Appeal considered 'fitness for purpose' and 'merchantability' (now 'satisfactory quality') separately. It considered that the material – in this case specified slag – was perfectly adequate (merchantable) for many applications but not for the application in question; that is, it was not fit for its intended purpose.

The court also said that the express terms as to 'fitness for purpose' agreed by a contractor in a contract create inescapable obligations. Where terms have to be implied, much depends on how detailed the client's requirements are in terms of design, specification and drawings. For a contractor's design and build contract the client's requirements are likely to be more in outline and the contractor's obligations correspondingly more onerous.

The Supply of Goods and Services Act does not require professionals (that is, those who provide services) to give such warranties. They are required to use reasonable skill and care.

If a professional gives a warranty that completed works will be fit for their purpose and they are not, liability will arise even if reasonable skill and care were used and therefore the professional has not been negligent. The damages flowing from a breach of warranty (which are the cost of making the works fit for their purpose) are different from, and can be higher than, those for negligence (which are the reasonably foreseeable loss caused by the negligence). Further, a 'state of the art' defence would not be available. A professional could not argue, for example, that a particular piece of knowledge was not available at the time the design was prepared. Say, as has happened, a professional specified high alumina cement at a time when its unsuitability was not known, the professional would not have been negligent in preparing his specification as he would be entitled in any claim to say that at the time of specification no competent professional in his field would have known that this material was not suitable – the state of the art defence. However, a contractor would be liable for breach of a warranty for fitness for purpose, notwithstanding the state of knowledge in the industry, because the material he had supplied would neither have been 'satisfactory' nor reasonably fit for the purpose for which it was intended.

The expression 'warrants' indicates that the professional is promising that a condition will be fulfilled, and it is in this context that it also appears in non-standard conditions of engagement. For example:

> The professional warrants that all work performed on this project related to its design will conform to the drawings and specifications and be operable in accordance with the professional's design intent;

and

> The professional warrants that all services performed hereunder by itself, its employees or agents, will be performed by persons who are extraordinarily skilled in their profession and in accordance with the highest standards of workmanship in their field.

Care must be taken to ensure that such warranties are only given where appropriate. A warranty can also be an undertaking as to the truth of a statement, for example, 'the professional warrants reasonable skill and care has been used'.

Some non-standard conditions of engagement express as warranties terms that should be expressed as obligations, and can do so in relation to obligations already performed and those going forward. Warranties are only necessary in collateral warranties or third party rights where statements are to be made about services, etc. that have been performed for other people, which services have usually taken place before the warranty has been entered into. An appointment should only contain obligations that will apply from the date the appointment is entered into, unless the parties agree the appointment is to have retrospective effect. If, however, some of the obligations have not been complied with from the beginning because, for example, the professional was unaware of them, the wording of those obligations should be amended to make it clear that they apply from a certain date, such as the date the appointment was entered into.

The risks attached to making statements, whether in an appointment, in reports or to third parties, are considered in Chapter 14.

As to what could constitute a warranty for fitness for purpose: if the professional warrants or agrees to procure that:

> Any part of the works designed pursuant to this Agreement shall be fit for the purposes described in the tender documents or the contract or the written requirements of the Client.

It is immediately apparent that this is a warranty for fitness for purpose because of the inclusion of the phrase 'fit for the purposes'. The professional would need to look at the tender documents or the written requirements to see what the purposes are. Some could be very detailed. If ultimately the works were not fit for the purposes described in those documents or in the written requirements of the client, there would then be a breach of that obligation by the professional. It would not matter that the professional's design had been done properly and carefully and therefore that there has been no negligence.

Say the appointment contained the following obligation:

> The Project Manager shall take the initiative to secure delivery of the Project in accordance with the Project Brief. In doing so, the Project Manager shall achieve the Purpose of the Project.

If the project manager does not achieve the 'Purpose of the Project', however that was defined, he would be liable even though he had not been negligent. The 'Purpose' could well be things that the project was to achieve and thus the obligation would amount to a warranty for fitness for purpose.

Another way in which a professional can inadvertently agree to a warranty for fitness for purpose is if the professional agrees to comply with requirements or obligations in other documents and these are sufficiently defined so that any failure to achieve them amounts to a breach of a warranty for fitness for purpose. Matters in the client's brief or the employer's requirements could impose warranties for fitness for purpose and if the professional in the appointment agrees, for example, to comply with the objectives in the employer's requirements, then it is likely that he has agreed to a warranty for fitness for purpose.

The standard conditions of engagement of the ACE, RIBA, RICS and in PSC3 do not include any warranties for fitness for purpose, nor any statements or avoidance of doubt provisions that no obligations as to fitness for purposes are included.

The reason for this is that the professional bodies are concerned that their standard forms reflect the common law duties of professionals and the professional indemnity insurance generally available to them.

The professional indemnity insurance arrangements of professionals usually reflect the obligation imposed at common law; that is, to exercise reasonable skill and care. Thus, insurance that is on a negligence-only basis would not cover warranties for fitness for purpose. Insurance on a legal or civil liability basis might cover such warranties because, once accepted, they become a legal liability. However, these types of policies often specifically exclude or limit recovery for breach of warranties for fitness for purpose, as the risk is considered too high. Giving such a warranty could therefore mean the professional is not insured or is not entitled to a full indemnity from his insurers. The risks must be very carefully assessed before a professional agrees to give one.

Most clients will not wish their professionals to undertake obligations that are outside their professional indemnity insurance as this could mean the professional has few if any assets to meet a claim from his client and will agree to amend the wording in the appointment to seek to avoid this. Frequently this is done by adding 'the Consultant will use reasonable skill and care to' before the warranty for fitness for purpose. There is a doubt, however, as to whether this will achieve the desired result, i.e. that the professional will only be liable if he has failed to use reasonable skill and care, as is explained later in this section.

Another approach is for the client to agree that 'Nothing in the Appointment imposes or seeks to impose on the Consultant any warranty for fitness for purpose' or 'Nothing in the Appointment imposes any greater obligation on the Consultant than that of using the degree of skill and care set out in clause [ ]'. Either of these provisions may assist the professional if a claim is brought for a breach of a warranty for fitness for purpose as the courts may interpret an obligation that amounts to a warranty for fitness for purpose as subject to a provision that appears to be an overall qualification applying to all obligations undertaken by the consultant, even though the two are inherently inconsistent. Whether, if the first wording is used, the warranty for fitness for purpose would then be construed as a strict obligation, or whether the test would be that the professional would only be liable if he had been negligent, is uncertain. The second wording might, however, be successful in making the test that of whether the professional has used the requisite skill and care – see comments on the Charles Haswell case below. Such an approach by the courts cannot be assumed, however, as there is at present no case law on an appointment which contained either such provision.

The case of *Costain Ltd* v. *Charles Haswell & Partners (2009)* gives some indication as to how the courts may construe strict obligations (which in this particular case probably also amounted to a warranty for fitness for purpose). (Note: strict obligations in law are treated the same as warranties for fitness for purpose and are outcome based, the test being 'has the object been achieved?', not 'has the professional exercised reasonable skill and care in carrying out the obligation?'.)

The case concerned the design of foundations for two structures undertaken by Haswell for Costain for a new water treatment works and a sludge treatment plant and water pumping station.

The Haswell consultancy agreement had a clause (7.2) saying the professional would exercise all reasonable professional skill, care and diligence.

But it also had a further clause:

> 7.4 Any part of the works designed pursuant to this Agreement if constructed in accordance with such design, shall meet the requirements described in the Specification or reasonably to be inferred from the Tender Documents or the Contract or the written requirements of Costain and designed in accordance with good up to date engineering practice and with all applicable laws, by laws, codes or mandatory regulations and in all respects with the requirements of the Contract.

Haswell tried to argue that clause 7.4 should be read as being subject to clause 7.2 (reasonable skill and care). The judge said that 'it seems to me quite plain that clause 7.4 is adding something different from clause 7.2 . . . clause 7.2 is a general provision relating to all the services by the consultant as a professional man . . . By contrast clause 7.4 is limited to one particular part of the consultant's obligations *viz* the design of the permanent works'.

His view was that clause 7.4 was 'expressed in mandatory terms and in my opinion imposes strict liability in contrast with clause 7.2'.

He found nothing wrong in principle with a consultant giving express warranties which impose strict liability 'or a performance obligation such as that the finished building will be reasonably fit for a specified purpose', that is a warranty for fitness for purpose.

He also dismissed Haswell's arguments that Costain had no good commercial reason to impose strict liability on Haswell. Clause 7.4 was sufficiently clear so as to prevail over any lack of conformity (if there was any) between Costain's obligations to the client and those in the consultancy agreement.

So Haswell had 'to in effect ensure that, if any part of the works is constructed in accordance with Haswell's design, that part will meet the requirements described in the Specification including the requirements in relation to permitted settlement tolerances'.

It is apparent from what the judge said that very clear words would have been needed in the appointment to make a separate strict obligation or warranty for fitness for purpose subject to an overall qualification contained in a clause which sets out the professional's duty of care and to make it clear that it was only in the case of a failure to exercise the requisite degree of skill and care that liability would arise.

For the reasons given above, it is the author's view that where an appointment contains warranties for fitness for purpose and an overall provision to the effect that nothing imposes on the professional a warranty for fitness for purpose or a greater obligation than to use reasonable skill and care, these qualifications may not be effective. The latter qualification, however, may stand a greater chance of success than the former.

If 'an avoidance of doubt' provision to the same effect was included, this might well not be thought to be sufficiently clear a qualification to the separate contractual obligation. It would also not help a judge in trying to construe the agreement if such a provision were included in an appointment that did not in fact contain any warranties for fitness for purpose.

A statement to the effect that the professional does not accept any obligations of fitness for purpose is a weaker approach, since it states what one of the parties intended, and if it were included in an appointment which also contained clear, unambiguous obligations (such as a warranty for fitness for purpose) it is possible that the obligation would stand, as this would not be a sufficiently clear qualification to that obligation.

Clients may well respond to the argument that if they insist on professionals signing up to warranties for fitness for purpose there is a danger there will be no or insufficient professional indemnity insurance to meet any claim by agreeing to remove or re-phrase the obligation so that it is not a warranty for fitness for purpose and states what the professional can in fact do.

Further, a board directive that no appointments can contain warranties for fitness for purpose can also be helpful when negotiating on warranties for fitness for purpose as any client who continues to insist on warranties for fitness for purpose will know they are going to need board approval, which they may not get (or not without considerable delay).

It is a commercial decision for the professional as to whether or not specific warranties for fitness for purpose are accepted or if the professional requires a qualification which could (but not certainly) be effective to remove the effects of having accepted a warranty for fitness for purpose.

The safest course is not to accept any obligation that could amount to a warranty for fitness for purpose. As has been said, all such obligations should be amended to say what the professional can and will do, properly qualified if necessary. Some could be properly expressed by saying that the professional will have 'due regard' to the brief, the client's requirements, etc. – see Section 3.2 – or that he will carry out the services 'with a view to' seeing that something is achieved.

4.3. Absolute or strict obligations

Non-standard conditions of engagement often contain absolute or strict obligations, for example

(*a*) the professional shall ensure the most efficient and cost-effective solution
(*b*) the professional shall ensure that the finished product meets/complies with the requirements of the Client in all aspects
(*c*) the professional shall comply with the client's requirements
(*d*) the professional's site staff shall procure that construction is in accordance with the plans and drawings.

The law may take a literal approach to such obligations – if the professional fails to comply with the obligation he will be in breach of contract regardless of whether or not he was negligent and/or was prevented from compliance for reasons beyond his control.

The simple rule to remember is that if a professional says he will do something then he has to do it. This is why obligations in appointments and the professional's services should be described in terms of what he can do with appropriate qualifications.

The effect of undertaking absolute or strict obligations is much the same in law as if a professional had signed a warranty for fitness for purpose – see Section 4.2.

'Ensure' means to guarantee and is an absolute performance obligation. 'Secure' has a similar meaning. These words were considered by the Court of Appeal (*John Mowlem & Company* v. *Eagle Star and Others (1995)*) who upheld the judge's view that the words 'ensure' and 'secure' in obligations to 'ensure the regular and diligent progress' of the works and to 'secure the completion of the same' in a management contract meant exactly what they said and could not be watered down to an obligation to do no more than use best endeavours. As a result, Mowlem was not entitled to an extension of time for delays that were caused by sub-contractors beyond Mowlem's control.

Other words and phrases can be used to create absolute or strict obligations, for example, 'comply' or 'procure'.

'Procure' imposes an obligation that the professional will ensure something happens. In the example above, the professional's site staff cannot 'procure' that construction is in accordance with the plans and drawings because this depends on the contractor and/or sub-contractors over whom the professional has no control. There is no direct contract with them, nor are they the professional's employees. Such an obligation should not be undertaken.

The same is true of promises to 'comply' with something. This should be undertaken only if the professional is certain that compliance is possible, having checked the documents in which such a requirement could be contained if necessary.

It may well be the case that compliance will depend on other parties such as the contractor or other consultants carrying out their obligations, and/or on factors outside the professional's control such as the obtaining of planning permission or the consent of a third party. Also, the documents referred to, such as the employer's requirements or the client's brief, could contain specific requirements which are to be achieved and so, by accepting an obligation to comply with them, the professional could also be signing up to a warranty for fitness for purpose.

None of this means the professional can disregard the client's brief, requirements or budget, etc. He has to have regard to these as part of his professional duties. There is no problem, for example, in agreeing to have 'due regard' to various matters or requirements, provided they are relevant to the services and the professional does in fact take account of them when preparing the design or carrying out the services – see Section 3.2. Such obligations could also be qualified by 'use reasonable endeavours to', provided this is in relation to something the professional can do.

Duties in relation to the work of the contractor can give rise to similar problems. A duty to assess, for example, whether the contractor's proposed materials are 'a satisfactory equal' to those specified is an unnecessary absolute – the criterion should be 'satisfactory equivalent'. A technical understanding of what is 'equal' might be that two types of material possess similar performance characteristics and can fulfil the same function for the required length of time, but with no inference of being identical in all respects. However, 'equal' could be interpreted as meaning material of exactly the same size, characteristics and so on without any difference at all. The word 'equivalent' does not give rise to similar confusion.

Sometimes obligations to 'prove', 'show' or 'ensure' something are linked to an obligation to investigate and/or obtain samples. A professional can never be certain that investigations and sampling are indicative of the 'whole' and thus no agreement should be given to any clause in a professional appointment that suggests otherwise, without adding an appropriate caveat.

Sometimes non-standard conditions of engagement will qualify strict or absolute obligations by reference to the duty of care, for example, 'the professional will use reasonable skill and care to ensure that . . .'. It will depend on the wording in each case as to whether the professional will only be held by a court to be liable if he has failed to use reasonable skill and care, or whether the court will decide the professional undertook to ensure something happened. In the latter case, the professional could be in breach even if what is intended by the parties is that the professional would only be liable if, and to the extent that, negligence by the professional had arisen – see discussion of this in Section 4.2.

CIC, for example, deals with the brief by introducing this qualification:

> 3–2 In performing the Services and within the scope of the Consultant's role, the Consultant shall exercise the Standard (i.e. the higher duty of skill and care) to see that:

3–2.1 the Client Brief (as updated by agreement between the parties from time to time) is complied with, provided that this shall not preclude the exercise of skill and judgement by the Consultant;
3–2.2 the Project Brief is consistent with the Client Brief (as updated by agreement between the parties from time to time); and
3–2.3 the Services are performed in accordance with the Project Brief.

The words 'to see that' may well create a strict obligation.

As has been said, the best way to deal with warranties for fitness for purpose and strict or absolute obligations is to amend them to say what the professional can do and will do, properly qualified if necessary – see last paragraph of Section 4.2.

The ACE Agreements do not impose strict obligations on professionals.

RIBA 2010 has no provisions that could amount to warranties for fitness for purpose but it is not clear whether by saying in clause 2.1 that the professional shall use the requisite skill and care 'in performing the Services and discharging all the obligations under this clause 2' the professional is only to be liable if he has failed to do so or whether these are separate obligations which have to be discharged in addition to using reasonable skill and care. The strict obligations in the RIBA 2010, however, do not cause the same problems as those in the 2007 Conditions.

RICS 2011 has a number of strict obligations such as 'The Consultant complies with the Client's Brief' and 'The Consultant complies with the Statutory Requirements' but in the case of the former it is qualified by 'exercising the Requisite Standard'. This raises the same question: is the professional only to be liable if he fails through negligence to comply with the brief but any failure to comply with the statutory requirements will result in strict liability?

PSC3 probably does contain some strict obligations. These are more difficult to identify because PSC3 is written in the present tense. The wording 'The Consultant provides' must surely mean he is obliged to do so. Does 'The Consultant . . . does the work so that the Condition stated for each Key Date is met by the Key Date' impose a strict obligation? As PSC3 also uses 'may', it is assumed that these other provisions are strict obligations. Most of these obligations relate to the procedures for communication, early warnings, cooperation, holding/attending meetings, and operating quality management systems and notices.

The obligations in professional appointments should be expressed clearly to avoid such ambiguities.

The dangers of professionals taking on absolute obligations, or indeed any obligation more onerous than to exercise the reasonable skill and care expected of their profession, is well illustrated by a recent case in the Court of Appeal (*Platform Funding Limited* v. *Bank of Scotland (2008)*). A firm of surveyors received standard instructions to value a property for the purpose of a loan. The surveyor carried out the survey, prepared a report in the usual way, and a loan was made secured by a mortgage on the property. The borrower, however, defaulted and the property was sold by the lenders for less than the sum they had advanced. The lenders sued the surveyors for the unrecovered balance.

The case against the surveyor was not that he had failed to carry out the valuation with reasonable skill and care or that he had been negligent in any other way. The simple fact was that the surveyor had surveyed the wrong property. It had been to the right street, but the houses were not numbered. He had got in touch with the borrower, who had directed him to and shown him around a house on the road, but unknown to the surveyor it was not the one that he had been instructed to survey. The one that he was supposed to be surveying was, in fact, only half built.

Whether or not the surveyor could be blamed for this situation is, for present purposes, beside the point, because no allegation was made that he had been negligent. The case was quite simple – the surveyor had undertaken to survey a particular property and had not done so; he was therefore in breach of contract. The question for the court was simply this – was the obligation to undertake the survey an absolute obligation? If it was, there would be no defence.

On this question – of whether the obligation was absolute – the Court of Appeal, consisting of three judges, was divided. All three agreed that the valuation itself was required to be carried out subject to the obligation to exercise reasonable skill and care, that this was the normal obligation for a professional person and that clear words or circumstances would be necessary in order to conclude that there was a stricter obligation. However, the majority distinguished between the actual obligation to do the survey and the manner in which the survey was done, and held that the obligation to do the survey was an absolute obligation. The surveyor having failed to survey the right property was therefore found liable. The third judge, however, held that the obligation to exercise reasonable skill and care applied to everything that the surveyor did, which included locating the correct property.

The circumstances of this case were, of course, unusual, although similar situations have been before the courts. However, the general point that arises from these cases is that the professionals in question, by accepting (though probably without realising it) an absolute obligation, were thereby accepting a risk which no one would think should have fallen on them, unless they had actually been negligent in failing to prevent that risk from eventuating.

4.4. Compliance with statutory requirements and the introduction of Eurocodes

An obligation to comply with statutory requirements is a strict obligation and should not be undertaken for the reasons given in the preceding section. Many bespoke appointments widen this obligation to cover all legislation, statutory instruments, codes of practice and British Standards. More recent versions can also refer to Eurocodes.

An obligation expressed in this way also ignores the fact that a professional may, exercising his professional judgement, decide to use one and not the other of conflicting standards or provide a solution which is the technical equivalent of a particular standard.

There is no need for a separate obligation in relation to compliance with statutory and other requirements, as consideration and adoption of such requirements would be subject to the duty to use reasonable skill and care in any event. If, however, an express obligation is accepted then it ought to be expressed as follows:

> The professional will use reasonable skill and care in complying with the relevant [standards].

The Eurocodes were adopted on 1 April 2010 to replace British structural design standards and also civil design standards. This has resulted in two particular concerns. The first is whether the structural Eurocodes could impose a warranty for fitness for purpose and the second is the practical problems that have resulted from their introduction.

The wording in EN 1990 (basis of structural design) published in 2002 as later amended states:

> A structure shall be designed and executed in such a way that it will, during its intended life, with appropriate degrees of reliability and in an economic way, sustain all actions and influences likely to occur during execution and use and meet the specified serviceability requirements of a structure or structural element.

It is thought this could impose an obligation equivalent to a warranty for fitness for purpose on designers, because although there is a separate requirement to use reasonable skill and care this is not expressed as an overriding provision; in fact it is the quoted provision that is expressed to have precedence.

At the time of writing it is thought that a need for a change is recognised but this could well take some years.

As to their implementation, they are to be used on all publicly funded projects above the minimum financial thresholds (or a technical equivalent is to be used). It is thought that there is a good knowledge of Eurocodes in the civil sector but an apparent reluctance to use them otherwise. This could be because of cost, a lack of training/education, the lack of promotion and the fact that they will not be incorporated into building regulations until 2013 at the earliest. Some errors have been found in them and it is anticipated more will be discovered as they become more widely used. (If a professional is aware of these errors he clearly should not be following Eurocodes even if there is an obligation to do so.)

Working to British Standards will become a problem as these have been withdrawn and the BSI will no longer update them. By 2013 all British Standards will be outside the five-year revision cycle. Finally, it is not intended at present that the Eurocodes will be maintained.

The current suite of Eurocodes will be maintained until 2013, but there will be no fundamental changes. There will then be a process of review, development and updating, following which it is anticipate that a new generation of Eurocodes will be published from 2015. Until 2013, therefore, it would be prudent for the client and the construction team to clarify from the outset which standards are to apply.

4.5. Compliance with third party agreements and 'back to back' clauses

Clients will often enter into agreements with funders, purchasers, tenants and other people with interests in the proposed development, and these agreements may well impose on the client wide obligations in relation to the development, such as meeting the conditions for the planning consent, how the development itself is to be carried out, what it is to achieve, how the fit-out works are to be done, timetables for completion of some or all of it and possibly penalties in the form of liquidated damages or other payments if the objectives are not met. Clients will as a result want to ensure that their professionals and the contractor will carry out their services and the works in such a way that all these requirements are met and the client does not face claims from these third parties. However, instead of working out what is required from the construction team to achieve this and putting this into their respective appointments, clients often refer to these third party agreements and then require the professional not to do anything which will put them in breach of the client's obligations under those agreements. If phrased in these terms, this is a strict obligation of which the professional could be in breach even if he has not been negligent or for reasons outside his control. For example, he may need information from others before he can meet a timetable for a certain aspect of his design which his client has agreed to give a tenant by a certain date, and he gets that information too late to be able to do so.

Even if the wording is amended so that the professional is relieved of the obligation to comply if he is prevented from doing so for reasons outside his control, or it is only negligent acts or omissions of his that can give rise to liability under these sorts of clauses, these provisions give rise to considerable problems.

First, the third party agreements can be very widely drawn and some may not be in existence at the time the appointment is entered into, for example, agreements that the client may enter into with purchasers/tenants in the future may be within the definition of 'third party agreements'. It is necessary to split them into two groups – those that exist at the time of the appointment and those that may be entered into later. In respect of the first group, the professional will be able to look at them and judge whether or not he will be able to meet those of the client's obligations which affect his services and/or what effect such compliance may have on his services and/or the fee. The professional and the client will then be able to discuss and agree what alterations if any are needed to the obligations in the appointment, the fee or the services.

It is obviously not possible to do this in respect of third party agreements that are not yet in existence and many non-standard conditions of engagement fail to make this distinction. The professional cannot say in advance whether or not he will be able to comply with them if he does not know what the client has signed up to. The time for compliance may have already passed by the time the professional sees it, and if so he would already be in breach. Also, without knowing what the obligations are, the professional does not know what effect complying with them will have on the services and/or the fee. The most, therefore, that can be said in relation to these later agreements is that the professional will have due regard to them, which will take into account the time at which he was given them, but this will not help with any significant effect this obligation may have on the professional's services or the fee, unless the appointment also provides for that – see also Section 3.2.

If a professional is told of commitments the client has entered into and/or is given documents to enable him to take these commitments into account, it is likely that the professional's general duty to take reasonable skill and care will extend to having proper regard to these matters when performing the services.

The ACE Agreements, RIBA 2010 and PSC3 have no provisions concerning third party agreements.

RICS 2011 has the following:

> 3.6 The Consultant performs the Services exercising the Requisite Standard so that no breach by the Consultant of this Appointment causes or contributes to any breach by the Client of any term of the Building Contract and/or any Client Contract that has been notified to the Consultant by the Client, unless the Consultant is prevented from doing so by circumstances outside its reasonable control. Provided that:
>
> (*a*) the Consultant is not required to comply with this Clause to the extent that any term of the Building Contract and/or any Client Contract notified to the Consultant after the date of this Appointment exposes the Consultant to any liability in excess of that which is reasonably foreseeable by the Consultant at the date of this Appointment; and
>
> (*b*) (subject to clause 3.6(a)) where any term of the Building Contract and/or any Client Contract notified to the Consultant after the date of this Appointment requires the Consultant to perform Additional Services in order to comply with this Clause, the Consultant is entitled to an Additional Services Fee in accordance with this Appointment.

This contains a strict obligation to perform the services so that no breach of the professional puts the client in breach of terms in other agreements, qualified by a reference to 'the Requisite Standard', for the effect of which see Section 4.2. It is also qualified by 'unless the Consultant is prevented from doing so by circumstances outside its reasonable control'.

This may be enough to relieve the professional from strict compliance – this will depend on the nature of the terms with which the client has to comply and whether something outside the professional's control is the only reason why the professional could fail to comply with this obligation which may not be the case. 'Client Contracts' are 'the particulars of all contracts between the Client and third parties, which the Client notifies to the Consultant, relating to the financing or re-financing of the Project, the disposal and/or letting of any interest in the Project, the provision of any enabling works for the Project, or the giving of consent or permission for the carrying out of the Project'. This is very wide indeed and could result in the professional having to read and consider the client's obligations in a large number of very lengthy documents.

The way the RICS deals with the time at which the professional is given the agreements is by providing that where they have been notified to the professional after the date of the appointment, the professional is not required to comply with them if it 'exposes the Consultant to any liability in excess of that which is reasonably foreseeable by the Consultant at the date of the Appointment'. (CIC has a similar approach to the professional's obligations in relation to third party agreements.) This could be something that is very difficult to demonstrate. Most professionals will be aware of the general type of obligations the client could enter into with funders, purchaser and tenants, etc., as well as other professionals and contractors, and therefore in broad terms be aware of the sort of the liability the client could incur if they are not met. All these would, therefore, be 'reasonably foreseeable'. This provision might therefore only catch exceptional liabilities, for example, if the client had to pay a purchaser a substantial amount of liquidated damages if the development was completed late.

The RICS recognises that compliance with these client obligations could cause the professional to incur additional fees (sub-clause (b)), but the definition of 'Additional Services' as 'any services relating to the Project which are not identified as Basic Services' would restrict payment to where these services are in fact additional and not, for example, if some element of design or other work (i.e. an existing service) had to be re-done or modified unless that was specifically excluded from 'Basic Services'.

Many appointments anticipate that a design and build contractor will carry out the works and that the professional will be novated to the contractor. If so, obligations to comply with third party agreements need further consideration.

First, once novated the professional is employed by the contractor and he will no longer be in a position where he can continue to perform his services so as to comply with the client's obligations under these agreements. His obligations will be owed to the contractor and the contractor's interests may conflict with the client's. The professional may as a result be unable to or instructed not to perform his services so that such obligations are met. Thus, this obligation should not continue post-novation.

Secondly, it is sometimes provided that post-novation the professional will continue to have an obligation not to put the client in breach of his obligations in relation to the building contract. If under the appointment terms 'the client' is now 'the contractor', such an obligation will now relate to not putting the contractor in breach of his obligations to the client under the building contract.

The obligations of the contractor to the client under the building contract are very different obligations from those of the client and could also include warranties for fitness for purpose. If a professional accepts an obligation not to put the contractor in breach of the terms of the building contract, he could find he has signed up to warranties for fitness for purpose which could take him outside his professional indemnity insurance – see Section 4.2. All the other points made above about whether the professional can undertake such an obligation apply but with more force, bearing in mind that the building contract is unlikely to be in existence at the time the appointment is signed.

The clauses which contain obligations not to put someone in breach of obligations in another contract are often included in what are called 'back to back' clauses. There they can be coupled with an obligation to take on all the obligations in that other contract which relates to the professional's services. Their objective is to ensure that all the obligations under the first contract are passed on in exactly the same form to the professional so that there is no 'gap' between the obligations undertaken by the first party and those which are being passed to the professional.

Such clauses are also found in sub-consultancy agreements and this is discussed further in Sections 12.4 and 12.5.

4.6. Compliance with the client's instructions

Non-standard conditions of engagement can include a strict obligation that the professional complies with the client's instructions. As this is a strict obligation the professional will be in breach if he does not comply, but what if the instruction is to do something which in the professional's judgement is unwise or unsafe? What if compliance means the professional has to carry out his services in a totally different way to that which he envisaged, which is very disruptive or results in him incurring additional costs that he cannot recover?

Why is such an obligation needed at all? After all, professionals are not likely to disregard their clients' instructions if they are proper instructions.

ACE Agreement 1 has no such obligation.

RIBA 2010 provides:

> 3.5 The Client (or the Lead Consultant, or other consultant designated by the Client) may issue reasonable instructions to the Architect.

This is acceptable and most clients will alter the obligation to 'reasonable instructions' once the difficulty of accepting strict compliance with instructions is pointed out. This would then make it a fair objective test.

Some clients and the RICS have opted for 'lawful' instructions. RICS 2011 wording is:

> 3.9 The Consultant complies, so far as reasonable practicable, with all lawful instructions given to it by the Client . . .

Although 'lawful' would mean a professional would not have to comply with an instruction which would involve the professional breaking health and safety law or other statutory requirements, it would not relieve the professional from compliance with an instruction which has serious commercial effects for the professional or was an instruction to do something about which the professional had serious concerns. The 'as far as reasonably practicable' is of some limited assistance but query whether this would cover the situation where the instruction could be carried out by the professional but at considerable cost.

What about the situation where the professional has to take a view and act independently of the client, for example, when he is the architect or contract administrator under one of the JCT forms of building contract or the engineer under the ICE Conditions of Contract or the employer's representative? The client may be entitled to give an instruction or even a lawful instruction which directs the professional to do something which would mean he is not acting independently but at the client's direction. If so, the professional would be in breach if he disregarded that instruction but he would also be in breach of his duty to act independently. If the client's instruction had to be 'reasonable' he could refuse.

RICS 2011 has recognised this and added to clause 3.9:

> . . . provided that, where it (the consultant) is required to exercise discretion in assessing matters between the Client and the Contractor, it does so fairly and impartially.

RIBA 2010 does not mention this situation.

ACE Agreement 1 says:

> F2.7 If in the performance of the Services the Consultant has a discretion exercisable as between the Client and a Contractor or Sub-Contractor the Consultant shall exercise that discretion fairly.

PSC3 contains several instances where the employer (or indeed the consultant) can instruct the other to do things – such as attend risk-reduction meetings.

The general provisions in PSC3 about client instructions are as follows:

> 20.2 The *Employer* may give an instruction to the *Consultant* which changes the Scope or a Key Date. After Completion, an instruction is given only if it is necessary to Provide the Services.

20.3 The *Employer* does not give an instruction to the *Consultant* which would require him to act in a way that was outside his professional code of conduct.

There are therefore several limits to the sort of instruction the client can give to the professional under PSC3. These are comparable with the sort of instructions a client can give to a contractor under a construction contract. The 'Scope' is basically the services the professional is to provide and 'Completion' is when the professional has done all the work which the 'Scope' states he is to do and has corrected 'Defects' or if this is not in the 'Scope', when the professional has done all the work necessary for the client to use the services and for others to do their work. It is difficult to envisage when an instruction 'necessary to Provide the Services' can be given after completion.

An instruction to change the 'Scope' and to change a Key Date are compensation events which entitle the professional to additional time and/or money. The proviso which allows the professional not to comply with an instruction is slightly oddly worded. The codes of conduct produced by professional institutions do not always deal with matters concerning professional judgement and what may or may not constitute negligence in any particular circumstance. It might have been expected to have referred to 'professional judgement'.

4.7. Best endeavours/reasonable endeavours

'Best endeavours' is an onerous obligation.

The court (*Sheffield District Railway Company* v. *Great Central Railway Company (1911)*) has described 'best endeavours' as 'broadly speaking [to] leave no stone unturned' but that 'the limits of reason must (not) be overstepped with regard to the cost . . . ' In other words, considerable effort including the expenditure of money, whether recoverable from others or not, must be expended.

The court *(IBM UK Ltd* v. *Rockware Glass Ltd (1980)*) has also stated that 'best endeavours' required non-owner land purchasers to 'take all those steps in their power which are capable of providing the desired results . . . being steps which a prudent, determined and reasonable owner, acting in his own interests and desiring to achieve that result, would take'.

More recently the courts (*Midland Land Reclamation Limited and Leicestershire County Council* v. *Warren Energy Limited (1997)* unreported) have confirmed that an obligation to use best endeavours imposes an obligation to do what can reasonably be done in the circumstances. Thus, a company should take action which, having regard to costs and the degree of difficulty, is commercially practicable. It is not required to take action which would lead to its financial ruin or undermine its commercial standing and goodwill. It must also incur such expenditure as is reasonable in taking such action and act in the interests of the company.

It has also been held that an obligation to use best endeavours may impose an obligation to litigate or appeal a decision, although not if the action was doomed to failure or was unreasonable in all the circumstances. It can be overridden by other duties – for example, an obligation or direction to use best endeavours to pass a resolution did not require the defendant to give bad advice to shareholders nor to continue to recommend the resolution once it ceased to be in the company's interests.

The case of *Rhodia International Holdings Limited and Another* v. *Huntsman International LLP (2007)* said 'best endeavours' probably requires a party to exhaust all of a number of reasonable courses which could be taken in a given situation to achieve a particular aim.

Professionals, therefore, should be careful to assess what might be involved in terms of time, additional costs, and the actions that might have to be taken before undertaking an obligation to use 'best endeavours' in relation to a particular activity.

What of 'reasonable endeavours', and is there a distinction between this and 'all reasonable endeavours'? The former involves the person under the obligation to use reasonable endeavours to balance 'the weight of their contractual obligation' to the other party against 'all relevant commercial considerations' such as the obligations he may have to third parties, its reputation and the cost of the relevant course of action. These considerations are primarily subjective, based on the circumstances of the person who is under the obligation, and he is not required to sacrifice his commercial interests. The Rhodia case said that 'reasonable endeavours' probably requires the party to take only one reasonable course in a given situation to achieve a particular aim.

The courts have said 'all reasonable endeavours' is probably a middle position, somewhere between the two, implying something more than reasonable endeavours but less than best endeavours. It has been likened to 'do your best'. The Rhodia case, however, equated 'all reasonable endeavours' with 'best endeavours'. Both these obligations could also require the expenditure of money but more so in the case of 'best endeavours'.

In *Jet2.com Limited* v. *Blackpool Airport Limited (2011)* the commercial court considered an 'endeavours' clause in a contract between the operators of an airport and a low-cost airline. The clause read:

> Jet2.com and BAL will co-operate together and use their best endeavours to promote Jet2.com's low cost service from BA (i.e. Blackpool Airport) and BAL will use all reasonable endeavours to provide a cost base that will facilitate Jet2.com's low cost pricing.

In this particular case, there was clearly considered to be a potentially significant difference between 'reasonable endeavours' and 'all reasonable endeavours' and the court noted that the latter was closer to 'best endeavours', the most stringent of the three, than to 'reasonable endeavours', the least stringent. 'All reasonable endeavours' in this particular context, required the airport to act, in effect, against its commercial interest, by allowing the airline to use the airport outside its normal operating hours, at a time when the airport was making substantial operating losses.

It is likely that any consideration of whether any 'endeavours' clause has been complied with will depend on the particular facts. It would be prudent, though, if a professional who is under such an obligation kept a record of the steps he had taken to try to discharge it.

4.8. Timescale for professional services

If nothing is said in an appointment about the time for performing services, the law implies a term that they will be carried out within a reasonable time.

Non-standard conditions of engagement may include a specific requirement (that is, a strict or absolute obligation – see Section 4.3) for the completion of professional services within a certain timescale. This should only be agreed to where it can definitely be met, as a failure to comply can lead to a client (who could be a contractor) incurring losses which are claimable from the professional. This means that the services required of the professional must be precise, that resources can be guaranteed and that there is no liability for the actions (e.g., late supply of information, gaining approvals, etc.) of either the client itself or third parties which may not be under the direct control of the professional. An example of when an unqualified obligation to meet a timetable can be accepted is where the professional agrees to provide the client with a report by a certain date. Provided meeting this is within his control and dependent on his own resources and is not dependent on obtaining information from others the professional should be able to meet this obligation. If the professional fails to do so, he will be in breach of this obligation and liable to the client in damages. Where the professional's ability to comply with the programme or fixed timescale is dependent on others or matters beyond his control, the obligation should be appropriately qualified.

Where the professional has direct control of a third party, for example, a sub-consultant, the professional will be responsible to the client for any default and delays caused by that third party. It may be possible to pass on claims from the client to the third party, particularly if there is a proper appointment in place. However, the client's claim will be against the professional, which will inevitably involve it in additional time and probably not wholly recoverable costs. Delays can be caused by a failure to perform when required, or by other events (e.g., errors in drawings).

A client's late delivery of information (whether this is to be supplied by the client or the client's other professionals or contractors), instructions or approvals to a professional can result in delays to the professional's timetable. Generally, it will be implied into a contract that such matters will be provided in a reasonable time and so as not to delay the professional. It is better, however, that this situation is dealt with expressly in the professional services agreement.

Extracts from the standard conditions of engagement dealing with the client's obligation to give decisions, information, etc. are set out in Section 6.1.

If the obligation to comply with a programme is to be included in the appointment, the appointment should also cover all the matters which may delay the professional (and thus relieve the professional of this obligation) and state, if appropriate, how they should be dealt with. Formal provisions for claiming extensions of time are becoming more usual in non-standard professional services agreements. PSC3 has such provisions. They need to be considered very carefully – do they cover all the possible circumstances which may arise, do they give a right to additional costs (and, if so, who decides those costs) as well as time, and can the decision later be opened up and reviewed? What are the consequences if there is failure to meet the required date or dates? If the professional has to pay liquidated damages, this is probably outside his professional indemnity insurance – see Section 15.6. PSC3 provides for certain events to be 'compensation events', giving an entitlement to additional time and money if applicable, and an option clause can be included whereby liquidated damages will be payable if there is a failure to meet the completion date – see also Chapter 9.

ACE Agreement 1 sets out the professional's duty in relation to the programme as follows:

> F2.5 All requests to the Client by the Consultant for information, assistance or decisions shall be made in a timely fashion. Subject always to conditions beyond the Consultant's reasonable control (including acts or omissions of the Client, any Lead Consultant if the Consultant is not so appointed, any Other Consultants or third parties), the Consultant shall use reasonable endeavours to perform the Services in accordance with the programme set out in Part D: *The Programme for the Services* or with any programme agreed with the Consultant from time to time.

RIBA 2010 does not have any specific obligation in relation to the timetable.

RICS 2011 provides:

> 3.7 The Consultant performs the Services and provides the Documents so as to comply with the Programme, unless the Consultant is prevented from doing so by circumstances outside its reasonable control.

Most clients accept that professionals cannot comply with a strict obligation in relation to the programme, but the qualifications that may be attached to such an obligation are not always as general as 'beyond its reasonable control' and cannot be accepted unless they do in fact cover all the eventualities that may prevent the professional from complying.

4.9. Acceleration of the services

Non-standard conditions of engagement can include an obligation on the professional to accelerate his services if so required by the client. Often this will carry no right to additional payment and therefore this would have to be at the professional's expense. This may be acceptable if the professional has fallen behind the programme through his own default, but not if this is due to others or events outside his control. The professional should be entitled to payment in such circumstances.

Sometimes the client has the right to 'decide' whether or not acceleration is due to the professional's default – this should be subject to a right to challenge the decision if it is wrong, or the decision should be 'reasonable' so that it can be challenged if it is not.

Sometimes there are provisions whereby the client can ask the professional for an estimate of costs and for other information where he is considering a genuine acceleration of the works, for example, to bring forward practical completion. There should be a formal timetable for this and clarity as to the basis on which the client and the professional will go forward if agreed. If the professional is to be entitled to the costs of providing the estimates and information whether or not there is acceleration, this should also be made clear.

The standard conditions of engagement do not deal with acceleration, and any requests for acceleration under those forms of contract would therefore be subject to the professional's agreement.

4.10. Statements concerning reliance

A client will sometimes insert a statement into an appointment that it has or will rely or exclusively rely on the professional in respect of the project or any other matters.

Such wording is not needed in order to establish a duty of care because the appointment already does this. Its origins lie in the law of tort where reliance on a third party needs to be established as part of determining whether or not a duty of care in tort is owed by that third party.

It can also carry a risk to the professional because it could mean that the client or client's staff or other parties engaged by the client (e.g., to check the professional's work) do not have to take responsibility for their own decisions or for information provided, or that the professional will remain responsible for a particular aspect of the services notwithstanding that the client has taken separate professional advice on this matter and has relied on that advice and not on advice from the professional.

The responsibility for decisions, information and advice in these circumstances should be determined by reference to the facts at the time – the professional should not agree in advance that reliance in all cases or for all matters has been placed on the professional, or it would not be possible later to argue that it has not.

If such statements cannot be deleted it might be possible to amend them so that 'reliance in respect of the project' is substituted by 'reliance only in respect of those matters for which the professional is responsible and to the extent of that responsibility'. The same principles apply if the appointment contains any statement that third parties will rely on the professional. This should be deleted as again it is trying to put in place a relationship between the professional and the third party where the third party will be able to sue the professional in tort.

The principles for this are discussed more fully in Section 4.1 and Chapter 14.

4.11. Deleterious/prohibited materials

Deleterious/prohibited materials clauses are now almost always included in non-standard professional services agreements. A professional is usually required to state that certain listed materials will not form part of the specification. A typical clause (as in CIC) is as follows:

> 6–1.2 The Consultant shall exercise the Standard [i.e. the higher duty of skill and care] to see that, unless authorised by the Client in writing (or, where such authorisation is given orally, confirmed by the Consultant to the Client in writing), it shall not specify or authorise for use in the Project any materials which are generally known within the Consultant's profession at the time of specification or authorisation (as applicable) to be deleterious or hazardous to health or safety or the durability of the Project in the particular circumstances in which they are proposed to be used.
>
> 6–1.3 In complying with §6–1.2 above, the Consultant shall have due regard (where applicable) to the guidelines contained in the edition of the publication *Good practice in the selection of construction materials* (Ove Arup & Partners) current at the time of specification or authorisation.

[Note: the Ove Arup publication has now been replaced by the British Council for Offices' publication *Good practice in the selection of construction materials 2011*.]

A professional already has a duty 'to exercise reasonable skill and care', and this will extend to the specifying of materials, so such a clause is not necessary. It is not included in the ACE Agreements.

RIBA 2010 refers in the Schedule of Role Specifications to designers having 'due regard' to the Ove Arup publication (Sheehan and Ove Arup and Partners, 1997) but adds 'unless authorised by the Client in writing to specify or authorise products or materials generally known within the Designer's profession at the time of specification or authorisation to be deleterious or hazardous to health and safety or to affect the durability of the construction works'. The drafting appears to have gone awry – it is thought this wording is intending to state that the professional should not specify such products unless authorised in writing.

RICS 2011 says in relation to deleterious materials:

> 4.5 Where the Appendix states that this clause is applicable, the Consultant exercising the Requisite Standard:

(*a*) does not specify for use in connection with the Project any materials which by their nature or application contravene any British Standard or EU equivalent current at the time of specification or which are otherwise generally known within the construction and engineering industry at the time of specification to be harmful to health and safety or to the durability of the works in the particular circumstances in which they are specified for use; and

(*b*) insofar as is reasonably practicable and having regard to the nature and extent of the Services sees that such materials are not used in the construction of those parts of the Project to which the Services relate.

This is another example of a strict obligation qualified by the reference to the duty of care – see Sections 4.2 and 4.3. It is wider than many such clauses as it refers to 'any British Standard or EU equivalent' but it is confined to the time of specification.

The obligation to see that such materials are not used during construction relates to 'the extent of the Services' which it is assumed is a reference to the duties of the professional on site (but does not deal with the situation where he may not visit the site at all) and confines it to the parts of the project to which the services relate.

A professional may, therefore, be prepared to accept an express obligation in relation to deleterious materials, provided the description of the materials (which can be set out in a list) is acceptable (i.e. they are materials he does not wish to specify) and provided the obligation is properly drafted. Some are very widely drawn. For example, references prohibiting the specification of materials 'generally known to be deleterious' are unspecific and could be open to argument. The obligation should also be qualified by stating the time that such knowledge is relevant (that is, at the time of specification) and to whom this was known (that is, the professional concerned). If the time of knowledge is the time of 'use', this would mean the professional will have to check his specifications, etc. at the time the material or product is being used on site, which could be a long time after he has prepared the specification and/or at a time when he has no duties in relation to the project.

Phrases such as 'materials used will comply with all applicable codes' or 'British Standards' are often inappropriate, because not every building material is covered by a British Standard and different national standards might conflict. If a reference to a British (or any other) Standard remains, it should refer to the Standard relevant at the time the specification was prepared. See also Section 4.4 on Eurocodes.

The obligation is often extended so that it reads 'ensuring' or 'seeing that' such materials are not used in the construction of the project. Such an obligation cannot be undertaken where there is no duty of supervision or inspection. Even if there is such a duty and even if the obligation is limited to those parts of the project with which the professional is concerned, it should still not be accepted without amendment. It is an absolute obligation and therefore creates risks for the professional – see Section 4.3. A professional simply will not be able to ensure no such materials are being used and in any event it is the contractor's obligation to supply materials of satisfactory quality and suitable for their purpose – see Section 4.2. A professional may be able (depending on the nature of appointment) to undertake that the client will be notified immediately if the professional becomes aware that any deleterious/prohibited materials are being used.

An additional problem can arise where deleterious/prohibited materials have not been referred to or listed in the appointment between the client and the professional, but are mentioned for the first time in a collateral warranty. It would be unfair to expect a professional to warrant that certain materials have not been included in any specification when such prohibitions were not notified to him at the time of preparing that specification. In such circumstances it would be entirely reasonable to require the deleterious/prohibited material clause in the collateral warranty to be deleted.

Some manufacturers were upset by the description 'deleterious materials' and threatened legal action on the basis that, while the materials in question may not be suitable in all applications, they are not inherently deleterious. As a result, it is preferable to refer to 'prohibited materials'.

4.12. Key personnel

Proposals for professional services agreements often require details to be provided of the particular individuals who will be involved, on the understanding that those individuals would be made available if the proposal is accepted. There are then detailed provisions in the appointment itself dealing with those individuals. These need to be checked to ensure they are practicable. How can a commitment be made to keep key personnel on a project when they might become ill, relocate, leave, become engaged on other projects, be promoted and so on?

Where possible the number of key personnel named should be kept to a minimum and otherwise only a general obligation included to provide 'an appropriately qualified/ experienced member of staff' instead of providing names.

Key personnel provisions should be qualified appropriately, so that if, for example, the client has a right to consent to any removal or replacement, this right must be qualified by 'such consent not to be unreasonably withheld or delayed' if the professional's internal arrangements are not to be disrupted. There are sometimes provisions that the client can insist on the removal of an employee; if so, the cause must be stipulated and the client's opinion be qualified by 'reasonable' so that it is not possible to act in an arbitrary manner.

The naming of particular individuals may make it more likely that the client will sue that individual direct in tort as well as suing the professional firm in contract if that individual has been negligent. In order to establish a cause of action the client needs to show either that the individual committed the negligent act or made the omission complained of, or that he had sufficient management and control over the people who did.

Neither ACE Agreement 1 nor RIBA 2010 has provisions relating to key personnel.

RICS 2011 deals with personnel as follows:

> 7 The Consultant nominates the key personnel set out in the Appendix as the persons primarily responsible within its organisation for the carrying out of the Services for so long as they remain in the employment of or (if applicable) partners of the Consultant. The Consultant ensures (unless it is prevented from doing so by circumstances outside its reasonable control) that such persons devote such time and attention to the Services as is necessary for them properly to fulfil their duties. Such persons are not changed other than by written agreement between the parties (such agreement not to be unreasonably withheld or delayed).

It is curious that the professional should be obliged to undertake a direct obligation to 'ensure' these key people devote the necessary time and attention to these services – this would allow the client to complain and ultimately to sue for breach of contract if this was not the case and could result in unwarranted interference by the client with how the professional carries out the services. Sometimes clients have genuine concerns as to the amount of time a key director or other person spends on his project, and include an obligation that such a person devotes all his time to the project. Such obligations should only be entered into in exceptional circumstances and even then it cannot be to devote 'all his time' as this would be impossible and/or impractical. The wording in the RICS clause allowing for a change in such personnel would be appropriate for these circumstances.

Requiring a written agreement each time a key person is changed could also be impractical.

PSC3 provides in relation to key personnel:

> 22.1 The *Consultant* either employs each *key person* named to do the job for him stated in the Contract Data or employs a replacement person who has been accepted by the *Employer*. The *Consultant* submits the name, relevant qualifications and experience of a proposed replacement person to the *Employer* for acceptance. A reason for not accepting that person is that his relevant qualifications and experience are not as good as those of the person who is to be replaced.

It is not clear whether this is the only reason the client can have for rejecting an individual. It is implied because if the client withholds an acceptance for a reason not stated in the contract this is a compensation event. If he rejects the person for any other reason this will entitle the professional to additional time and/or money. The period within which the client has to give his acceptance is 'the period for reply', which will be set out in the contract data.

> 22.2 The *Employer* may, having stated his reasons, instruct the *Consultant* to remove a person employed by the *Consultant*. The *Consultant* then arranges that, after one day, the person has no further connection with the work included in this contract.

Again, it is not clear what can constitute 'reasons' on which the client can instruct the removal of a professional's employee. Does there have to be more than one? Do they have to be reasonable? The one-day period after which that person is to have 'no further connection with the work included in this contract' is mandatory but may be difficult to achieve in practice. A proper hand-over could take longer than that, and it is difficult to see why the client would object to that.

Professional Services Agreements
ISBN 978-0-7277-4158-5

http://dx.doi.org/10.1680/psa.41585.061

Chapter 5
The services

5.1. Definition of tasks

A clear definition of the tasks involved in a professional appointment is essential in order to describe correctly what is being undertaken and to help make clear the respective responsibilities of all parties to the project. They will also be the duties against which a professional is judged in any later claim for breach of contract or negligence. If a professional has agreed to carry out something specifically, but has in fact done something else, that will be a breach of the professional's obligations under the appointment. Those tasks, if extensive or complicated, may need to be listed on an attached appendix to the letter of appointment and should be referred to in its text. It goes without saying that a professional should only agree to tasks that he can undertake, and deal with in the timescale required.

Other clarifications may be necessary. If terms such as 'as required' or 'as necessary' are used to describe the extent of any particular services, there could be a major difference of opinion as to what this was meant to encompass. 'Supervision' gives rise to particular difficulties and therefore is dealt with separately – see Section 5.6. Even if the professional's tasks and role are perfectly clear between professional and client at the time of negotiation, over the period of a contract the individuals involved may no longer be available to deal with any question concerning these. Third parties who later become involved, such as funders, purchasers and tenants, will only be able to refer to the duties contained in the appointments attached to or referred to in their warranties. The same would be true of any client, contractor or funder who later takes over by way of novation or via rights allowing it to step into the place of the original client. The statement of tasks should therefore be definitive and exhaustive. It would not be sufficient for the client and professional to agree that they understand the wording to mean something specific when the wording seems quite clearly to indicate something different or more extensive. The professional should not rely on a statement by the client's representative to the effect that 'you'll never be held to that' or 'we won't expect you to do all that' because the representative may not be around when the written obligation is later insisted upon. The professional would then be left to try to prove a variation by consent.

It may be necessary to state expressly any limitations that have been placed on a task or that something has been specifically excluded, again so that both parties are clear as to what is to be done. If services can only be performed after information or assistance is obtained from third parties, this should be stated.

Generally, the role of each type of professional is self-evident, for example, a structural engineer will provide structural designs and a quantity surveyor will undertake measurements and pricing. However, in some circumstances engineers might undertake the measurement of quantities, architects might provide some minor design and detailing, and so on.

It is, therefore, essential clearly to define in writing the respective tasks and services to be provided, wherever there can be confusion. Non-standard conditions of engagement often do not list comprehensively the normal services to be provided and optimistically do not anticipate what might constitute additional services for which additional payment should be made to the professional.

Lack of clarity has led to disputes as to what the professional is responsible for. For example, who is to check the elements of fire-proofing in a building (usually an architect but a structural engineer is typically involved in the fire-proofing of structural steelwork), flat-roof detailing and so on? Another area which should be clarified is the extent to which a professional has responsibility for designs carried out by specialist sub-contractors – see also Section 12.2.

An area that may need specific clarification is the role of a project manager. The role of a project manager can be a source of confusion for clients.

The words used for the services should be descriptive and not prescriptive. If the latter are used, for example, by adopting 'ensure', 'comply' or 'procure', a strict obligation will be created and also possibly a warranty for fitness for purpose – see Sections 4.2 and 4.3.

5.2. The standard conditions of engagement

ACE Agreement 1 provides for the professional to describe the services to be performed either by attaching one or more of the services schedules produced by the ACE or by attaching a schedule agreed by the parties. ACE Agreement 1 makes no provision for additional services as such. If the ACE services schedules are used, the professional amends them as appropriate to show which services are or are not to be performed within the agreed fee, including those that in earlier versions of the schedules would have been additional services and are now described as 'Other Services to be performed by the Consultant'. The professional can also indicate whether he will arrange for any services to be performed by others. The ACE services schedules have been published in predominantly the same form for many years and many professionals still use them without substantial amendment.

The provisions for limiting liability in ACE Agreement 1 presuppose that the extent of the professional's responsibility for advising on pollution and contamination matters, asbestos matters, and terrorism is made clear – see Section 16.9. If the professional is not advising on any or all of these matters, this too should be clear because in such a case it is anticipated that liability will be excluded.

There is also provision in the particulars of agreement for the professional to describe the particular parts of the project in respect of which the services are to be performed – defined as 'The Works'.

In the earlier versions of the ACE Agreements, lists of the possible elements comprising 'The Works' could be amended or completed but it is thought that these may now be out of date, although they could be used as a starting point. It is obviously useful and could be crucial in the event of a claim if a professional, for example, is to be responsible for 'earthworks and excavation', 'structures in timber or reinforced plastic', 'above ground drainage' or 'fuel gas incoming supply and distribution'.

The Particulars also provide space for the professional to set out for which parts of the project he will not be providing services, which can be helpful in clarifying matters for the client.

RIBA 2010 has a services part to its Schedule which is divided into Role Specifications and Design Services, and other services and additional services can be added. It also includes the RIBA Outline Plan of Work prepared in 2007 and amended in November 2008 giving the various work stages and the OGC's Gateways.

The specified roles are for project manager, lead consultant, CDM (construction, design and management) coordinator, cost consultant, contract administrator, employer's agent, lead designer, architect as designer, landscape designer, civil and structural engineer as designer, and building services engineer as designer.

It is not clear if these roles are to be descriptive – it can hardly impose obligations on or set out services to be performed by third parties – but if the professional were to take on any of these roles it is assumed the professional is to perform the services listed later on in the schedule in respect of each 'role', although they are referred to as 'activities'.

The Role Specification for designers has wide design services for all designers and includes having due regard to the Ove Arup Good Practice publication (Sheehan and Ove Arup and Partners, 1997) – see Section 4.11. There are also some general site activities if the designer is to be employed during the construction stage including 'Inspection generally of the progress and quality of the Relevant Design as built', in other words the design for which the designer is responsible.

The Role Specifications finish by listing some 'services' the architectural designer, structural designer and building services designer services 'include' and there is a statement as to what elements the specialist designer's services relate.

Part 2 of the services part of the Schedule then lists the 'Design Services' divided, as before, into the work stages.

It is anticipated that the services will be deleted if not required, and some are put in italics because these may need to be moved to suit project requirements.

Part 3 of the services part of the Schedule has a tick box list of other site building and related services that may be provided.

Finally, there are some 'Special Services' which are 'only required if the need arises during the commission, which may be instructed as additional Services'.

RICS 2011 'Basic Services' are to be identified in Schedule 1. The professional indicates which of the services apply – building surveyor (broken down into various headings), CDM coordinator, employer's agent, project manager, quantity surveyor or other, and attaches the relevant services schedule produced by the RICS. It is expected that the professional will tick the box adjacent to the particular services in the services schedule that are to be basic services. Additional services can also be indicated.

PSC3 provides no services schedules. The parties are to describe the services in 'the Scope' and the contract data states which documents contain the scope – see Section 9.2.

The CIC has produced a comprehensive schedule of services called 'The CIC Scope of Services' to accompany the CIC Consultants' Contract, although it can be used separately. It is for use on major projects. It comprises a list of tasks which are (or may be) required on all projects. From these lists, parties can draw up schedules of services for the appointment of consultants, specialists and contractors by allocating the tasks to whoever is to undertake them. It provides a full menu of services and a clear framework for agreeing the level of services required and the responsibilities of each party.

A series of tables details the tasks to be undertaken as part of the 'definition process'. The term 'definition' is used rather than 'design'. It includes management, coordination, costing, programming, health and safety and so on.

The definition services will be undertaken by a range of personnel, depending on the contractual arrangements and responsibilities and the skills required for a particular project. The parties need to allocate the tasks to whichever participant is to undertake them.

The stages match well with international systems, those adopted by many leading developers, the BPF and the RIBA Plan of Work.

The design tasks are divided into architectural design, civil and structural engineering design and building services design. The contractor will also be involved in a level of defining what is to be constructed under the construction contract, for example, preparing manufacturing and installation drawings (as a minimum), and may be involved at an earlier stage to advise on constructability. A contractor appointed under a design and build contract will be fully involved in design.

There are further roles in the definition process: cost consultancy (some tasks will typically be undertaken by a quantity surveyor) and a health and safety consultancy (including the role of a CDM coordinator).

Set out as separate roles are two important functions: project lead and design lead. In brief, the project lead role is to lead and manage the team. The design lead role is to lead and coordinate those undertaking design.

There is one further vital role, and that is the role of the client, acting either himself or through a representative. The tables set out the input required from the client as part of the definition process, whether the client be a developer or end-user or a design and build contractor.

Alongside definition services, there are review services, which involve reviewing definition work undertaken by others: for example, consultants engaged by an end-user client to review definition services undertaken by a design and build contractor.

The specific scope schedule provides a further level of detail, allowing responsibility for the definition of 'components' to be allocated between parties. So the potential for confusion over responsibility for underground drainage or sprinkler systems should really be a thing of the past as the elements and the stages for which design services are required are defined and allocated. A further benefit of having the details in a clear framework is that it provides for change management.

5.3. Approving, checking, reviewing, etc. services

It is important to realise that if a professional undertakes to approve, review, comment on, examine or otherwise check someone else's work and documents, etc., the professional will incur some responsibility for those tasks. The extent of that responsibility will depend on what was required to be undertaken as well as what was actually undertaken.

For example, if a professional undertakes to check another consultant's or sub-contractor's design and fails to find an error, he can become jointly liable with that other consultant or sub-contractor – see Section 12.7 on joint and several liability. It will not matter if that area of design was not within the professional's expertise. If, however, the purpose of the check is set out and correctly describes what the professional is to do, for example, 'check that the design information . . . is consistent and integrated where necessary', he can only incur liability if he fails to make that particular check or fails to do so carefully. If the design check is to be comprehensive – reviewing every document and detail, for example, where a professional is being replaced – the duty should state this. It will then follow that the professional will become liable for all of that third party's design, unless any limitation or exclusion of liability applies. If it is a check of major elements of a structure, again this should be stated. The same is true for any other type of review or approval – always state what the exact nature and extent of the review is and its purposes and the purpose of any approval.

An example in the ACE schedule of services G(a) for civil and structural engineering is 'Examine detailed designs . . . submitted by Contractors or Sub-Contractors for the Works or parts thereof, in respect of conformity with the Consultant's design and in particular in respect of general dimensions, structural adequacy of members and connections and compliance with performance criteria', and in G(b) for mechanical and electrical engineering: 'Examine the proposals of Contractors or Sub-Contractors for carrying out commissioning procedures and performance testing in relation to the Works (i.e. that part of the project in connection with which the professional is to perform his services). Comment to the Client on any requirements of these proposals affecting the programme for the Works'.

The RIBA design services have 'Reviewing design information from contractors or specialists to establish whether that information can be coordinated and integrated with other project information'.

5.4. Review of shop drawings

FIDIC (International Federation of Consulting Engineers) makes particular reference to approval with regard to shop drawings. It clearly identifies that approval of shop drawings is, in fact, only a review process and the word 'approve' should be avoided since it recognises outsiders (presumably including clients and perhaps those with English as a second language) generally take approval to mean 'unqualified acceptance'. It should therefore be excluded from a professional services agreement. Instead, FIDIC states that the professional should state unambiguously what is required from a contractor with regard to shop drawing submissions and hence clarify his own duties to his client at the same time.

The following wording which can be incorporated into a shop drawing specification clause is generally in accordance with FIDIC advice. With simple amendments, it could equally be used within a professional services agreement:

> The Contractor shall design all connections, prepare all fabrication and erection detail drawings and calculations for the structural steelwork and shall submit [two] copies to the [Professional] for review at least [fourteen] days before being required. The detail drawings shall give all information necessary for the fabrication and erection of the component parts of the structural steelwork including the grade of steel, location, size and type of welds and the location, size, type and grade of bolts. The review is for checking general conformance with the requirements of the Contract Documents. The [Professional] shall, when appropriate, return one copy of the drawings or calculations with comments to the Contractor. The Contractor shall supply [two] amended copies of the drawings and calculations, again allowing [fourteen] days for further review by the [Professional].
> Review by the [Professional] shall not relieve the Contractor of responsibility for any error subsequently discovered in the detail drawings.

There is even less risk if the following is added:

> Corrections or comments made on shop drawings or calculations during this review will not relieve the Contractor from compliance with the requirements of the drawings and specifications. This check is only for review of general

> conformance with the design concept of the project and with the information given in the Contract Documents. The Contractor is responsible for confirming and correlating all quantities and dimensions; selecting fabrication processes and construction techniques; and performing work safely.

Having received the relevant information from a contractor, the professional should proceed with extreme care without using the words 'approval' or 'approved'. The following provides an example of the written notification that might be provided to a contractor:

> We herewith return a copy of each of the shop drawings and/or shop drawing calculation sheets listed below. These have been reviewed in accordance with Clause . . . of the Specification. Action, if any, is as noted.

DRAWING NO.	*CATEGORY*			
	1	2	3	4

CALCULATION SHEET NO.	*CATEGORY*			
	1	2	3	4

Note:
Category 1 – No comments.
Category 2 – Make corrections noted on the relevant copies and proceed without resubmitting.
Category 3 – Amend and resubmit.
Category 4 – Totally rejected – see endorsements (on drawings, calculations, separate letter).

The foregoing information could be (and frequently is) incorporated additionally and more briefly in the form of a *pro forma* stamp, with tick box, on each reviewed document.

5.5. 'As-built' drawings

Professional appointments can require the preparation and issue of 'As-built' drawings. However, a professional is not in a position to provide such drawings, as a professional's site staff are neither omnipresent nor able to measure and record in detail all of the works. There is a risk therefore that a client could rely upon incomplete or inaccurate information in an 'As-built' drawing for, say, extension or remedial works, and incur additional costs which the client then attempts to recover from the professional. Such drawings should be provided by the contractor.

'As-built' drawings are therefore best not referred to as such at all; nor should any obligation be undertaken in respect of them. There is no mention of them in ACE Agreement 1, nor in the services schedules to RICS 2011.

What the professional can supply as set out in the ACE services schedule is:

> On completion of the Works deliver upon request to the Client one copy of each of the final drawings supplied by the Consultant to Contractors for the purpose of constructing the Works.

5.6. 'Supervision', 'inspection' or 'monitoring' on site

The matter which perhaps most frequently seems to give rise to misunderstanding between a client and a professional is what is meant by 'supervision', 'inspection' or 'monitoring' of the contractor's works.

In practice it is never possible for a professional to 'supervise' or 'inspect' the works to the same degree and in the same manner as a contractor, however many site staff are provided, or to guarantee that works will be constructed in accordance with the drawings and specification. Further, the professional should not tell the contractor how to construct the works, what plant and labour to use, and so on. Indeed, it would be dangerous for a professional to attempt to do so because the professional or the client would then be responsible for those instructions and their consequences, and could become liable in respect of them. (In some circumstances, where the professional may have to give instructions to a contractor, he should be aware of the risks in doing so.) The responsibility for the works should always rest with the contractor.

The case of *McGlinn* v. *Waltham Contractors Ltd & Others (2007)* commented on the inspection duties of an architect. It is also a classic example of what architects fear – not only was the contractor in administration, leaving only the architect and other professionals to be sued, but the works had come to a halt before handover, at which stage a snagging list would have been prepared and many of the defects put right by the contractor. Would the architects be liable for not having had these sorts of defects rectified earlier?

On inspection generally, the judge said the frequency and duration of inspections should be tailored to the nature of the works, so that the important elements of the works were inspected at the appropriate stages. In particular, it was not good enough for an inspector to religiously carry out an inspection either before or after fortnightly or monthly site meetings and not otherwise.

The fact that defective work has been carried out, and covered up, between inspections will not necessarily therefore be a defence for an inspector who should have become aware of the defects by being present at the right time. The inspector should also time their visits so that covering up of work does not have to be delayed pending their inspection. If an element of the work is to be repeated throughout parts of the building, the inspector should ensure that they inspect that work at an early stage so as to form a view of the contractor's ability to carry it out.

However, the judge was clear that the inspecting professional was not required to go into every matter in detail, and it is almost inevitable that some defects will escape their notice. The approach of automatically blaming the inspecting professional for every builder's defect is misconceived.

As for leaving defects to be rectified under the snagging process, the judge drew a distinction between temporary disconformities, where the work has not yet been finished and where it could be premature and counter-productive for the inspector to interfere, and defective work which has been completed. In the latter case, the inspector should usually require the work to be rectified as the works progress.

The judge also had some comments on what the inspecting professional should do when they notice a defect. Generally, it is not sufficient simply to point it out to the contractor, or even to write to the contractor without any follow-up. They must take all possible steps to ensure that the defective item is replaced. The alternative would be to make a deduction from the contractor's valuations.

If the inspector does not take the appropriate action at the appropriate time, he may well be liable for the cost of rectification if, for whatever reason, it is not subsequently possible to require the contractor to rectify the work and the cost cannot be recovered from it.

The architects were not held liable for not noticing defects that only became apparent after the element of work in question (the laying of roof slates) had been completed. The architects' duty required them to inspect this element of the work when it commenced and when it was completed. They did not have a duty to re-inspect it before the snagging stage, which of course was never reached.

It is necessary to show that the inspector has been negligent (unless their contract imposes a stricter duty). In the case of mechanical and electrical works, there were a number of items that the experts agreed 'might ideally' have been identified on periodic inspection. The judge held that this did not form the basis for a finding of negligence.

In deciding whether an ordinarily competent architect would have picked up any particular defect, the judge said at one point that he had endeavoured not to lose touch with reality.

The perils of a general obligation to supervise are also exemplified by the following example (*McKenzie* v. *Potts and Dobson Chapman (1995)*).

A contractor built a house and engaged an architect who agreed to supervise. The floors cracked after the property was purchased and it transpired that the contractor had (*inter alia*) used some clay backfill instead of hardcore. Both the contractor and the architect were sued. The architect had visited site in accordance with the agreement, but had not been there when the contractor had put in the backfill. The contractor had covered the clay backfill with sand. The architect accepted the builder's assurance that the appropriate materials had been used and so did not uncover it to check. He was found jointly liable with the contractor. Even worse, the judge found the architect 40% liable to the contractor's 60%. On appeal to the Court of Appeal against the apportionment, the Court found that although some other judges might have taken a more sympathetic view of the degree of culpability of the architect, it was within the spectrum of discretion of the trial judge to hold the architect 40% to blame and so the Court of Appeal would not interfere.

Section 12.7 discusses joint and several liability in more detail.

In order to avoid the misunderstanding 'supervision' and perhaps 'inspection' can cause, those words are excluded from the ACE services schedules, the more correct description of 'monitor' or 'monitoring' being used instead, combined with a statement as to the purpose of such monitoring:

> to monitor that the Works are being executed generally in accordance with the contract documents and with good engineering practice.

There is a reference to inspecting the works but this is in relation to defects as follows:

> Inspect the Works on completion and, in conjunction with any Site Staff, record any defects.

RIBA 2010 services include for the contract administrator:

> Administering the terms of the building contract including reviewing the contractor's progress and the progress of construction works . . .

It is appropriate to note at this point that the words 'monitor' and 'monitoring' have their own implications where there is a requirement in a professional services appointment for a professional to monitor progress, performance, or cost.

RIBA 2010, for example, has under the role specification for the lead consultant 'Monitoring the work of the consultants'. This could be quite an onerous task. The project manager's monitoring duties are even wider: 'Monitoring performance and activity, cost planning and control, programme and progress, and reporting to Client.'

The question of site staff is, of course, linked to 'inspection' or 'monitoring'. Both the ACE and RIBA standard conditions of engagement include provisions whereby the professional recommends the appointment of site staff if appropriate and agrees details, scope, payment and so forth, as well as the number of site visits with the client. Both also deal with the responsibilities of the contractor (or any sub-contractors), notwithstanding the provision of the professional's site staff so that the respective responsibilities are made clear.

ACE Agreement 1 concerning site staff and visits to site provides:

F4.1 If in the opinion of the Consultant the execution of the Works warrants full time or part time Site Staff to be deployed at any stage the Client shall not unreasonably withhold consent to the employment and/or deployment of such reasonably qualified technical and clerical Site Staff as the Consultant shall advise. The Client and the Consultant shall discuss, agree and confirm in writing in advance of such deployment the number and levels of staff to be deployed to Site, the duration of such deployments, the frequency of occasional visits, the duties to be performed by Site Staff and whether such Site Staff shall be employed by the Consultant or by the Client.

F4.2 The Client shall procure that the contracts of employment of Site Staff employed by the Client empower the Consultant to issue instructions to such staff in relation to the Works and shall stipulate that staff so employed shall in no circumstances take or act upon instructions in connection with the Works other than those given by the Consultant.

F4.3 The Consultant shall not be responsible for any failure on the part of any Site Staff employed other than by the Consultant to comply with any instructions given by the Consultant.

F4.4 The Client shall be responsible for the cost and provision of such local office accommodation, furniture, telephones and facsimile apparatus and other office equipment, protective clothing and transport on Site as shall reasonably be required for the use of Site Staff and for the reasonable running costs of such accommodation and other facilities, including those of stationery, telephone and facsimile charges and postage.

RIBA 2010 concerning site visits and site staff provides:

2.3 The Architect shall inform the Client upon becoming aware of

2.3.1 a need to appoint Other Persons, other than those named in the Project Data, to perform work or services in connection with the Project . . .

3.8 Where work or services, other than those to be performed by the Architect, are required, the Client shall appoint and pay Other Persons under separate agreements to perform such work, services or products . . . Such Other Persons shall include site inspectors or clerks of works.

The Client shall confirm in writing to the Architect the services to be performed by Other Persons, their disciplines and the expected duration of their employment.

A client will sometimes attempt to dictate a professional's site staffing establishment. However, it is the professional's duty to recommend to a client the resources necessary to administer the contract. If there is nothing in the appointment concerning the terms on which site staff are to be employed, it is prudent for the professional to set this out specifically to the client and obtain the client's express agreement. All the matters set out in one or other of the standard conditions of engagement mentioned previously should be included.

Proposed staff should not be named initially if possible, unless at that early stage individuals can be firmly committed to the future work. It is usually better to refer to them by grade (for example, resident engineer, assistant site engineer, quantity surveyor or architectural assistant) and to give an indication of qualifications and experience for the most senior staff.

After the recommended level of site presence is agreed, the names (with relevant short CVs) of proposed staff can be provided with salaries and budget estimates of costs, as well as the professional's conditions of employment for site staff, on which budget costs have been based.

Inspection or 'supervision' by professionals appointed by design and build contractors is discussed in Section 5.8.

5.7. The duty to warn/report on deficiencies or unsafe designs or practices of others/continuing duties

Whether a professional is under a duty to warn about some particular matter will normally depend on the professional's role on the project and the terms of the professional's engagement (once it is established, of course, what the matter in question is).

In some situations, the existence of a duty to warn will be so obvious that the question would not be asked. The question would usually arise where circumstances have arisen that are a little out of the ordinary and with which the professional may not be directly involved.

The question of whether a professional has a duty to warn about deficiencies in the performance of other professionals arose in a case in 1996 (*Chesham Properties Ltd* v. *Bucknall Austin Project Management Services Ltd & Ors*). The case involved allegations of negligence and breach of contract in connection with a construction development. Among other things, it was alleged against the project manager, the architects, the structural engineers and the quantity surveyors that each of them had a duty to advise the client of deficiencies in the performance of the others (and indeed of deficiencies in their own performance). This question was considered in relation each of them in turn, and involved consideration of each one's terms of engagement.

In the view of the judge, the project manager was plainly under such a duty, on the true construction of its contract. The key provision, as the judge saw it, was an obligation in its contract:

> to implement and maintain all monitoring procedures including the performance of the contractor on-site, *and all consultants* (emphasis added).

Monitoring, in such a context, said the judge, cannot sensibly be confined to passive observation. It must include reporting to the principal on the performance being monitored by reference to the standards that should be achieved.

At the time when the architects were appointed, it was not then contemplated that a separate project manager would be appointed, and indeed the architects took on some project management responsibilities themselves. In that context, the judge held that it was clear beyond doubt that there could be no duty incumbent on the architects to report any deficiencies in the performance of the project manager (after the latter was eventually appointed). The remaining question therefore was whether the architects had such a duty in relation to the performance of the structural engineers or the quantity surveyors.

The appointment of the architects included the following term from the RIBA booklet at the time:

> Provide management from inception to completion; . . . appoint and coordinate consultants, construction managers, agents and contractors; monitor time, cost and agreed targets; monitor progress of the works; . . .

The judge held that, although the situation was not as clear-cut as in the case of the project manager, on the balance of probabilities, there was a duty on the architects to report on deficiencies in the performance of the other consultants (i.e. the structural engineers and the quantity surveyors). He referred to the express obligation to monitor as part of general management responsibilities, and although what was to be monitored was not explicitly the performance of the other consultants, there is not much point in reporting on failures to meet time and cost targets or to maintain due progress of the works if the reports do not attempt to identify the causes of those failures.

The structural engineers seem to have been appointed on standard conditions of the Association of Consulting Engineers (now the Association for Consultancy and Engineering). The key terms relied on (in support of a duty to report on the deficiencies of others) were:

> 6 Normal services
> 6.5 General. Without prejudice to the preceding provisions of this clause, the Consulting Engineer shall from time to time as may be necessary advise the Client as to the need for the Client to be provided with additional services in accordance with clause 7.
> 7 Additional services not included in normal services.
> 7.2(g) Checking and advising upon any part of the Project not designed by the Consulting Engineer.
> 7.3(b) Legal, financial and other professional services.

The judge commented that the terms of clauses 7.2(g) and 7.3(b) were no doubt very wide, but in his view they must, with their heading of 'additional services', be read as supplementing and being ancillary to the 'normal services' in clause 6 rather than giving the structural engineers *carte blanche* to advise general roving inquiries by lawyers and others into the conduct of the other professionals, still less imposing any duty to do so. Therefore, the judge considered, on the balance of probabilities, that there was no duty on the engineers to report on deficiencies in the performance of the architects or the quantity surveyors, although he did not conclude that the allegation was obviously unsustainable. (The existence of any duty to report on deficiencies in the performance of the project manager was dismissed, as in the case of the architects.)

There was no written document covering the appointment of the quantity surveyors. It was assumed that the quantity surveyors were appointed to carry out the normal functions of quantity surveyors: to prepare the bill of quantities, carry out the functions of quantity surveyor under the building contract, ascertain the value of the contractors' work as executed, and prepare monthly statements and an estimate of the contractors' final account.

The judge found that it was quite impossible to imply, from this, terms imposing a duty to report on the deficiencies of others; it was clear beyond doubt that such a duty did not exist.

The judgement was on a trial of preliminary issues, and the judge emphasised that, where he did find a duty to exist, he could make no findings at the time on the precise ambit of the duty. In any event, the judge's findings were obviously particular to the precise circumstances and, in particular, the contractual terms in question.

Nonetheless, it is worth noting that the circumstances generally do not seem to have been that unusual, and the contractual terms fairly standard. In the case of the quantity surveyors, it was a matter of considering any implied terms in the absence of express terms.

The main lesson to take from the case is the potential importance of the express terms of the professional's appointment.

The question of whether there is a duty to warn of deficiencies in the performance of contractors may also sometimes arise. Where a professional has inspection duties and the deficiencies concern work designed by him, the duty may be obvious, assuming that the professional is aware of the deficiencies. A problem may arise where the work in question is, or might be thought to be, outside the remit of the professional.

The point arose in a recent case – *Hart Investments Ltd* v. *Fidler (2006)* – where part of a building collapsed during excavation works due to lack of propping of the underpinning. The structural engineer had in fact prepared drawings and a method statement with details of temporary support of the underpinning, so it was difficult for him to argue that he had no duty to be concerned with this aspect of the works on his visits to site, and the main issue at trial was whether he in fact knew of the lack of propping – the court held that he did.

However, the recorder trying the case also had this to say on the duty of an engineer generally in relation to temporary works:

> If an engineer employed by an owner in respect of permanent works observes a state of temporary works which is dangerous and causing immediate peril to the permanent works in respect of which he is employed, he is obliged to take such steps as are open to him to obviate that danger. It seems to me that that follows, partly as a matter of common sense, but also because the engineer is, after all, instructed in relation to the permanent works as a whole. It would appear strange if he is under a duty to take such steps as he can to see that they survive for say, the next 25 years, or whatever the design life of the building is, but is not obliged to take any steps to warn of an immediate danger to those works caused by an imperilling act by the contractor.

It may always be more likely that a duty to warn will be held to exist where there is a physical danger – see also later in this section. However, it is worth noting that the recorder considered that the position was analogous to a solicitor's negligence case, where it was held to be the solicitors' duty to point out a plain and obvious danger, although the danger in question involved matters on which the solicitors had not been instructed to advise. As one would expect, the danger in question in the solicitor's case was a financial, not a physical, danger.

The recorder's comments might indicate that there is a duty to warn of any circumstances of which a professional becomes aware which might affect adversely the accomplishment of his own work. The point may sometimes, of course, be academic, because a professional in such a position might well want to take some action anyway, regardless of any duty to do so.

What about a professional's obligations in respect of unsafe designs specified by their client? In the McGlinn case the architect was found liable for the unsafe designs specified by its client for a residential dwelling.

One involved gaps in the stone balustrades on the terraces. These were wider than permitted by the building bye-laws. The architect said that when the client was warned that this was potentially dangerous for children, the client made it clear that children would not be visiting the property and that therefore this was not a concern.

Another issue concerned the external stairs, for which a handrail was not provided. This was because the client did not want one, as the client, at the trial, accepted.

In the case of the gaps in the balustrades, the judge held that the architects should have been tougher with the client, and in the case of the missing handrail the judge said that there was no evidence that the grave danger posed was ever communicated to the client. The architects were also held liable for the balustrade being too low – this had been done because the client wanted to have a good view of the sea.

Curiously, the judge does not seem to have addressed the question of whether it would have made a difference if the architect had been 'tougher' with the client or given sufficient warnings on any of these matters.

Maybe the client in this case would have agreed to the designs being changed: that is the implication of the judge's findings that the architect was liable.

Nonetheless, such instances do raise the question of what happens if a professional has a client that insists on the implementation of a design that is unsafe despite having being warned about it.

Again, the implication of the judgement is that, if the designer has been sufficiently tough with the client, he will not be liable – at least, not to the client. But how tough does the designer have to be? And what about liability to a third party who is injured many years later, after the client has sold the property?

Giving a warning to the client may not be considered a sufficient discharge of the duty that a designer owes to subsequent occupiers that suffer personal injury as a result of a negligent design, particularly if the client is not under a duty to such people.

And what about criminal liability under the Health and Safety at Work Act or the CDM regulations, for example? The CDM regulations have now been extended to cover designs of buildings for use as a workplace, in other words, designs that, for such buildings, could give rise to defects in the completed works.

Some would say that designers required to prepare designs that might be considered unsafe, either because of the way they would have to be constructed or because of the actual results, should simply decline to do so and may have to walk away from the job.

As the case described illustrates, however, the relevant element of the design may be quite minor, and walking out could seem rather an extreme reaction.

It could also give rise to difficult contractual issues: is the client necessarily in breach of contract, thus entitling the designer to walk away? At the very least, the designer may have difficulty getting paid for the work done.

The terms of the designer's contract may assist if, for example, it is made clear that all instructions given by the client must be reasonable and that the client is under an obligation to comply with all safety regulations.

Perhaps the test will be this: has the designer given such clear and emphatic warnings to the client that it can fairly be said that the design in question will then, in reality, not be the designer's but the client's?

The CDM regulations define a designer as any person who prepares a design, whatever his or her nominal function on the project. A client that insists on an unsafe design, against advice, may arguably be treated as the real designer for the purposes of both civil and criminal liability. But the warning given to the client would have to amount to an effective disclaimer of responsibility.

What about continuing duties? What does a professional have to do if some new piece of knowledge, for example, comes to light after he has completed his services?

It was established in the Abbeystead case (*Eckersley* v. *Binnie and Partners (1988)*) that the engineers did not have a duty to review that continued beyond the time when the project had been completed. As a result, there was no duty to keep abreast of developing knowledge about methane after the project was finished and pass it on to the water authority responsible for operating the facility. The Court of Appeal said such a duty 'would seem to place startlingly onerous responsibilities upon professional organisations... It would seem to involve duties relating to work completed many years before the professional organisation is actually or constructively put on notice of fresh material.'

One of the judges did indicate that there may be examples involving dangers to life and health where some response by a professional man might be called for.

This approach was followed in a recent case concerning solicitors where the court had to consider whether solicitors had an ongoing duty to review the suitability of drafting amendments and/or drafting advice given to a contractor in the light of legislative and other amendments – *Shepherd Construction Limited* v. *Pinsent Mason LLP (2012)*. This was based on the concept of a 'single contract', that is, that there was one continuing retainer. This part of Shepherd's case was struck out. Akenhead J said it would be 'commercially and professionally worrying if professional people are to be held responsible for reviewing all previous advice or indeed services provided'. There would have to be very clear terms in order to establish such a contract and this would be even more true for architects or engineers than solicitors.

During construction the position is different. It is reasonably well established that a professional has a continuing duty to keep his design under review after it has been completed if he is supervising the implementation of his design and if something happens that should put him on notice that the design should be reviewed. This would encompass defects that become apparent during construction, such as cracking or a leaking basement, when the professional should inform the client as to what he knows about design defects as a possible cause of the problem. It could also extend to later knowledge that is acquired, for example, about a method of construction or a material.

It is also clear that there is a duty on professionals (and contractors) to warn the client if they become aware of conditions that might reasonably affect the safety or integrity of the works during the project.

Professionals should remember, however, that if they do review projects and make suggestions as to how the danger could be dealt with, they could become liable for that advice even if given gratuitously – see Chapter 14.

5.8. Services for a design and build contractor

As a professional can be directly employed by a contractor as a sub-contractor, it is necessary to ensure that the obligations and the services agreed to be provided are clearly stated. The contractor is likely to have a wide-ranging specification and it is important that the professional's brief is clearly defined and where necessary limited exactly to that which it is prepared to undertake. It must not be left vague or ambiguous, especially if it is a fixed fee appointment.

Particular attention needs to be paid to any requirement to inspect the contractor's work. A contractor's supervisory work is assumed to be that necessary to enable the contractor to meet its obligations under the building contract, which may be a 'fitness for purpose' obligation – see Section 4.2. 'Inspection' or 'supervision' by a professional on behalf of a contractor may be seen by an unsophisticated third party (such as a contractor's employer) to be part of the contractor's supervision obligations. To agree to 'inspect' or 'supervise' could involve a professional in any claim that is made relating to its work. If the professional is found to be concurrently liable with the contractor for the loss suffered by the employer, and the contractor has gone out of business, the professional could be liable for 100% of the damages, for example, see *McKenzie* v. *Potts and Dobson Chapman (1995)* discussed in Section 5.6. This is on the basis that each is responsible for the whole of the employer's loss.

Even where the main design and build construction conditions of contract limit a contractor's design obligations to 'skill and care', the site works will almost certainly still need to be 'fit for purpose' and all associated supervision and inspection by, or on behalf of, the contractor will relate to those 'fitness for purpose' obligations. In addition, a contractor's supervision (as opposed to a professional's 'inspection') is outside a professional's normal business and is not usually covered by professional indemnity insurance.

Occasionally for commercial reasons (e.g., a valued contractor/client relationship or generous terms of payment) it might be considered appropriate to accept a restricted site services role, involving such duties as dealing with site queries regarding design matters and some attendance at progress meetings for administrative purposes during the course of the works.

As discussed earlier, 'supervision' by a professional is not an appropriate word, and this is particularly so in a design and build appointment. The design and build contractor is the professional's employer, and one does not supervise (that is, control) one's employer.

Frequently, design and build contractors wish to restrict the services provided during the construction phase in order to minimise the professional's fees.

A professional engaged by a contractor will often visit site only to

(*a*) obtain/confirm information (e.g., ground conditions) related to the professional's design services (*b*) attend site meetings which are reasonably requested, in order to clarify written information previously provided by the professional to the contractor relating to the works.

For its own protection relating to any provision of site services, the professional should

(*a*) keep a detailed record of all events related to any site visits
(*b*) not comment on any aspect of construction on site (except obviously hazardous construction, and then in writing).

Care must be taken in any report given to a contractor – see Chapter 14.

There is a need for the professional to clarify, define and limit his task and related services. A general reference to 'ensure compliance with the specification in the contractor's main construction contract' (which could include, for example, all works to ensure obtaining planning consents, and temporary works design) is a strict obligation and is not acceptable – see Section 4.3. It is also not achievable in practice as compliance depends on factors outside the professional's control.

Responsibility for the totality of the main 'construction contract' must not be accepted. Obligations should be limited solely to those contained in the professional services agreement. The contractor should state which main contract clauses are proposed to be specifically applicable to the professional so that they can be reviewed – see also Section 4.5.

If professional services include the production of a specification (or the supply of a standard specification), which is to relate to any portion of the contractor's work, the specification should be checked to see that it does not place an obligation on the professional to 'inspect' or 'approve' work on site.

Limit any interpretation by the professional of the results of any investigations to its effect on the professional's design and not to the contractor's method of construction.

Claims against professionals from contractors are more likely because of the following.

(*a*) Design and build contractors usually have to commit themselves to fixed prices. If a professional's error or miscalculation results in the design and build contractor incurring additional expense, the contractor is likely to claim those costs from the professional. If a professional is providing services so that the design and build contractor can price a tender, it is essential that suitable qualifications and caveats are stated regarding information or materials provided – for example, the possibility that trial pits or samples may not be representative of site conditions as a whole. If less detail or a less rigorous investigation are being provided than would normally be recommended because the professional is working on a speculative basis, this should be recorded in the professional's appointment or in the relevant report.

(*b*) The contractor is more likely to claim against a professional because of the latter's professional indemnity insurance, than to claim against a sub-contractor with, possibly, a design responsibility.

5.9. Contractors' obligations

As already discussed (see Chapter 4) contractors' obligations are very different from professionals' obligations (e.g., in relation to absolute or strict obligations, warranties for fitness for purpose, and duties of supervision). As a general rule, such obligations should not be taken on by a professional. Further, the insurance arrangements for contractors are very different and a professional's professional indemnity insurance should be checked before such obligations are assumed. It may be possible for the professional to arrange for a contractor to carry out work as agent for the client – see Chapter 13.

A detailed study of a contractor's obligations is outside the scope of this text.

5.10. Hired-in plant operators

Where hired-in plant is required to enable a professional's agreed tasks to be carried out, it will probably be by means of an agreement with a hire plant company using the company's own terms or some nationally agreed standard form. The agreement should, wherever possible, be arranged by a professional acting as an agent for a client – see Section 2.2. Where this is not possible, and the professional has a direct contract with the hire company, the professional should be aware that under some conditions of hire it can become responsible for additional risks such as

(*a*) loss (e.g., theft) or damage to the plant
(*b*) consequential losses (e.g., loss of hire charges) of the owner/supplier while the plant is not available because of loss or damage
(*c*) damage or injury to third parties arising out of the use of the plant
(*d*) the actions of operators/drivers, even if wages are paid by the owner/supplier
(*e*) ensuring the plant is legal if using public roads.

Any relevant insurance held by the owner/supplier should be checked and, where there are gaps in cover, further insurance agreed or arranged. This may require a new insurance policy or extensions to existing policies. In addition to insurances, road-worthiness and taxation certificates must be valid in order for the plant to be used on public roads, and the professional should clarify who is responsible for these.

5.11. Provision of Services Regulations 2009

The Provision of Services Regulations 2009 came into force on 28 December 2009 and require the majority of private sector businesses providing services in the UK to make certain information available to clients and to deal with client complaints promptly. They also prohibit discrimination against individual clients in the provision of services on the basis of place of residence.

The Regulations apply to services provided by engineers, architects and surveyors.

Certain information must now be made available to clients at the outset of the contract. As far as the services are concerned, this information is the price of the services, where the price is pre-determined by the business; the main features of the services, if not already apparent from the context; the general terms and conditions, if any, that the professional uses; and the existence of contractual terms, if any, that the professional uses concerning the competent courts (e.g., that the English courts have jurisdiction) or the law applicable to the contract (e.g., that it is governed by English law).

If the professional provides clients with an information document with a detailed description of his services, he must include in it the following information.

(*a*) If the professional carries out other activities that are directly linked to the services in question, details of this and the measure taken to avoid conflicts of interest.
(*b*) If the professional is subject to a code of conduct, trade association, or professional body that has non-judicial dispute resolution procedures, details of this and how to access information about it.

The professional must supply the following information if asked (although the professional may choose to make this information available in all cases).

(*a*) If the professional does not have pre-determined prices, the method for calculating the price (or a sufficiently detailed estimate).
(*b*) If the professional is carrying out a regulated profession (e.g., as an architect), a reference to the professional rules applicable in the EEA (European Economic Area) state in which the professional is established and how to access these rules.
(*c*) If the professional carries on other activities which are directly linked to the services in question, details of this and the measures taken to avoid conflicts of interest. That information should be included in any information document in which the professional gives a detailed description of his services.
(*d*) Any codes of conduct to which the professional is subject and the websites from which these codes are available, specifying the language version available (e.g., CIBSE's code of conduct is available at www.cibse.org).

The professional can make information available to a client using any one of four methods.

(*a*) Providing clients with such information on the professional's own initiative.
(*b*) Making it easily available to the client at the place where the professional provides the services, or where the contract for the services is included, for example, at the professional's premises.
(*c*) Making it easily available to the client electronically (e.g., by providing the exact address of where the information can be found on the professional's website).
(*d*) Including it in any information documents the professional supplies to the client, which set out a detailed description of the services the professional provides.

There are also requirements under these Regulations concerning professional indemnity insurance – see Section 15.14.

5.12. Variations and the power to omit

Parties to a contract are always free to agree variations to that contract. An agreement to vary must be supported by consideration but this is usually found because both parties will normally have given up existing rights and conferred new benefits on the other. Some non-standard conditions of engagement provide for this to be done formally, for example, any variation has to be in writing signed by both parties. This could be impractical. During a project all sorts of variations can be agreed, both to the obligations undertaken and to the services.

Because consent would otherwise be needed, clients often reserve the right to instruct variations and/or changes and/or additional services, sometimes through a pre-agreed procedure and sometimes without the consent of the professional being required. Usually the fee basis on which any varied additional work is to take place is also set out in the contract.

There are limits, however, as to the extent of the variations that can be instructed.

The cases on this topic mainly concern contractors but the same principles from those cases set out below would also apply to a professional's appointment.

What constitutes a variation is a matter for contractual definition. The approach of standard forms varies, reflecting the type of work and method of payment which they envisage.

The carrying out of 'additional' work not mentioned in the contract will not constitute a variation if the work is included within the professional's overall obligation to complete.

There is no variation unless the professional is required to carry out work additional to that necessarily included in the contract.

There is no implied power to order variations. In the absence of any express provision for variation in the contract, the client will be unable unilaterally to vary the works.

If the client has no express power to order a variation then the professional will be under no obligation to accede to a request by the client to vary the works.

A professional will not always be entitled to payment for variations. The professional will have to show that on a true construction of the contract, there is an express or implied right to payment for the variation in question. That is

(*a*) the work needs to be outside the scope of the work the professional has already agreed to carry out for the contract sum
(*b*) the client needs to have expressly or impliedly agreed to pay for it.

If the professional can show he has a right to payment, the valuation of variations will either be carried out by reference to a mechanism contained within the contract, or on the broader basis of a reasonable valuation of the work (*quantum meruit*) – see Section 3.9. The contract mechanism may provide for a reasonable valuation in default of applicable rates – see Section 7.8.

If the work falls outside the scope of the contract the client may be liable to pay the professional if a promise to pay can be found.

Where additional or varied works are instructed which are 'so peculiar, so unexpected, and so different from what any person reckoned or calculated upon', they will not be capable of being a variation to the works (*Thorn* v. *London Corpn (1876)*). If carried out, the varied works will be the subject of an entirely separate agreement under which the professional is entitled to payment on a *quantum meruit* or fair valuation basis. Alternatively, it will be open to the professional to simply refuse to carry out works which are outside the scope of the original agreement.

The court will look at the specific wording of a contract and each case will be decided on a case by case basis. The court will also look at the conduct of the parties and the surrounding circumstances at the time the works were varied to construe whether or not there was a separate agreement. This may include correspondence or agreements between one of the parties and relevant third parties.

In *Sir Lindsay Parkinson Ltd* v. *Commissioner of Crown Works and Public Buildings (1949)*, the contractors agreed to build an ordnance factory for an estimated cost of £3.5 million, with provision for a further £300 000 of extras. The extra work ordered by the employer was so extensive that the total cost of the factory was almost £7 million.

The Court of Appeal held that by virtue of an implied term, the employer was not entitled under the contract to order work materially in excess of the agreed sum: 'the additional work called for and executed later was something outside the contemplation of the parties . . . the contractors executed it at the request of the [employer] and they are entitled to be paid a reasonable profit or remuneration in respect of it'.

If work is 'outside the contemplation of the parties' then the professional may be entitled to be paid on a *quantum meruit* basis.

See also Section 3.9 for the situation where the original price may be recalculated on a *quantum meruit* basis.

What about a right to omit or reduce the original services agreed? Again, there is no right to do this unless such a right is included in the contract, although of course the parties can agree to this as a variation. It is becoming increasingly common to find this right or power in non-standard conditions of engagement, although it has long been included in construction contracts together with a basis for the adjustment of the contract price.

Depending on the construction of the contract, the right to omit may not extend to the ordering of variations. It may also not give the client the right to omit part of the work from the contract with the object of giving it or similar work in substitution to another contractor and it may not permit the client to remove all the work.

Professional Services Agreements
ISBN 978-0-7277-4158-5

http://dx.doi.org/10.1680/psa.41585.085

Chapter 6
Client provisions

6.1. Client's obligations to provide information and decisions, etc.

In addition to the obligation to pay (and payment is dealt with in Chapter 7), non-standard conditions of engagement should also deal with the client's other obligations. These are to give information and decisions, approvals, assistance, instructions and decisions, etc. It is surprising how frequently non-standard conditions of engagement omit any such obligations, with the result that the only express obligation of the client will be to pay the professional. These other obligations would often be implied into the contract as an implied term, except where the appointment has an 'entire agreement' clause – see Sections 3.6 and 3.7. However, it is better to have this included specifically, to avoid uncertainty and to remind the client what he has to do. It may also be necessary to provide for the particular requirements of the commission such as to set out the facilities which the client will be providing or the requirements necessary for access. There may also be a need for specific provisions concerning the client's project manager or representatives.

In ACE Agreement 1 the client obligations are as follows:

Information needed by the Consultant

F3.2 The Client shall supply to the Consultant, without charge and in such time so as not to delay or disrupt the performance by the Consultant of the Services, all necessary and relevant data and information (including details of the services to be performed by any Lead Consultant if the Consultant is not so appointed and any Other Consultants) in the possession of the Client, the Client's agents, servants, any Lead Consultant if the Consultant is not so appointed, any Other Consultants or any Contractors.

Assistance

F3.3 The Client shall give, and shall procure that the Client's agents, servants, any Lead Consultant if the Consultant is not so appointed, any Other Consultants and any Contractors, give such assistance as shall reasonably be required by the Consultant in the performance of the Services.

Decisions

F3.4 The Client shall give the Client's decisions, instructions, consents or approvals on or to all matters properly referred to the Client by the

> Consultant in such reasonable time so as not to delay or disrupt the performance of the Services by the Consultant.

These provisions summarise the client's obligations correctly – that is they set out the obligations that a court would probably imply if the appointment were silent on these matters. The professional is going to need all this from the client if he is going to be able to perform his services properly.

Some non-standard conditions of engagement provide that the professional will only be given information if he specifically requests it from the client. This presupposes that the professional knows what information the client or his other advisers and consultants have, which is not necessarily the case. Also, it runs counter to the client obligations under the CDM Regulations to provide certain information about the site, etc. to its designers – see Chapter 23. Such clauses can go on to provide that any requests for information must be made not too long before or too close to the time when it is required. This seems unnecessarily pedantic.

ACE Agreement 1 also includes provisions concerning the authority of the client's representatives and the responsibility of contractors. The latter is in the following terms:

> **Responsibility of Contractors**
> F3.6 The Client shall require that any Contractor executes and/or manages the execution of the Project and/or the Works in accordance with the terms of the contract between the Client and any such Contractor.

This, however, does little more than make it clear that the contractor is responsible for what he does and the client has duties under his contract with him.

The final client obligation in ACE Agreement 1 deals with the professional acting as agent for the client and is commented on in Section 13.10.

RIBA 2010 deals with the client's obligations in relation to information and decisions as follows:

> 3.2 The Client shall supply the initial statement of the Client's requirements and shall advise the relative priorities of the Client's requirements, the Brief, the Construction Cost and the Timetable.
> 3.3 The Client shall supply, free of charge, all the information in the Client's possession, or reasonably obtainable, which is necessary for the proper and timely performance of the Services and the Architect is entitled to rely on such information.
> 3.4 The Client shall give decisions and approvals and shall take such actions necessary for the proper and timely performance of the Services.

Again, this is very similar to any implied obligations on the client. The statement that the professional is entitled to rely on information provided to him could concern clients, as it may seem to relieve the professional of any responsibility for having to exercise any independent judgement of his own as to the qualify of the information that is being given to him – see Section 6.2.

RIBA 2010 also has a sub-clause in clause 3 which deals with the authority of the client's representative. Clause 3 also deals with the client's obligation to instruct and pay for the making of planning applications and other consents, to appoint and pay other parties for the other work and services required and confirm this to the professional. The client also has to procure the legal advice necessary for any dispute and there are confidentiality obligations.

Clause 3 also has statements as to who the client is to hold responsible for what. He is to hold the contractor responsible for the management and operational methods necessary for the proper carrying out and completion of the construction works. There are also acknowledgements that the professional does not warrant the 'competence, performance, work, services, products or solvency of any such Other Persons', that is those to be appointed by the client. This is a little curious as this would not necessarily be the case. The professional might incur some responsibility to the client if he had considered and advised on the appropriate people to be appointed: in making those enquiries and recommendations he is expected to use reasonable skill and care. This provision could relieve him of that responsibility.

There is also an acknowledgement that the professional does not warrant that planning permission or other approval will be granted, or compliance with the construction cost and/or the timetable, which may need to be reviewed as a result of matters such as variations.

Other than in relation to the statement about competency, etc., these do no more than make clear to the client who is responsible for what and gives the necessary warnings about planning consents, etc. and the lack of certainty of the initial construction cost and programme.

RICS 2011 contains more specific obligations on the client which reflect the services their members provide and what they need to do:

8.1 The Client provides the Consultant with any relevant and necessary information which is in its possession or control by the date specified in the Programme or (if no date is specified) in such reasonable time so as not to delay or disrupt the performance by the Consultant of the Services.

8.2 The Client gives any decisions, approvals, consents, or instructions required within a reasonable time, having regard to the Programme.

8.3 The Client gives, and procures (so far as reasonably practicable), that the other members of the Professional Team give, such assistance to the Consultant as is reasonably required by the Consultant for the performance of the Services.

8.7 The Client appoints the Professional Team and the Contractor in sufficient time and on terms that allow the Project to be successfully delivered and so as to secure compliance with all relevant Statutory Requirements (including the CDM Regulations).

8.8 Where the Services require the Consultant to issue or give certificates or statements under the Building Contract and/or any Client Contract, the Client procures, so far as reasonably practicable, that the Professional Team provides the Consultant with such information or confirmation as is necessary for it (acting reasonably) to issue such certificates or statements.

Again, this does little more than set out the general obligations of any client and recognises that it is the client who appointed the other professionals and the contractor and it is therefore only the client who can ensure those parties comply with their contractual obligations.

PSC3 has a client obligation:

20.1 The *Employer* provides information and things which this contract requires him to provide in accordance with the Accepted Programme.

This is a strict obligation on the client and 'things' refers to all the notices, decisions, acceptances, replies and other communications the client has to give. The other client obligation in the PSC3 relates to the giving of instructions – see Section 4.6.

6.2. Accuracy of information provided by the client

Clients seem to have become exercised in recent years about the nature of the information they give to their professionals and whether they should have any responsibility for it. This has not arisen out of any case on the subject and may reflect their lawyers' desire to afford their clients the utmost protection. In the context of a construction project, however, it could have some undesirable and unexpected consequences for both client and professional.

ACE Agreement 1 does not say anything about the status of the information provided by the client.

RIBA 2010 has the statement about the professional's reliance on that information commented on in Section 6.1.

RICS 2011 says:

8.4 The Client does not warrant the accuracy of any information provided by it to the Consultant save for the following:

(*a*) the Client's Brief; and

(*b*) any specialist reports and surveys provided by the Client upon which it is reasonable for the Consultant to rely in performing the Services.

8.5 The Consultant notifies the Client upon becoming aware that any of the information referred to in Clause 8.4 is incomplete or contains any inaccuracies or discrepancies.

CIC has a similar provision as to which information can be assumed by the professional to be accurate – the client brief, any other information provided by the client from its own knowledge, and any specialist reports and surveys provided by the client and on which it is reasonable for the professional to rely in performing the services. It also says in terms that the professional is not entitled to assume 'the accuracy of any other information provided by the Client or on its behalf'.

A statement that the client does not 'warranty the accuracy' of any information means not only that the client does not know whether the information is accurate, but puts in question the extent that the professional can rely on it as being accurate when he is preparing his designs. This may in fact be no different from the situation where nothing is said about the status of the information given to the professional. He should use his professional judgement to assess the value of the information including discussing it with the client or others. If it is important, however, that a piece of information provided by the client is accurate: such a provision might well mean that the professional has to do more in relation to checking it before relying on it than would otherwise be the case.

If nothing is said about the status of information provided to the professional, the usual situation would be that the professional in carrying out his services has to take a view as to whether the information was correct. If there was something obviously wrong or omitted then he would be expected to point this out, as with any ambiguity, inconsistency, discrepancy, etc. which could have a bearing on what he is doing. As to whether he should rely on the information provided, again he would need to consider whether, in his professional judgement, he could do so without further enquiry. Take a site investigation report – if it was recently prepared by a reputable firm for the client and contained information which the professional might reasonably expect, then it might be reasonable for him to rely on it. If it contained something which would put a reasonably competent professional of the same discipline on enquiry, then he would be expected to raise it or question the information.

The RICS seems to be saying that in the case of the brief the professional can assume everything in it is accurate, but by introducing 'upon which it is reasonable for the Consultant to rely in performing the Services' in (b) the professional has to take a view as to the extent that he can rely on specialist reports and surveys.

These clauses in these standard conditions of engagement are surprising – one would have thought that the client would want his professionals to exercise their professional skill and judgement when considering the matters in the brief or any other information he is given. And in the case of the CIC clause how does the professional know if the information is 'provided from the Client's knowledge'?

The wording also begs the question as to why it is necessary to say anything about the 'specialist reports or surveys' if the test as to whether or not they are accurate is bound up with whether or not it is reasonable for the professional to rely on them. This wording would mean, however, that anything else the client gives the professional is not accurate and possibly therefore that the professional cannot rely on it.

Perhaps it is intended that the words 'provided by the Client' means the client's own information and not information from other professionals or other information, but if that is the case why bring in 'specialist reports and surveys'?

Some clauses in non-standard conditions of engagement go wider than this and say the client does not warrant the accuracy of any information whatsoever.

As to the obligation to notify the client if the information is incomplete or contains any inaccuracies or discrepancies as in RICS 2011 and CIC, nothing is said as to what the client has to do in these circumstances or how the matter is to be resolved.

These sorts of clauses should be deleted as the common law has a perfectly good way of dealing with the many different sorts of information that can be given to the professional and the extent of reliance the professional should place on it.

If the client is concerned about the status and accuracy of any information provided to the professionals, the best thing to do would be to discuss this with them and seek their views as to what is needed, how reliable information could be obtained, etc.

6.3. Nothing the client does relieves the professional

It is common to find in non-standard conditions of engagement provisions which say in effect that no approval, comment, decision or instruction by the client will relieve (or exclude or limit) the professional of his liability and/or any of his obligations under his appointment. Such clauses ought to be resisted as they can reduce the professional's rights and/or defences.

First, as far as liability is concerned, the common law has rules which would determine whether or not the professional's liability has been affected by the conduct of the client (or others on the client's behalf). That conduct may amount to contributory negligence which can only be ascertained on the actual facts in question. If the client is contributorily negligent, there is no possible justification for excluding the usual apportionment under the Contribution Act – see Sections 12.7 and 12.10.

The other possibility is that the client's conduct, whether negligent or not, could constitute a break in the chain of causation which would absolve the professional of liability altogether. This is just another way of saying the professional's negligence has not caused the damage in question. This would arise, for example, when the client decides to rely on somebody else or refuses to take the professional's advice. Again, in these circumstances, why should the professional still be held liable? A professional should not be required to say in advance what the effect of the client's behaviour should be.

Secondly, it makes no sense to say that no decision or instruction of the client will relieve the professional of his obligations under the appointment. Circumstances change during a project and something the client required of the professional at the start may not in fact be required, or the client may require something different from what was originally requested – say a concept design only and not a detailed design of a particular element. To that extent, therefore, the original 'obligation' has been changed and the client should not be able to revert later to the original obligation and insist on its compliance, as this sort of clause might well entitle him to do. If there is any variation to the original obligations, this should be determined on the usual basis – is it a variation and has it been agreed by the other party?

Professional Services Agreements
ISBN 978-0-7277-4158-5

http://dx.doi.org/10.1680/psa.41585.093

Chapter 7
Payment of fees

7.1. Offers, estimates and quotations

It is important when making a fee proposal for a professional to distinguish between whether an offer is being made to do the work for a certain amount, or whether it is an estimate only which is therefore subject to review and possible amendment. If it is an estimate, it must be clearly stated as such. For example:

> As discussed you will appreciate that it is not possible for us to predict the level of fees and disbursements that we would incur in undertaking the above Services but for budgetary purposes we currently estimate a total in the order of £xxx inclusive of fees and disbursements but exclusive of VAT.

or:

> Our fees and disbursements will of course reflect the actual services undertaken but you may wish to allocate domestic budgets now for those services. We advise for that purpose that we currently estimate a total of £xxx (fees plus disbursements) excluding VAT. Should the Services develop so as to need revision of this estimate we shall advise you.

In the second example, it is imperative that if the estimate is going to be exceeded the client is informed promptly or it may be able, quite properly, to argue that, as no revision was indicated, the estimate has become the fee payable.

There is no principle in contract law that a document describing itself merely as an estimate cannot be treated as an offer to contract (*Croshaw* v. *Pritchard (1899)*). The acceptance of an estimate can create a binding contract.

It would, of course, be open to the professional to argue that he did not intend to be bound by the estimate, or that the estimate was not an offer as it was only a statement of price to be used as the basis for further negotiations. The distinction between an estimate and an offer can be a fine one, and the individual circumstances must be considered.

If an offer or quote for a specified sum is given in relation to a specified piece of work, and that is accepted, the agreement is binding and no revision to the fee will be permitted unless the offer itself makes clear the circumstances in which that will happen. It is also useful in those circumstances to set out how fees will then be charged – for example, on a specified hourly rate.

If any fee proposal is to be related to the 'Cost of the Works' or 'the Cost Plan', always check the definition of 'the Works' or 'the Cost Plan' and the provisions as to what is to happen if the cost of the works or cost plan is exceeded, and that the final settling up provisions are satisfactory.

7.2. Lump sums and fixed fees

The principles in relation to offers are particularly important where a client wishes to pay a professional on a lump sum or fixed fee basis. Further, and subject to the provisions of the Construction Act 1996 relating to instalments (see Section 7.6), unless payment intervals are also agreed a client can insist on the services being completed before the lump sum fee is paid to a professional for services rendered. If these stage payments are based on, say, a regular percentage of the total being paid monthly, there is a danger that those accumulative percentages will lag behind the actual amount and costs of input of the professional. If the client should then cease trading the professional would make a disproportionate loss. The court (*Bosky* v. *Shearwater Property Holdings (1992)*) held that a quantity surveyor was entitled only to that part of the agreed lump sum remuneration which had accrued at the date of termination, rather than the value of work carried out at that date. A better solution therefore is to agree stage payments based on the actual input of the professional at the end of each stage. This can be particularly important where the client has a right to terminate at any time – see Chapter 8 for this and also a discussion of what is payable on termination and suspension.

7.3. Disbursements

The basis on which disbursements and expenses are to be paid and whether or not they are included in the fee should be made clear.

7.4. Conditional payment and speculative work

Clauses relating to payment need to be reviewed to see whether any condition has been placed on payment, for example, 'subject to the satisfactory performance by the professional of the services and its other obligations the client shall pay . . .'. Such clauses would allow the client to withhold payment if the client believes the services to be unsatisfactory. This would be a subjective test. If a professional is prepared to accept such a condition it would be better to make the test objective, that is, 'if in the reasonable opinion of the client the performance by the professional of the services is satisfactory the client shall pay . . .'

There have been some non-standard conditions of engagement which make the payment of fees conditional on the signing of collateral warranties. This would clearly be unacceptable if some or all of those warranties could be signed after the completion of the services – is the professional to receive no fee until the last warranty is signed?

Especial care needs to be taken with the terms on which any speculative work is done or conditional fees are undertaken. For example, if work is undertaken in order to obtain planning permission, is the professional to be paid whether or not the planning application is successful or only if the planning permission is obtained? If the latter, is it clear what constitutes 'obtained'? Is the professional to do all the work in connection with any appeal within the original fee, regardless of how long this takes or how far the appeal is taken? Is there to be a success fee? If so, is it clear when it is to be paid and have all contingencies been covered?

If payment is dependent upon 'funding', for example, is it clear what will trigger the payment? Is it the signing of the funding agreement? Is it when any or all conditions attached to the funding have been fulfilled, even though these may not affect the professional, or is it when monies are released to the client? Does the fee agreed reflect the length of time the professional may have to wait to be paid?

Similar considerations apply to competition work, for example, are the fees only payable if the competition is won? What if the scheme is not then implemented?

It is perfectly possible for a professional to undertake work on a speculative basis and take the risk of never being paid for that work. This could happen, for example, when a contractor asks the professional to do some design work to assist in his tender and it is agreed that if the contractor fails to win the tender the professional gets nothing. If it is agreed, however, that if the contractor wins the professional will be employed, care will need to be taken in formulating that post-tender agreement – the essential terms of that appointment, including the services to be undertaken and the fee to be paid, would need to be agreed at the time of the original contract or there is a danger that this will only be an 'agreement to agree' which would not be enforceable.

The terms of the licence of any copyright work produced by the professional in these circumstances also need particular consideration. Is the client to have a licence in any event, only on the payment of fees, or only if the project goes ahead (see Chapter 10)?

7.5. Conditions precedent to payment

Some rights to payment or additional payment can be subject to conditions precedent – that is the right to payment does not arise unless the condition precedent is satisfied. If clients are very concerned about the delivery of collateral warranties, for example, they can make payment of the instalment then duedependent on their delivery. More usually, conditions precedent to payment are based on the giving of notices or the sort of invoice or documentation that has to be submitted before the client has an obligation to pay. Where additional payment is concerned, the obligation to pay can be dependent on the professional having given prior written notice that additional fees are going to be incurred and that he is about to perform additional services and sometimes it is also provided that the basis of payment has to be agreed before the obligation to pay arises. The courts will give effect to conditions precedent provided they are clearly worded and thus if a professional fails to comply with the precondition there will be no right to payment, however much additional work he has carried out, or however much additional cost has been incurred.

7.6. Payment of fees by instalments

It is assumed that a professional can ascertain the appropriate fees required for the scope of the work concerned, so that aspect is not considered further in this text. The sequence of payments needs to be considered also; that is, if the provision of services extends over several months or years a professional will inevitably require periodic payments for work done to date or when certain landmark points have been achieved. Non-standard conditions of engagement may have no provision for this, or their provisions may not be adequate. Such matters need to be agreed in the appointment.

Dealing with the effect of delay on instalments or stage payments is always difficult but any provisions concerning this should be fair. If fee payment is related to stages, there needs to be some provision to protect the professional if a stage becomes very delayed, especially if this is due to default by others. This will be of less importance if there is an entitlement to additional fee payment for the costs of delay and disruption.

ACE Agreement 1 provides for payment in instalments by reference to a schedule or at weekly, monthly or quarterly intervals or by references to stages of the works.

RIBA 2010 provides that the intervals are to be of not less than one month or as specified in the schedule of fees.

RICS 2011 clause 9.3 provides for instalments to be set out on the dates or on completion of the activities set out in a schedule, but if nothing is said monthly instalments apply.

PSC3 deals with payment in a different manner from these standard forms – see Chapter 9.

7.7. The Construction Act 1996 and the Local Democracy, Economic Development and Construction Act 2009

The Construction Act 1996 which applies to 'construction contracts' – see Section 25.3 – provided for certain matters in relation to payment to be included in such contracts. For the most part. professional services agreements are construction contracts. The Local Democracy, Economic Development and Construction Act 2009 has amended some of the payment provisions in the Construction Act but the Construction Act 1996 provisions still apply to contracts entered into prior to 1 October 2011 in England and Wales and 1 November 2011 in Scotland.

The following sections, where appropriate, describe the provisions of the Construction Act 1996 and indicate whether or not they have been amended by the Local Democracy, Economic Development and Construction Act 2009. Section 7.19 describes the amendments made by the Local Democracy, Economic Development and Construction Act 2009 and Section 7.20 the revisions that have been made to the standard conditions of engagement as a result.

7.8. Instalments under the Construction Act 1996

Section 109 of the Act provides the following.

(*a*) A party is entitled to payment by instalments unless the contract specifies the duration of the work is to be less than 45 days or the parties agree the duration is estimated to be less than 45 days.
(*b*) The parties are free to agree the amounts of the payments and the intervals at which, or circumstances in which, they become due.
(*c*) In the absence of such agreement, the relevant provisions of the Scheme for Construction Contracts (England and Wales) Regulations 1998 (the Scheme) apply – see Section 25.3.

However, the use of the Scheme by adoption or default should be avoided if possible because the Scheme has been drawn up with contractors, not professionals, in mind. The basis of calculation is by reference to aggregate amounts over periods and relates to the 'value of any work performed' plus 'an amount equal to the value of any materials manufactured on site or brought onto site' plus 'any other amount or sum which the contract specifies'. It is not clear what the 'value of any work' means where professional services are concerned or how that relates to an agreed fixed sum or hourly basis of payment, etc.

It would be much safer, therefore, for the parties to agree the intervals and the amount to be paid at each interval or a mechanism for deciding the same.

The instalment period prescribed by the Scheme is 28 days – which may prove helpful in negotiations with clients. There is nothing, however, in the Act itself which would stop a client from attempting to impose longer (or very long) instalment periods or instalments intervals that do not reflect the work done at any particular stage. The difficulty then would be in deciding whether or not to agree to the client's proposal or let the Scheme apply. The Scheme would, however, automatically bring with it the method of calculating the amount due at each interval.

The Local Democracy, Economic Development and Construction Act 2009 does not make any amendments to these provisions.

7.9. Additional fees

It is surprising how many non-standard conditions of engagement contain neither provisions for additional fees nor a basis of payment for additional services. The terms should generally include a specified list of additional services, and the basis of fees for those additional services, and provide for additional payment for such matters as changes by the client, or contractor changes accepted by the client, and for delays or for additional work caused by reasons beyond the professional's control. It may also be appropriate to have a provision allowing for the annual or other revision of hourly rates if the project is to go on for a long period. Again, the basis of such additional payment should also be set out.

Such matters are variously covered by the standard conditions of engagement as follows.

ACE Agreement 1 no longer provides for additional services as such – the services schedule sets out only the services that the professional is to perform which can include some or all of those listed under 'Other Services' (i.e. optional services). Professionals should therefore advise clients before they undertake any additional services, agree with the client that they are additional, and agree on a fee basis therefore before starting on them, as they are outside the scope of Agreement 1. In the earlier ACE Agreements it was anticipated that there would be a list of additional services the client could ask the professional to perform and if he did so the professional had to provide them.

ACE Agreement 1 therefore only deals with the circumstances in which the professional could become entitled to additional payment, and provides as follows:

> F5.11 If the Consultant has to carry out additional work and/or suffers disruption in the performance of the Services because:
> (*i*) the Project or the Works or Brief is or are varied by the Client; or
> (*ii*) of any delay by the Client in fulfilling the Client's obligations or in taking any other step necessary for the execution of the Project or the Works; or
> (*iii*) the Consultant is delayed by others (or by events which were not reasonably foreseeable); or
> (*iv*) the Project or the Works are damaged or destroyed; or
> (*v*) of other reasons beyond the control of the Consultant
> the Client shall make an additional payment to the Consultant in respect of the additional work carried out and additional resources employed (unless and to the extent that the additional work has been occasioned by the failure of the Consultant to exercise reasonable skill, care and diligence) and/or the disruption suffered. The additional payment shall be calculated (unless otherwise agreed) on a time basis as set out in Part E: *The Schedule of Fees* with payment by instalments in accordance with Part E: *The Schedule of Fees* save that such instalments shall start at the next instalment date so provided. The Consultant shall advise the Client when the Consultant becomes aware that any such additional work is required or disruption shall be suffered and shall if so requested by the Client give an initial estimate of the additional payment likely to be incurred. Where payment for such additional work or disruption is to be in the form of lump sums, these lump sums and the intervals at which instalments shall be paid and the amounts of each instalment should be agreed prior to the additional work being commenced. For the avoidance of doubt it is hereby agreed that if the Consultant carries out any work which subsequently becomes redundant the Client shall (unless otherwise agreed) pay the Consultant therefor on a time basis as set out in Part E: *The Schedule of Fees*.

RIBA 2010 has a provision for payment where the professional has performed 'Other Services' specified in the services schedule but not included in the basic fee.

These are, it is assumed, the 'Other Services' which are to be set out in Part 3 of the Schedule and which have to be deleted if they are not required. It is thought that this is meant to be a list of additional services that the professional could provide if asked. These 'Other Services' comprise some sites, buildings and related services. There are then some special services also listed in the Schedule under 'Other Services' which 'are only required if the need arises during the commission and which can be instructed as additional Services'. It is not clear whether these are also 'Other Services' for the purposes of this clause. There is also a reference to additional services in the Schedule and in clause 5.1 but not here.

Fees for 'Other Services' can either be a lump sum, or sums, or time charges, or other agreed methods. It is assumed this should be agreed at the time the 'Other Services' are instructed, but nothing is said about this and the fees and expenses part of the Schedule seems to anticipate time charges plus expenses only for such services.

There is a provision for index linking of all lump sums and of time charges every 12 months. Clause 5.7 says 'Every 12 months, lump sums complying with clause 5.5, less any amounts previously claimed, and rates for time charges shall be revised in accordance with changes in the Average Earnings Index . . .'. It is not clear why this should be 'less any amounts previously claimed'. Is it trying to make clear that it is only the lump sums yet to be paid and hourly rates going forward that are to be index linked?

Rates for mileage and printing are also to be revised in accordance with changes in the Consumer Price Index.

This clause then goes on to provide that the 12 month period commences 'on the anniversary of the Effective Date of this Agreement specified in the Project Data or, where there are lump sums calculated by reference to the construction cost, the period commences on the date of calculation of the lump sums'. As the date on which the lump sum is to be calculated varies, it is not clear when the sum gets indexed. These provisions are not at all clear.

As to an adjustment to the fee, RIBA 2010 provides:

> 5.8 The Basic Fee:
> 5.8.1 shall be adjusted including due allowance for any loss and/or expense if:
> (*a*) material changes are made to the Brief and/or the Construction Cost and/or the Timetable; and/or
> (*b*) the Services are varied by agreement.
> 5.8.2 shall not be adjusted for any reduction of the Construction Cost arising solely from deflationary market conditions not prevailing at the Effective Date. The Basic Fee shall continue to be based on the Construction Cost current prior to the date of such reduction.

This says the basic fee has to be adjusted 'including due allowance for any loss and/or expense' if (a) any material changes are made to the brief and/or the construction cost and/or the timetable and/or (b) the services are varied by agreement. It does not say on what basis the basic fee is to be adjusted; neither does it say that the adjusted fee becomes the basic fee and payable as such by the client thereafter.

This is not a satisfactory provision. It appears to apply to all bases of fees. It is difficult to see why this should apply to time charges or any fees fixed by reference to percentages of the Construction Cost. In the 2007 Conditions and SFA/99, it only applied to pure lump sums and in SFA/99 the adjustment was treated as additional fees. The 2007 Conditions did not say how the adjustment was to be made.

What is the 'loss and expense' that is anticipated? Is this the professional's if his services are reduced by a material change or variation? Why are not material changes dealt with under the additional fees clause 5.9?

Clause 5.8.2 says there is to be no adjustment for any reduction in the Construction Cost which arises solely from deflationary market conditions not prevailing at the effective date. The basic fee continues to be based on the Construction Cost current prior to the date of such reduction. This seems to provide that no fees fixed by reference to a percentage of the Construction Cost are to be reduced if there are deflationary market conditions which reduce that cost after the effective date. It is not clear how the basic fee can continue to be based on the Construction Cost prior to the date of such reduction, when other provisions could require estimates to be made at a later date which would presumably allow for such reductions. Nor is it clear how the final Construction Cost gets calculated.

As to additional fees, RIBA 2010 provides:

> 5.9 Where the Architect for reasons beyond the Architect's reasonable control incurs extra work or loss and expense for which the Architect would not otherwise be remunerated, the Architect shall be entitled to additional fees calculated on a time basis as set out in clause 5.6 unless otherwise agreed. Matters in relation to which the Architect shall be entitled to additional fees include but are not limited to circumstances where:
>
> 5.9.1 the Architect is required to vary any item of work commenced or completed pursuant to this Agreement or to provide a new design after the Client has authorised the Architect to develop an approved design;
>
> 5.9.2 the nature of the Project requires that substantial parts of the design cannot be completed or must be specified provisionally or approximately before construction commences;
>
> 5.9.3 performance of the Services is delayed, disrupted or prolonged; and
>
> 5.9.4 the cost of any work, installation or equipment for which the Architect performs Services is omitted from or not included in the Construction Cost.
>
> This clause 5.9 shall not apply where any adjustment under clause 5.8.1 applies to the same events.

> The Architect shall inform the Client on becoming aware that this clause 5.9 will apply. This clause 5.9 shall not apply where any change or extra work or expense arises from a breach of this Agreement by the Architect.

This clause gives the professional an entitlement to additional fees where for reasons beyond his reasonable control he incurs extra work or loss and expense for which he would not otherwise be remunerated. These fees are to be paid on a time basis unless otherwise agreed. The circumstances where he is entitled to an additional fee include those where he is required to vary any item of work commenced or completed pursuant to the agreement, or to provide a new design after the client has authorised him to develop an approved design; or the nature of the project requires that substantial parts of the design cannot be completed or must be specified provisionally or approximately before construction commences.

Clause 5.9.3 also says that the professional is entitled to additional fees where the performance of the services is delayed, disrupted or prolonged or the cost of any work, installation or equipment for which the professional performs services is omitted from or not included in the Construction Cost. It then says that clause 5.9 does not apply where there is any adjustment under clause 5.8.1 for the same event. In the case of clause 5.9 the professional has to inform the client on becoming aware that the clause will apply – this is not the case where there is to be an adjustment to the fee under clause 5.8.

Finally, clause 5.9 says that no additional payment will apply where any change or extra work or expense arises from a breach of the agreement by the professional. (Note: 'Change' has been added in the 2010 Conditions.) Again, this does not seem to apply to an adjustment to the fee under clause 5.8.

Both the ACE and the RIBA say that the professional has to advise or inform the client on becoming aware that additional work will be required. This is a contractual obligation and the professional will be in breach if he does not do so, but it does not amount to a condition precedent because the right to payment has not been made conditional upon such notice being given. If the professional does not notify, or is late in notifying, the client, this will, however, be a breach of a contractual obligation and the client may be able to demonstrate that this has caused him additional cost or expense which he would not otherwise have incurred, and that he is entitled to set this off against the amount payable to the professional in respect of additional fees. Equally, if informing the client late has meant the additional fees are higher, the client may only have to pay the additional fees that would have been payable if he had been informed in time.

RICS 2011 provides:

> 10.1 If at any time it is apparent to the Consultant that Additional Services are likely to be required, it notifies the Client of this as soon as reasonably practicable after identifying the requirement and submits a proposal for Additional Services to the Client.

10.2 The Consultant performs such Additional Services as it is instructed in writing to perform by the Client provided they are reasonably capable of being performed by the Consultant. The Consultant may, in the case of an emergency, perform Additional Services without such an instruction, but not otherwise.

10.3 If the Consultant is instructed by the Client to perform Additional Services, it gives the Client, within a reasonable time of the Client's request, a written estimate of:

(*a*) the cost of providing such Additional Services, taking into account any reduction in work or savings which might result; and

(*b*) the effect (if any) of such Additional Services on the Programme.

10.4 Wherever practicable, the parties agree the Additional Services Fee and the timing of its payment before Additional Services are performed. If the Client instructs the Consultant under clause 10.2 to perform Additional Services notwithstanding that an Additional Services Fee is not agreed, or if the Consultant performs Additional Services in the case of emergency, the Additional Services Fee is based on the time-charges set out in Part D of Schedule 2.

'Additional Services' comprise any services relating to the project which are not identified as basic services and listed in the schedule – they are therefore traditional additional services.

There is also a right to an adjustment to the fee under clause 9.11 if 'the performance of the Basic Services is materially delayed and/or disrupted as a result of a change in the scope, size, complexity or duration of the Project or for any cause outside the Consultant's reasonable control'. This restricts the right to additional payment to delay or disruption only. Notices of the intention to claim must be given as soon as reasonably practicable together with a written estimate of the adjustment to the fee and programme. There is no adjustment to the fee to the extent that the delay or disruption is due to any negligence or breach by the professional.

The RICS provisions are unusual in that there is a mechanism to allow the agreed additional fees to be varied further, as follows:

10.5 The Additional Services Fee is adjusted if the performance of the Additional Services is materially delayed and/or disrupted as a result of a change in the scope, size, complexity or duration of the Project or for any other cause outside the Consultant's reasonable control.

10.6 If the Consultant intends to make a claim for an adjustment of the Additional Services Fee under clause 10.5, it gives the Client:

(*a*) notice of such intention as soon as reasonably practicable after it becomes aware of any material delay and/or disruption to the Additional Services; and

(*b*) a written estimate of the proposed adjustment to the Additional Services Fee and of the likely effect on the performance by the Consultant of the Services and on the Programme.

Wherever practicable, the parties agree the adjustment to the Additional Services Fee and the timing of its payment. If the adjustment to the Additional Services Fee is not agreed, such adjustment is based on the time-charges set out in Part D of Schedule 2. Unless otherwise provided in the Appendix, such time-charges are inclusive of all disbursements and expenses other than the disbursements and expenses referred to in Part C of Schedule 2.

10.7 The Consultant is not entitled to an adjustment to the Additional Services Fee to the extent that the delay and/or disruption of the Additional Services is due to any negligence of, or breach of this Appointment by, the Consultant.

This last provision and the equivalent provision applying to fee adjustments allows for a proportionate reduction to be made to the Additional Services Fee to reflect the extent to which the professional's negligence or breach has caused or contributed to the delay or disruption.

7.10. Date for payment of fees

Non-standard conditions of engagement may make no provision, or no adequate provision, for the period within which a professional's fees must be paid. This is obviously unacceptable and needs to be negotiated to the professional's satisfaction, even where the agreement is a construction contract and therefore the Construction Act will apply.

7.11. Due date and final date for payment of fees under the Construction Act 1996

Where the Construction Act 1996 applies Section 110 provides that any 'construction contract' must include the following.

(*a*) An adequate mechanism for determining what payments become due and when.
(*b*) A final date for payment of any sum which becomes due.
(*c*) The parties are free to agree the period between the date on which a sum becomes due and the final date for payment.

It does not necessarily follow that providing for intervals for payment such as the first of each month, also means that the payment is due on that first day of the month. It is more usual, and probably better practice, for the appointment to state when an amount is due. Prior to the amendments made as a result of the Local Democracy, Economic Development and Construction Act ACE Agreement 1 provided that 'Payments shall become due for payment on submission of the Consultant's invoice therefor' and RIBA 2010 stated that 'Payment shall become due to the Architect on the date of issue of the Architect's account' and ACE Agreement 1 provided for the final date for payment to be 28 days thereafter and RIBA 2010 for 28 days from the issue of the relevant account. RICS 2008 provided for payment to become due on the date the client received each invoice and the final date for payment to be 28 days thereafter or at a time to be stated in the Appendix.

The Scheme for Construction Contracts is incorporated if no mechanism is included in a 'construction contract' for establishing the due date and the final date. It provides a rather complicated formula for establishing those dates.

(*a*) The 'due date' is the later of 'the expiry of seven days following the relevant period', that is, the 28-day period prescribed by the Scheme for instalment periods (if the 'construction contract' fails to provide for this) or the instalment period fixed by the appointment itself or 'the making of a claim by the payee'. An invoice or an application for payment would qualify as 'making a claim', provided it contains the information prescribed by the Scheme. 'Making a claim' is 'a written notice given by the party carrying out the work under a construction contract to the other party specifying the amount of any payment or payments which he considers to be due and the basis on which it is, or they are, calculated'.

(*b*) The 'final date for payment' is 17 days thereafter. This makes a total of 24 days.

(*c*) If the contract is so short that there are no instalments, or where it is the last payment, the Scheme provides that payment becomes due on the expiry of 30 days following completion of the work or the making of a claim by the payee, whichever is the later, and the final date for payment is again 17 days thereafter.

No alterations to these provisions are contained in the Local Democracy, Economic Development and Construction Act.

7.12. Interest payments and Late Payment of Commercial Debts (Interest) Act 1998

At common law, interest is not payable on late payment of debts unless there is an express agreement to that effect or an agreement can be implied from a course of dealing. Most non-standard conditions of engagement make no provision for interest. Various statutes, however, provide that interest may be claimed on outstanding payments in any proceedings.

Section 35A of the Senior Courts Act (formerly the Supreme Court Act 1981) provides that simple interest is payable on debts or damages in the High Court, and Section 69 of the County Courts Act 1984 makes similar provisions in relation to county court debts or damages. Section 35A of the Senior Courts Act gives the court power to award interest at such rate 'as the court thinks fit or as the rules of the court may provide' on all or part of a debt between the date on which the cause of action arose and the date of the judgement. Such interest only applies if a claim form has been issued.

Where no rate of interest is fixed, there is no rule as to the amount of interest that will be allowed by the court, and the rate may vary according to the practice of the court and the circumstances of the case. Long ago, the courts allowed a rate of 5% for commercial transactions and in other cases interest at a rate of 3% to 4%. Nowadays the courts award interest at more commercial rates.

The Late Payment of Commercial Debts (Interest) Act 1998 provided that from 1 November 1998 there was to be a statutory right to interest on late payments. Initially, the Act only applied to debts owed by a large business or the public sector to a small business supplier. Businesses include those of professionals. The Act applied to any debt resulting from an agreement made between two businesses for the supply of goods or services. A large business is one that employs more than 50 people and a small business is one that employs 50 people or fewer. There are complicated rules for calculating the number of employees which is outside the scope of this text.

From 1 November 2000, the provisions covered debts owed to small businesses by all other businesses and the public sector. From 1 November 2002, the provisions covered debts owed to all businesses, and the public sector can use the Act against all other businesses and the public sector.

The rate of interest to be paid is the 'reference rate' plus a percentage (currently 8%). The reference rate is fixed for six-month periods: 1 January to 30 June and 1 July to 31 December each year. In August 2002 (the first six-month period), it was 4%. For the period 1 July to 31 December 2011 it was 0.5%. The 8% is then added to the relevant reference rate. It may be claimed from the date the debt becomes overdue (that is, the first day after any agreed credit period has ended or, if none, 30 days after the performance of services or after notice of the debt has been given, whichever is the later) to the date of judgement. Thereafter, interest accrues at the current statutory rate of interest. If, however, a contractual rate of interest has been agreed, that will apply instead, provided it is 'substantial'. Sometimes non-standard conditions of engagement require the professional to agree that a certain (and often fairly low) rate of interest is 'substantial'. Such a rate might well not be so considered by the courts. It is yet to be seen whether this is an effective bar to the professional being able to argue that the agreed rate of interest is not 'substantial' – the author's view is that it should not necessarily be so – 'substantial' being a statutory requirement.

ACE Agreement 1 provides for interest to be paid on amounts unpaid after the final date for payment in accordance with the provisions of the Late Payment Act. RIBA 2010 provides for interest at a daily rate equivalent to 8% over the dealing rate of the Bank of England current at the date a payment becomes overdue. RICS 2011 provides for interest to be paid on a daily basis at the rate of 4% above base rate and this is acknowledged as constituting a 'substantial remedy' for the purposes of the Act.

7.13. Payment of undisputed portions of invoices

Where the contract is a 'construction contract' and Construction Act applies, the provisions concerning notices of withholding described in Section 7.16 will apply and no monies can be withheld unless such a notice has been given. Where the Local Democracy, Economic Development and Construction Act applies, the provisions concerning notices of intention to pay less apply – see Section 7.19 – and no monies can be withheld unless such a notice has been given.

Non-standard professional services agreements which are not 'construction contracts' should contain provision for the payment of non-disputed portions of invoiced fees. This should state that the undisputed portion must be paid (in accordance with the agreed time for payment provisions), and also that interest is due on the payment of disputed amounts later agreed and paid.

ACE Agreement 1, RIBA 2010 and RICS 2011 all incorporate some or all of the payment provisions of the Construction Act 1996 as amended, whether or not the appointment is a 'construction contract' and the earlier versions of these agreements included some or all of the Construction Act 1996 payment privisions. None of them now contain express provisions concerning disputed invoices although earlier versions of the RIBA Agreements stated that the client shall not delay payment of any disputed part of the account.

7.14. Notice of payment under the Construction Act 1996

Where the Construction Act 1996 applies, Section 110(2) provides:

> Every construction contract shall provide for the giving of notice by a party not later than five days after the date on which payment becomes due from him under the contract, or would have become due if:
> (*a*) the other party has carried out his obligations under the contract and
> (*b*) no set-off or abatement was permitted by reference to any sum claimed to be done under one or more other contracts specifying the amount (if any) of the payment made or proposed to be made and the basis on which that amount was calculated.

If a 'construction contract' does not contain such a provision the relevant provisions of the Scheme apply. The Scheme simply repeats the foregoing provisions.

The reason for the rather convoluted provision about set-off is to stop payers arguing, when payment is due, that no notice was necessary either because nothing is due because the work or services are, for example, defective, or because under the contract (or any other contract between the same parties) monies/damages were owed by the payee to the payer. The payer must give a notice which specifies the amount, if any, of the payment made or proposed to be made and the basis on which that payment has been calculated. The notice therefore is needed even if payment is to be made in full or if no payment will be made. (Set-off is also discussed in Section 7.15.)

Failure to give the notice of paying is a breach of contract (although it is difficult to see what loss a payee suffers in the absence of such a notice if the amount asked for was received before, or on, the final date for payment).

The five-day notice of payment period in Section 110(2) is the only mandatory period stipulated in the Act. The parties cannot alter it, but they can arrange 'the due date' so that it minimises administrative inconvenience.

Although it is not necessary to set out the requirement for giving notice of payment, because if omitted it will be imported into every 'construction contract' by the Scheme. Earlier versions of the ACE, RIBA and RICS Agreements all incorporated it expressly.

It should be noted that if any appointment document contains these notice provisions even though the appointment is not a 'construction contract' within the Act, the client has to give the notice of payment (and, if necessary, notice of withholding – see Section 7.16) because these will then be express terms of the contract.

These notice provisions have been altered by the Local Democracy, Economic Regeneration and Construction Act – see Section 7.19.

7.15. Set-off generally

The common law has complex and sophisticated rules as to what can and cannot be withheld or set-off against monies due, such as professional fees.

The courts have recognised, for example, that a defendant to a claim for payment has the right to reduce or eliminate the amount claimed by using any debts which it is owed by the claiming party.

Legal set-off is a defence to a court action, not a cross-claim, but a cross-claim can form the basis of a set off. Legal set-off applies only if both claims are readily ascertainable and are due at the date of the action. The cross-claim must be for a monetary amount or a liquidated debt, and that amount must be quantifiable at the time the right of set-off is claimed. There can be no legal set-off where the cross-claim is for an amount that is not readily ascertainable, for example, damages, but there can be equitable set-off.

There is no requirement that the original claim and the cross-claim are linked to the same transaction, so amounts due under one contract could be set off against amounts due under another contract. However, the claims must be between the same parties 'in the same right'. This last requirement merely means, for example, that a personal claim could not be used to set off a contractual claim.

Legal set-off can be excluded by agreement between the parties in their contract.

There are also equitable rights to set-off in relation to the same contract – for example, a client has an equitable right to set-off a claim for negligent work by a professional against a claim for fees made by that professional. The circumstances have to be such that the court would have considered that the defendant was entitled to be protected against the plaintiff's claim.

The defendant's cross-claim need not be an ascertained amount, but it must be so closely connected with the claimant's claim that it would be unfair not to allow the set off. This is now stated as arising if the cross-claim flows out of and is inseparably connected with the dealings and transactions which gave rise to the claim.

However, it is not thought that in equity the amounts have to arise out of the same contract. Identical contracts on identical terms for the same project between the same parties should be sufficiently connected to permit the set-off in equity of amounts due under one such contract against an amount due under another such contract.

It is not necessary, therefore, for an appointment to have specific provisions concerning set-off. If it does, the provisions normally give a client wider rights than the common law, and so need to be considered carefully. Also, some contracts or collateral warranties expressly prohibit any set-off by the professional. This too would need to be considered carefully – see Section 21.11 in relation to the prohibition of set-off and counterclaims in non-standard collateral warranties.

In the case of *Axa Sun Life Services plc* v. *Campbell Martin Limited (2011)*, the Court of Appeal considered the following clause excluding the right of set-off:

> All monies payable by you to us under this Agreement shall be paid in full without any deduction or withholding other than as required by law and you will not (and renounce any right you may have to) assert any credit, set-off or counterclaim against us to justify withholding payment of any such Money or amount in whole or in part.

The Court held that there was no basis for this clause not taking effect in accordance with its terms (so it would effectively exclude the right of set-off). However, it was subject to the requirement of reasonableness in the Unfair Contract Terms Act 1977 (as decided in a previous case on a materially identical term) as the contract in this case was subject to the Unfair Contract Terms Act being on one of the party's standard business terms – see Section 16.1.

Equitable set-off can be used outside the context of legal proceedings in response to a liability to pay money otherwise due and in response to the exercise of rights by the other party. For example, if equitable set-off applies, a right to terminate which is contingent on non-payment cannot be exercised except insofar as the non-payment exceeds the cross-claim. For equitable set-off to arise, the claim relied upon does not have to be valid; it just needs to be made reasonably and in good faith.

Neither ACE Agreement 1 nor RICS 2011 contains any reference to set-off.

RIBA 2010, however, in clause 5.16, has a prohibition of all rights of set-off.

This is relevant to the Construction Act provisions concerning notices of withholding referred to in the next section (and to the notice of intention to pay less provisions of the Local Democracy, Economic Development and Construction Act), and this clause and its effect are therefore discussed in that section.

7.16. Notice of intention of withholding payment under the Construction Act 1996

Where the Construction Act 1996 applies Section 111 provides:

> (1) A party to a construction contract may not withhold payment after the final date for payment of a sum due under the contract unless he has given an effective notice of intention to withhold payment.

(2) To be effective such a notice must specify:
  (*a*) the amount proposed to be withheld and the ground for withholding payment, or
  (*b*) if there is more than one ground, each ground and the amount attributable to it,
  (*c*) and must be given not later than the prescribed period before the final date for payment.

(3) The parties are free to agree what that prescribed period is to be.

As usual, if the appointment does not set out what the prescribed period is to be, the Scheme applies. That stipulates that the prescribed period is not later than 'seven days before the final date for payment determined either in accordance with the construction contract or, if there is no such provision, in accordance with the provisions of the Scheme'.

Section 111 of the Act also states the notice of payment can be used as the notice of withholding (although it would have to be given earlier under Section 110 if the Scheme applied) if it contains the information prescribed in Section 111.

If there is a dispute as to the amount withheld or the grounds for withholding, the matter can be referred to adjudication. The Act then provides for a final date for payment once the adjudicator has decided what amount should be paid. It is 'seven days from the date of the decision or the date which apart from the notice would have been the final date for payment, whichever is the later'.

These notice provisions have been altered by the Local Democracy, Economic Development and Construction Act. Earlier versions of the ACE, RIBA and RICS Agreements all incorporated the notice of withholding provisions of the Act, although they did not all use the exact wording of the Act.

The RIBA 2010 clause concerning set-off is as follows:

5.16 The Client shall not withhold any amount due to the Architect under this agreement unless the amount has been agreed with the Architect or has been decided by any tribunal to which the matter is referred as not being due to the Architect.
All rights of set-off at common law or in equity which the Client would otherwise be entitled to exercise are expressly excluded.

This is not a reasonable provision and it also is against the spirit of the Construction Act. The Construction Act excludes rights of set-off where the payer has failed to serve a notice of withholding payment. Does the wording in RIBA 2010 mean the client must get a decision from a tribunal before withholding payment (unless that withholding has been agreed by the professional)? It would appear so. What about amounts that may be due but which the client wishes to withhold on account of negligent work?

In any event, the client could use the notice of payment provision to say what he is proposing to pay and then that is the amount that is due, not the amount in the professional's invoice. The timetable for this is quite tight, however.

Also, it might be argued that clause 5.16 is only concerned with what is due and does not affect the right to withhold on other grounds.

Finally, to what is 'otherwise' referring?

This is the provision that was requested by the ACA in order to protect professionals against clients who withheld payment for no very good reason. However, in the case of a 'construction contract', the intention of the Construction Act and any withholding provision is to prevent this, or at least make it much more difficult. The notice of withholding payment has to set out the ground for doing so (which may include a claim for negligence), or each separate ground if there is more than one, and the notice has to be given within a fixed time of an invoice being rendered. Notices that do not comply should be swiftly rejected by adjudicators and courts.

Of course, it may still be possible for potentially spurious claims to comply with the requirements of the notice. However, realistically there is nothing that can be done about that except to challenge the notice.

The Construction Act also includes a right to suspend performance if the payer fails to pay by the final date for payment and has not served an effective notice of withholding – that is, one in the correct form and served within the time prescribed. This is dealt with in Section 8.8.

7.17. Pay when paid clauses under the Construction Act 1996

The Construction Act severely restricts the rights of parties under 'construction contracts' to make any payments conditional on the receipt of monies from a third party.

Section 113(1) of the Act provides:

> A provision making payment under a construction contract conditional on the payer receiving payment from a third person is ineffective, unless that third person, or any other person payment by whom is under the contract (directly or indirectly) a condition of payment by that third person, is insolvent.

This particularly affects contractors and professionals taking on sub-contractors and sub-consultants, when previously they had been able to employ them on a 'pay when paid' basis. The risk of the client not paying or paying late is with the contractor or the professional.

If a pay when paid clause is rendered ineffective by the Act, the Scheme implies into the appointment all the payment provisions of the Act – the right to payment in instalments, a due date and final date for payment, the notice provisions, as well as the mechanisms for determining the amount due.

The only time such a provision is allowed is when the third person upon whom the pay when paid provision relies is 'insolvent'. This is defined in the Act. The definition initially attracted some attention from commentators who suggested the definition was incomplete. The Enterprise Act 2002 (Insolvency) Order 2003 amended the definition to add 'entering into administration within Schedule B1 to the Insolvency Act'. The definition should still be checked, however, by anyone seeking to take advantage of Section 113 to ensure the particular circumstances of the ultimate payer do match those described in the definition.

There is also the difficulty of how this might be expressed in an appointment by a professional of a sub-consultant. Payments to the sub-consultant during the contract cannot be on a pay when paid basis but there could be a general provision relieving the professional from the obligation to make future payments immediately the client is insolvent within the meaning of the Act. If the professional's contract with the client contains a provision for termination if one of the parties is insolvent and if the sub-contract or sub-consultancy appointment contains a provision that that appointment terminates if the professional's contract with the client terminates, this too will be sufficient to relieve the payer of any obligation to make future payments. What has to be paid up to that date will be governed by the contract with the sub-contractor or sub-consultant. (See also Sections 12.3 to 12.5 with reference to sub-consultancy appointments.)

These provisions concerning pay when paid clauses have not been altered by the Local Democracy, Economic Development and Construction Act.

7.18. Pay when paid clauses

Non-standard conditions of engagement may contain 'pay when paid' clauses, that is, the professional's client will pay the professional's fees only after the client has received them himself from whoever is paying the fees. However, a professional will not want to be dependent for payment of fees on a third party with whom there is probably no direct contract, and who might become embroiled in a dispute with the professional's client, become insolvent, or otherwise be a slack or difficult payer.

Pay when paid clauses should thus be rejected except for the strongest commercial reasons.

If accepted, pay when paid clauses must be considered carefully and amended to reduce the risks to the professional as far as possible if he is the payee. For example, the professional might be able to persuade the client to accept a requirement that the client will apply promptly for any payment due to the professional, will pay it over within a short time of receipt and will give access to documents or records so the professional can verify that those requirements are being followed. An obligation for the client to make an application to its funder for any additional payment due to the professional and to recover it on the professional's behalf might also be required.

As a result of the provisions in the Construction Act described in the preceding section, 'pay when paid' clauses have been outlawed by the Construction Act for all 'construction contracts' and so 'pay when paid' provisions are most likely now to be found in sub-consultancy agreements for overseas work.

Other mechanisms can also be incorporated in non-standard conditions of engagement which could significantly delay payment to the professional. For example, payment could be dependent on the services having been 'satisfactory' and then the right to payment (and thus when that payment becomes due under the appointment) does not arise until that is the case. Payment could depend on the opinion of a third party (for example, the project manager) as to whether the invoice is correct, or the professional could be treated as contractors are – obliged to put in an application for payment and it is only when this application is certified as correct by a third party or the client himself that the payment is due under the appointment. These latter provisions, known as 'pay when certified', have been the subject of criticism by commentators on the Construction Act – see Section 7.19 on the proposed amendments to the payment provisions to the Construction Act in this regard.

7.19. Amendments to the payment provisions of the Construction Act 1996 by the Local Democracy, Economic Development and Construction Act 2009

The Local Democracy, Economic Development and Construction Act 2009 (the amended Construction Act) has been passed and was brought into force in England and Wales on 1 October 2011 and in Scotland on 1 November 2011 ('the day appointed'). Separate statutory instruments in these countries were necessary to do this.

It applies to construction contracts entered into after 'the day appointed'. This means that, for some time, some appointments will be subject to the old regime and some to the new regime.

There are also some consequential amendments to the Scheme.

The amendments are complex and the following description is not exhaustive.

7.19.1 Instalments – when payments become due and the amount

Section 110 of the Construction Act says that every 'construction contract' shall provide an adequate mechanism for determining what payments become due and when – see Section 7.11.

The amended Construction Act says that it is not an 'adequate mechanism' where payment is made conditional on the performance of obligations (other than the obligation to pay) under another contract or on the 'decision' by any person as to whether obligations (other than the obligation to pay) under another contract have been performed. So a payer will not be able to stipulate, for example, to the payee that the payee only becomes entitled to an interim payment once the payer has performed such of the works including the payee's sub-contract works under the payer's contract with his employer or that the payee will only become entitled to payment when the architect/contract administrator/engineer decides that the payer is entitled to such payment.

There is an exception to this, thus making it an 'adequate mechanism' for the purposes of the amended Construction Act, and this is where the 'construction contract' is for the carrying out of construction operations by another person and the obligations referred to in the new sub-section are obligations on that other person to carry out those operations. This, it is thought, is the management contracting exception. Here the management contractor is employed to arrange for the construction operations themselves to be carried out by various works contractors and provide for him to certify payment if and when parts of the works have been completed. So it will be an 'adequate mechanism' if the management contractor can determine when payment is due.

Further, the amendments say it will not be an adequate mechanism where a 'construction contract' provides for the date on which a payment becomes due to be determined by reference to the giving to the payee of a notice which relates to what payments are due under the contract. This seems to be directed at contractual provisions which allow the client to say 'no payment is due (and therefore the whole mechanism for paying the professional does not start) until the client gives the professional a notice saying what payments are due'.

It is not known whether this is aimed at the 'pay when certified' type of provisions referred to in Section 7.18. Would such a provision be a notice 'which relates to what payments are due'? Or would it cover any mechanism whereby until the client decides and informs the professional as to what is due nothing is payable? If so, such provisions are very rare.

7.19.2 Payer/payee notices

The whole system for giving notices of payment and notices of withholding has been abolished by the amended Construction Act and replaced with something more complicated.

A 'construction contract' will have to contain a specific term that *either* the payer (or a specified person on behalf of the payer ('specified person')) has to give notice of payment ('the payer payment notice') to the payee not later than five days after the payment due date *or* the payee has to give notice of payment ('the payee payment notice') to the payer or 'specified person' not later than five days after the payment due date.

So far this is like the earlier notice of payment provisions except that the contract will provide who is to give the notice – the payer (or 'specified person') or the payee – and the timetable is the same – it has to be given not later than five days after 'the payment due date'. The payment due date is defined as 'the date provided for by the contract as the date on which payment is due'.

The section then describes what will constitute a 'payment notice'. Where the payer or the 'specified person' gives the notice, the notice has to specify the sum considered to be or to have been due at the payment due date and the basis on which it is calculated. Similarly in the case of the payee, the notice has to specify the sum that the payee considers to be or to have been due at the payment due date and the basis on which that sum is calculated. The notice has to be given even if the figure is zero.

What happens if the 'construction contract' fails to provide for either the payer (or specified person on his behalf) or the payee to give a payment notice? The new sub-section says as before that the Scheme will apply, and this makes the payer responsible.

7.19.3 No payer notice given

What happens if the payer (or the specified person on his behalf) was the one who had to give the 'payment notice' and fails to do so within the five days – what is to be paid in its absence? Section 110B of the amended Construction Act says in these circumstances the payee can give the notice, and provided it contains the right information so as to constitute a 'payment notice' then that amount becomes payable, but not by the original final date for payment, however, as the final date is postponed by the same number of days as the number of days that the payee gave notice after the date the payer could have given notice.

The amended Construction Act also says that if the contract has already permitted or required the payee *before* the time the payer is to give the 'payer notice' to notify the payer of the sum he considers due and the basis on which that is calculated and he has done so, then that will be a 'payee default notice' and the payee cannot give another such notice (i.e. he cannot change his mind and ask for more (or less)) under a new payee payment notice. The payee's invoice would qualify as a 'payee default notice' if the contract requires the payee to submit an invoice containing the information set out in the amended Construction Act before the payer gives a payer payment notice.

When is 'the final date for payment' in such circumstances? Nothing is said and so it is assumed that it will still be the original 'final date for payment'.

Is all this necessary for professionals' contracts? You can see why clients might still want to stipulate what is due to the contractor and make the certificate the 'payer notice' and it can also be seen why a fall-back provision is needed if the payer does not give a notice. But for professionals, the professional's invoice should be a sufficient notice if it sets out what is due and how it has been calculated, and as this notice can be given at any time up to five days after the due date, that should be sufficient. The professional would need to make the due date the date he submits the invoice (as in many non-standard conditions of engagement and the earlier ACE Agreements) rather than the date he is entitled to submit the invoice (e.g., the end of each calendar month) so as not to fall foul of the provision that the notice has to be given 'not later than 5 days after the payment due date'.

7.19.4 Requirement to pay the notified sum

Section 111(1) of the amended Construction Act provides that the payer has to pay the 'notified sum' by the final date for payment (calculated as before).

The 'notified sum' is either

(*a*) the sum in 'the payer payment notice'
(*b*) the sum in 'the payee payment notice'
(*c*) the sum in the notice given by the payee when the payer/'specified person' has not given a 'payer payment notice' – whether he actually gives the notice or relies on the fact that he has already give sufficient details for this to constitute a 'payee default notice'.

7.19.5 Paying less than the 'notified sum'

However, the payer/'specified person' can be relieved of the obligation to pay the 'notified sum' and can pay less if he gives the payee a notice of intention to pay less ('notice of intention to pay less') which specifies the sum the payer considers to be due *on the date the notice is served* and the basis on which that sum is calculated (again this sum can be zero) and gives that notice not later than the prescribed period before the final date for payment (which, as before, can be fixed by the parties or inserted by the Scheme). Further, that notice cannot be given before a 'payee payment notice' or a 'payee default notice' has been given where those are required to give a 'notified sum'.

There is nothing to stop a payer whose payer notice provides 'a notified sum' then giving a notice of intention to pay less, provided he does it at the right time and in the right form.

Section 119(6) of the amended Construction Act provides that the amount in the notice of intention to pay less becomes the amount to be paid in place of the 'notified sum' provided the notice gives the correct information and is given at the correct time (or it will not be a notice under subsection (2) of Section 111).

There, therefore, has to be a 'notified sum' before the 'intention to pay less notice' can be given. The sum in the 'notice of intention to pay less' if any is the sum that has to be paid by the final date.

There is some incentive for a payer not to stipulate too little in the 'payer payment notice' or 'notice of intention to pay less'. Sections 111(8) and (9) of the Act say where there are such notices but on adjudication (and in the case of adjudication only) the adjudicator decides that more should have been paid, then his decision shall be construed as requiring payment of the additional amount not later than seven days from the date of the decision or the date which, apart from the notice, would have been the final date for payment, whichever is the later.

It would not be easy to obtain an adjudicator's decision between the date any notice of payment has to be given and the final date for payment, as this would mean the final date for payment would have to be well over 28 days from the due date, in order to allow for the adjudication timetable. It would probably also be impossible to get such a decision between the time the 'notice of intention to pay less' has to be given and the final date for payment, because this is generally a couple of days; but it is not impossible that some clients may make this more than 28 days.

It is likely, however, that the 'later date' will be the date of the adjudicator's decision.

7.19.6 Insolvent payee

This is a new provision in the amended Construction Act dealing with insolvent payees rather than insolvent payers.

The payer does not have to pay 'the notified sum' (or the lesser sum included in a notice to pay less) by the final date where *the contract provides* that if the payee becomes insolvent the payer need not pay any sum due *and* the payee has become insolvent after the prescribed period before the final date for payment; that is, after the date by which the notice of intention to pay less has to be given.

This provides a very small window of opportunity for payers to be relieved of the obligation to pay, if the insolvency has to occur within the usual timetable for payment. This generally gives only a couple of days or so between giving notice of intention to pay less and the final date for payment. This timetable was designed to give the payer as long as possible to determine whether he is entitled to withhold any monies and to set out the basis and the amount. Now the payer might prefer a longer period during which he can wait to see if the payee has become insolvent so that he does not in fact have to pay.

'Insolvency' has the same meaning as in the Construction Act as amended in 2003 – see Section 7.17.

7.19.7 Suspension rights

Under the amended Construction Act, the right to suspend arises where 'the requirement in Section 111(1) is not complied with . . .': that is, 'The payer must pay the notified sum on or before the final date'. The 'notified sum' is that in the payer/payee default notice, subject to the provisions concerning the notice of intention to pay less. If such a notice has been given, Section 111(6) states that the requirement to pay the notified sum applies only in respect of the sum specified in the notice of intention to pay less.

That right to suspend subsists until the 'notified amount' is paid in full, and the same provisions as under the Construction Act apply – see Section 8.8 – about disregarding the time during which performance was suspended when calculating any time for the performance of any obligations under the contract: except that this has been amended to read 'any period during which performance is suspended in pursuance of *or in consequence of the exercise of* the right [to suspend]'. This could cover the time taken to demobilise, and also to remobilise.

This right still cannot be exercised without giving at least seven days notice.

The other changes to this particular provision by the amended Construction Act are as follows.

(*a*) The payee can suspend 'any or all of the obligations' – under the Construction Act it has to be all of the obligations. The professional therefore will be able to choose which obligations to suspend – it is the payee's right. No guidance is given as to which obligations are chosen or how they are chosen, but it is difficult to see why the professional would want to suspend only some of his obligations or services if the client is not paying him.

(*b*) The party in default is to be liable to pay the suspending party 'a reasonable amount in respect of costs and expenses reasonably incurred by that party as a result of the exercise of the right'. It is assumed this covers demobilisation and remobilisation costs and any other direct costs, for example, laying off people, disruption, redundancies, but no guidance is given. This is something that was not covered by the Construction Act. Again, this should act as an incentive on the payer to pay.

7.19.8 Will professionals still be able to suspend because clients forget to serve 'the intention to pay less notice'?

Where the obligation is on the payer/'specified person' to give the notice of payment, it is possible they could give this notice and then forget to serve the 'notice of intention to pay less'. It would, however, only be necessary for a payer to serve a 'notice of intention to pay less' if the notified sum in the notice of payment is greater than the amount the client thinks he is obliged to pay by the time the right to give 'notice of intention to pay less' arises. This is because there seems to be nothing to stop the 'payer payment notice' containing a sum less than the payee has asked for, for example, in his invoice. And thus the payer would only need to serve a 'notice of intention to pay less' if something had happened between the two dates which would reduce the amount payable further. As a consequence, the payee will only be able to suspend if the client forgets to serve the notice of intention to pay less and either pays less than the amount in the payer notice of payment or nothing at all.

However, where the payer/'specified person' fails to give the 'payer payment notice' the money does not have to be paid until the payee gives the 'payee default notice', because it does not become a 'notified sum' until then. This might alert the payer to the fact that he needs to serve the 'notice of intention to pay less' if he wants to pay less than the amount stipulated by the payee, but he could forget to do so, in which case the payee will be able to suspend if the payer fails to pay the full amount in the payee default notice.

However, if Section 110(B)(4) of the amended Construction Act applies and the client forgets to serve the notice of intention to pay less, then the professional will be able to suspend. This section relieves the payee of having to give a 'payee default notice' if 'the contract permits or requires the payee before the date on which the notice referred to in sub-section 1(A) is required [i.e. before the payer/"specified person" has to give the payer payment notice] to notify the payer/"specified person" of the sum the payee considers will become due on the payment due date and the basis on which that sum is calculated and the payee gives such notification in accordance with the contract: that notification is to be regarded [as sufficient compliance with the obligation to give a "payee default notice"]'.

So if, say, the contract required the professional to send an invoice with these sorts of details in it about the amount that is to be paid, and this had to be given before any obligation arose on the payer/'specified person' to give the payment notice (which it generally would), then it is possible the professional would not have to give the 'payee default notice' because his invoice would stand as sufficient notice for the purposes of sub-section (4), and then the payer might not remember to give either the 'payer payment notice' or the 'notice of intention to pay less', and the right to suspend could arise.

7.19.9 Pay when paid

Section 133 of the Construction Act – making payment conditional on the payer receiving payment from a third party 'ineffective' unless the third party is 'insolvent' – is unaltered.

7.20. Revisions to standard conditions of engagement as a result of the amendments

The approach adopted by the ACE, the RIBA, the CIC and the NEC in the amendments to the PSC3 contract is similar. The professional is to provide an invoice for each payment that contains the information needed to constitute a payee payment notice under the amended Construction Act. The date of receipt or date of the invoice is the 'payment due date'. The final date for payment is a specified number of days thereafter. The client then has to pay by the final date for payment the amount in the payee payment notice unless he gives the requisite notice of intention to pay less not later than a specified number of days before the final date for payment.

The RICS provides both for the payee's invoice to contain the requisite information and for a payer payment notice.

This means that the new procedures are substantially similar to the invoicing/notice of withholding procedures under the Construction Act.

For the first time, the ACE Agreements also include the statutory right to suspend if the client fails to pay the amount due by the final date for payment.

ACE Agreement 1 provides:

> F5.5 The Consultant shall send an invoice to the Client for each instalment of the fees and other sums payable under this Agreement. Each invoice shall specify:
> (i) the sum that the Consultant considers will become due on the payment due date under Clause F5.6, and
> (ii) the basis on which that sum is calculated.
>
> F5.6 Payment shall be due on the date the Client receives the Consultant's invoice.
>
> F5.7 The Consultant's invoice under clause F5.5 shall be the payment notice for the purposes of Section 110A(1) of the Housing Grants, Construction

and Regeneration Act 1996 as amended by Part 8 of the Local Democracy, Economic Development and Construction Act 2009.

F5.8 The final date for payment shall be 28 days after the payment due date.

F5.9 On or before the final date for payment, the Client shall pay to the Consultant either

(i) the sum stated as due in the Consultant's invoice issued under clause F5.5 ('the Notified Sum') or

(ii) the sum that the Client considers to be due as specified in any Pay Less Notice under clause F5.10.

F5.10 Not later than seven days before the final date for payment, the Client may give the Consultant a Notice that it intends to pay less than the Notified Sum (a 'Pay Less Notice'). Any Pay Less Notice shall specify:

(i) the sum that the Client considers to be due on the date the Notice is served; and

(ii) the basis on which that sum is calculated.

These clauses provide that the payment due date is the date of the receipt of the invoice – in earlier ACE Agreements it was the submission of the invoice – and that the final date for payment is 28 days thereafter as before. The notice of intention to pay less has to be given no later than seven days before the final date for payment, again as in the earlier ACE Agreements.

There is a slight lack of clarity in the drafting as to what the client has to pay by the final date for payment. It could be argued that he can pay either the amount in the professional's invoice or the amount in a pay less notice. It is likely, however, that in the event of a dispute this provision will be construed in a manner that is consistent with Section 111(6) of the amended Construction Act if the contract is a construction contract; that is, the client has to pay the amount in the invoice unless he has given a valid notice of intention to pay less within the prescribed timetable in which case he has to pay the amount if any in that notice. Alternatively, if the provision were held to be non-compliant with the amended Construction Act, Section 111(6) as described above would apply in its place, again if the contract is a construction contract.

In practice it is unlikely that the client will pay the greater sum (or that the professional will seek to argue that he should) and so it is not envisaged that this will cause any significant problems in practice and it is not thought that it is essential that these provisions are amended.

RIBA 2010 by its Amendment 1 of 1 October 2011 replaces clauses 5.14 and 5.15 with the following:

5.14 The Architect shall issue payment notices at the intervals specified in the schedule of Fees and expenses.

Each notice shall comprise the Architect's account setting out any accrued instalments of the fee and other amounts due, less any amounts previously paid, and stating the basis of calculation of the amount specified as due,

which shall be 'the notified sum'. The payment due date shall be the date of the Architect's payment notice. Instalments of fees shall be calculated on the Architect's reasonable estimate of the percentage of completion of the Services or stages or other services, or any other specified method.
The Client shall pay the notified sum within 14 days of the date of issue of the relevant notice (which shall be 'the final date for payment'), unless:

(*a*) The Architect has become or becomes insolvent;
(*b*) The Client issues a notice under clause 5.15.

Otherwise, the amount due and payable shall be the notified sum. The Client shall not delay payment of any undisputed part of the notified sum.
The Architect shall submit the final account for fees and any other amounts due when the Architect reasonably considers the Services have been completed.

5.15 If the Client intends to pay less than the notified sum, the Client shall give a written notice to the Architect not later than 5 days before the final date for payment specifying the amount that the Client considers to be due on the date the notice is served, the basis on which that sum is calculated and, if any sum is intended to be withheld, the ground for doing so or, if there is more than one ground, each ground and the amount attributable to it. The Client shall on or before the final date for payment make payment to the Architect of the amount, if any, specified in the written notice.
If the Client issues such a notice and the matter is referred to an adjudicator who decides that an additional sum is due, the Client shall pay that sum within 7 days of the decision.

The matters to be set out in the payee payment notice under clause 5.14 (i.e. in the professional's account) are not as expressed in the amended Construction Act. Section 110A(3) states the notice has to specify '(a) the sum that the payee considers to be or to have been due ate the payment due date in respect of the payment and (b) the basis on which that sum is calculated'. There could therefore be some inconsistency between this clause and the amended Construction Act and the professional would have to comply with them both if it is to be an effective notice.

It should be noted that the final date for payment is 14 days from the date of issue of the relevant notice. This is a reduction from the 28 days in the earlier RIBA agreements and imposes a tight timetable on the client. The notice of intention to pay less has to be given not later than five days before the final date for payment.

The amended Construction Act says the client is relieved of the obligation to pay the professional if the professional has become insolvent between the date when the notice of intention to pay less should have been given and the final date for payment. Clause 5.14(a) is wider than this in that it relieves the Client of the obligation to pay if the professional becomes insolvent. It does not say when, but it could be any time up to the final date for payment.

It is also not clear from this clause whether 'insolvent' is the same as 'insolvent' as defined by the Construction Act as amended in 2003. This clause may well be contrasted with the wording of clause 8.2.2 which refers to acts other than pure insolvency and so it could be thought this clause must be restricted to pure insolvency although that is not the definition under the amended Construction Act.

It can also seem unnecessary for this clause to include 'Otherwise the amount due and payable shall be the notified sum'. The first part of clause 5.14 already makes it clear that that notified sum has to be paid within 14 days unless either of the events then set out have happened. The clause then goes on to say that the client shall not delay payment of any undisputed part of the notified sum. This seems superfluous – either the client has to pay the notified sum, i.e. the amount in the professional's account/ payee payment notice or if less the amount if any in a valid notice of intention to pay less. This would presumably be the undisputed part.

Clause 5.15 not only includes the provisions required by Section 111 of the amended Construction Act but also refers to withholding of sums and the ground for doing so, or if there is more than one ground each ground and the amount attributable to it. This is not required by that section. This therefore is an additional contractual requirement to , 'grounds' for any sum withheld to which clients might object. As clause 5.16 is unamended – this places restrictions on the client's right to set off or withhold monies – it may be that it was thought necessary to retain the reference to withholding when dealing with the notice of intention to pay less. As to the effect of clause 5.16, see Section 7.16. The same comments apply to this new provision in clause 5.15.

Clause 5.15 then goes on to say, correctly, that the client shall on or before the final date of payment make payment to the professional of the amount if any specified in the written notice of intention to pay less. As with clause 5.14, it seems unnecessary for this clause then to provide that if no such notice is given the amount due and payable shall be the notified sum stated as due in the professional's account. Clause 5.14 says that 'the notified sum' is the sum stated in the professional's account and also provides for that sum to be paid by the final date for payment unless the professional is insolvent or a notice of intention to pay less has been given.

There is then the same provision as in clause 5.14 – that the client shall not delay payment of any undisputed part of the account – and the same comments apply as set out above in relation to clause 5.14.

The final part of clause 5.15 does not, in the author's view, make it quite clear that the 'additional sum' referred to is a sum that the adjudicator decides which is greater than the amount in the notice of intention to pay less. Also, under the amended Construction Act, the client has to pay that sum within seven days of the decision or the date which apart from the notice would have been the final date for payment. It is probable, however, that the second alternative will not apply because of the short timetable for payment under these amended clauses, but it might have been advisable to provide for it as set out in Section 111(9).

RICS 2011 provides:

9.6 The Consultant submits to the Client VAT invoices in respect of the Fee, the Additional Services Fee and disbursements and expenses on the dates or on completion of the activities set out in Part B of Schedule 2. All invoices are accompanied by such supporting document, records and receipts as are reasonably necessary for checking each invoice. The invoices and supporting documents, records and receipts (if any) specify the sum that the Consultant considers to be due to it on the payment due date in accordance with this Clause 9 and the basis on which that sum is calculated. The payment due date is the date the Client receives the relevant invoice (the 'due date'). The Client pays the Consultant the total amount of VAT properly chargeable on the supply of the Services.

9.7 The Client gives notice to the Consultant of the sum that the Client considers to have been due at the due date in respect of the payment and the basis on which that sum is calculated. Such notice is given not later than 5 days after the due date.

9.8 The Client, subject to any Pay Less Notice, pays the Consultant the sum referred to in the Client's payment notice under Clause 9.7 (or, if the client has not give notice under Clause 9.7, the sum stated in the invoice referred to in Clause 9.6) (the 'notified sum') on or before the final date for payment of the invoice. The final date for payment is 28 days after the due date or, if a different period is stated in the Appendix, within such period.

9.9 If the Client intends to pay less than the notified sum, it gives the Consultant notice of that intention not later than 7 days before the final date for payment by specifying the amount the Client considers to be due to the Consultant at the date the notice is given and the basis on which that sum is calculated (the 'Pay Less Notice'). Where a Pay Less Notice is given, the payment to be made on or before the final date for payment is not less than the amount stated as due in the Pay Less Notice.

9.10 If the Client fails to pay the notified sum (or, where a Pay Less Notice is given, the amount specified in that Notice) by the final date for payment, the Client pays to the Consultant simple interest on the unpaid amount for the period from the final date for payment until the date of actual payment, calculated on a daily basis at the rate of 4% above the Base Rate. It is acknowledged that this provision constitutes a substantial remedy for the purposes of Section 9(1) of the Late Payment of Commercial Debts (Interest) Act 1998.

The details to be included in the professional's invoice are those required by the amended Construction Act for a payee default notice and the payment due date is fixed by reference to receipt of the invoice with the final date 28 days thereafter. The difference from the ACE and the other agreements described in this section is that clause 9.7 requires the Client as payer to give the payer payment notice in the terms required by the amended Construction Act no later than five days after receipt of the invoice. This would mean, however, that if the client forgot to do so the professional's invoice would stand as the payee default notice (provided it did indeed include the requisite information) and there would be no extension to the final date for payment.

Clause 9.8 then makes the notified sum the sum in the client's payment notice or the professional's invoice (if the client's notice has not been given) which is in accordance with the provisions of the amended Construction Act.

Clause 9.9 then deals with the notice of intention to pay less, again in accordance with the provisions of the amended Construction Act, with the notice of intention to pay less to be given not later than seven days before the final date for payment.

CIC provides:

> 3-8.3 The Fee shall be paid by instalments, and the Consultant shall invoice the Client on the dates or on the occurrence of the events provided in §2-3. Each invoice shall show any accrued instalments and other amounts which the Consultant considers due at the payment due date and the basis on which those amounts have been calculated, plus value added tax, less amounts previously paid. It shall be supported by such documents, vouchers and receipts as the Client reasonably requires to verify it.
> 3-8.6 The following provisions apply to payment:
> 3-8.6.1 the due date for payment shall be the date on which the Client receives the invoice;
> 3-8.6.2 the invoice shall be the payment notice required by Section 110A(1) of the Act; and
> 3-8.6.3 the Client shall pay the Consultant the amount stated in the invoice or (if less) the amount specified in the Client's notice under §3-8.7 on or before the final date for payment, which shall be 28 days after the due date.
> 3-8.7 If the Client intends to pay less than the amount stated as due in an invoice, it shall give the Consultant notice of its intention to do so, stating the amount that it considers to be due to the Consultant on the date the notice is served and the basis on which such amount is calculated.
> 3-8.8 A notice under §3-8.7:
> 3-8.8.1 shall be the notice of intention to pay less referred to in Section 111(3) of the Act;
> 3-8.8.2 may not be given before the due date; and
> 3-8.8.3 must be given at least 7 days before the final date for payment.

The invoice requirements in clause 3-8.3 have been extended from those in the first edition so that the invoice also has to show the amount the professional considers due at the payment due date and the basis on which those amounts have been calculated. This means that the payee has been required to give the payment notice – it will be recalled that the amended Construction Act requires the contract to stipulate whether the payer or payee is to give the notice – and that the invoice will stand as the payee payment notice as it will contain all the information required by the amended Construction Act. Some additional information is also required: details of accrued instalments and other amounts that may be due, for example, additional fees.

Clause 3-8.6.1 is unchanged from the first edition. The due date for payment is the date on which the client receives the invoice. So if the professional sends his invoice in the correct form his payment notice will have been given 'no later than 5 days after the payment due date' as required by the amended Construction Act.

Clause 3-8.6.2 then goes on to state in terms that the invoice is the payment notice required by Section 110A(1) of the amended Construction Act and under clause 3-8.6.3 that the amount that the client has to pay by the final date for payment (28 days after the due date) is the amount in the payment notice or (if less) the amount in any notice of intention to pay less given by the client.

The notice of withholding provisions in clause 3-8.7 and 3-8.8 of the first edition have been replaced by the notice of intention to pay less requirements in the amended Construction Act and the client has to give this notice at least seven days before the final date for payment but not before the due date.

PSC3 brings in the amended payment provisions by the Amendments of September 2011 to Option Y(UK)2 which had the payment provisions in accordance with the Construction Act.

The relevant amendments are:

Page 41 Option Y(UK)2: Y2.2, third paragraph replaced with the following:
The *Consultant's* invoice is the notice of payment specifying the sum that the *Consultant* considers to be due at the payment due date (the notified sum). The *Consultant's* invoice states the basis on which the amount is calculated and includes details of the calculation in accordance with the contract.

Page 41 Option Y(UK)2: Y2.3 replaced with the following:
If the *Employer* intends to pay less than the notified sum, he notifies the *Consultant* of the amount which the *Employer* considers to be due not later than seven days (the prescribed period) before the final date for payment. The *Employer's* notification states the basis on which the amount is calculated and includes details of the calculation. A Party pays the notified sum unless he has notified his intention to pay less than the notified sum.

The date on which payment becomes due and the final date for payment have not been changed. Payment becomes due seven days after the date of the professional's invoice and the final date for payment is 14 days thereafter (although a difference period could be stated in the contract data).

The professional's invoice is to be the payee payment notice and is to specify the sum the professional considers to be due at the payment due date. This is called 'the notified sum'. The invoice has also to state the basis on which the amount is calculated and 'include details of the calculation in accordance with the contract'. This latter part is in addition to what is required by Section 110A(3) of the amended Construction Act and so professionals will need to note this as their usual invoices could be amended Construction Act-compliant, but not PSC3-compliant, if details as to the calculation required are set out in the contract and are not complied with.

The notice of intention to pay less has to be given not later than seven days before the final date for payment. This in effect gives the client seven days to decide whether or not he is going to serve such a notice. The notice has to notify the amount the client considers to be due and state the basis on which the amount is calculated and 'includes details of the calculation'. It is not made clear in the drafting that the amount considered to be due is to be that at the date of the notice is served so this again could lead to unintentional non-compliance with the amended Construction Act. The last part – '... includes details of the calculation' – is not required by the amended Construction Act and so the same comment applies as above.

The amendment then states 'A Party pays the notified sum unless he has notified his intention to pay less than the notified sum'.

It would have been clearer if the amendment had provided that the client had to pay the sum specified in the notice of intention to pay less rather than the notified sum as required by the amended Construction Act. There does not seem to be an obligation to pay this lesser sum, but as with ACE Agreement 1 it is likely that this provision will be construed in a manner that is consistent with the amended Construction Act.

Professional Services Agreements
ISBN 978-0-7277-4158-5

http://dx.doi.org/10.1680/psa.41585.127

Chapter 8
Termination and suspension

8.1. Termination generally

A professional services agreement should describe how the professional as well as the client is to terminate the agreement, what documents are to be delivered to the client, as well as stipulating the payment to be made in that event.

It needs to be noted, however, that any right to terminate has to be operated in accordance with its terms, for example, as to the length of any notice or the serving of any notice. Also, if it is termination for breach, there has to have been a breach, and likewise for insolvency or other stipulated cause. Otherwise the termination is unlawful and could be treated as a repudiatory breach by the other party – see Section 8.3.

There should also be an express provision to the effect that:

> the termination of the engagement shall be without prejudice to the rights and remedies of either party in relation to any negligent act or omission or breach of this Agreement [or 'accrued rights or claims'].

This is necessary to ensure that, notwithstanding termination, either party can bring a later claim, for example, for additional fees or for negligence. In the absence of such a provision, the termination of a contract would bring to an end all the parties' rights and obligations except accrued rights and obligations.

8.2. Client's right to terminate and payment on termination

Up until the mid-1980s the right of a client to terminate a professional's appointment for anything other than breach of contract or insolvency rarely appeared in non-standard conditions of engagement. Clients were prepared to employ their professionals on the basis of a 'whole appointment' – that is, for the whole project. The professional, as a result, could plan his work and the fee accordingly. This attitude was reflected in the standard conditions of engagement. For example, the ACE Conditions prior to the 1995 Edition were on the basis of a whole appointment with limited rights to terminate. This meant that if a client terminated the appointment for any reason not stipulated in the appointment, that contract had been terminated wrongfully, the breach could be 'repudiatory' – see next section – and the professional would be entitled to all the losses which flowed from that breach. These could include unavoidable expenses such as redundancy costs and loss of profit.

Non-standard conditions of engagement almost invariably contain an express right to the client to terminate at any time with or without a notice period. Standard conditions of engagement have tended to contain the same right.

Professionals need to consider whether a right to terminate at will with or without an agreed period of notice is acceptable, particularly in recessionary times when projects could be halted.

It should be noted, however, that it is thought that a client is not permitted to use a right to terminate at will so that he can employ another professional to carry out the same services, perhaps at a reduced fee, unless the contract sufficiently expressed his right to do so. This is because it is an implied term of such appointments (or indeed of building contracts) that the client will allow the professional to complete the work. However, if the reason for the termination is because the client has decided to abandon the project permanently or for the time being, or because he is bona fide dissatisfied with the professional's performance, this would be a legitimate exercise of the power. See also Section 5.12 on the power to omit.

Some recent cases have indicated that a right to terminate at will can only be exercised in good faith or for good cause. This depends on the circumstances of each case and advice would be needed to see if these principles would apply to any particular termination.

PSC3 has an express obligation of good faith/mutual trust and cooperation which could prevent a client from terminating capriciously or if the client seeks to use the threat of termination in order to try to re-negotiate the terms of the professional's appointment or fees.

If a right to terminate at will is accepted, the payment the professional is to receive in such circumstances becomes more important. There is authority to the effect that in such circumstances the professional will be entitled to full compensation, including loss of profit on the work performed up to termination, and other costs. However, it is more usual for the appointment to stipulate what is payable in these circumstances – and this is often very restricted. Generally, the professional will be entitled to that part of the fee that has accrued due up to the date of termination (or perhaps only up to the last interval on which any instalment is due) with no compensation for the fact that the professional's work has been disrupted, nor for staff possibly being reassigned or made redundant. Further, if the structure of the fee payments has not been considered carefully, the amount of work undertaken at the date of termination may exceed the fee payable.

There will often be an express stipulation that termination will not give rise to any claim for any damages against the client for loss of profit or other losses arising out of termination.

The client may also be given the express right to terminate if the professional is in breach of its obligations. This may be tempered by the inclusion of a time limit for putting things right. Such a right is unexceptionable, except perhaps if it is exercised when the breach is very minor and has no adverse effect on the client or the project. For this reason, the reference to breach is often qualified by it being required to be a 'material' breach.

Where the professional has been in breach the client may be entitled to claim for the losses that flow from and are a consequence of that breach. The common law has rules for assessing the damages payable if there has been a breach of contract, and any express provisions in appointments setting out what losses or damages the client will be entitled to in these circumstances will usually be intended to improve the client's position. Such provisions could also mean that the professional's professional indemnity insurers may not cover the full amount payable to the client because the indemnity provided by insurance may only extend to the damages properly recoverable in law or caused by the professional's negligence.

Some non-standard conditions of engagement state that if the appointment has been terminated because of the professional's breach, the breach will be deemed to be a 'repudiatory breach' or describes the damages the professional is to pay on the basis that the breach is repudiatory. Such a provision has to be deleted or the professional could have to pay more than would otherwise be the case – see next section.

If a client and a professional cannot agree whether a breach is in fact repudiatory, the courts will decide. This should not be prejudged in an appointment, and again professional indemnity insurers may not provide a full indemnity for the damages payable.

There may also be an express provision that the professional is not entitled to any outstanding fees when the appointment has been terminated for breach or for insolvency. Again, this needs to be considered carefully. The common law has established rules as to whether or not a breach is such that no fees or no outstanding fees should be paid. Any express provision about this could adversely alter those rules in favour of the client and so should be resisted.

Otherwise, the client's right to terminate should only generally be agreed where there is insolvency of a professional or acts leading up to its insolvency, such as the appointment of a receiver or an administrative receiver. The parties need to be clear as to what will constitute 'insolvency'.

8.3. Repudiatory breach

Repudiation occurs when a breach is so serious that it relieves the other party of the obligation to complete performance of the contract; that is, all outstanding obligations under the appointment come to an end.

This can be where one party so acts or so expresses himself as to show that he does not mean to accept the obligations under the contract any further.

In the client's case, this could be prolonged and serious failure to pay the fees due, and in the professional's case absenting himself from the project and/or not carrying out the services agreed. If the breach is 'repudiatory' it entitles the other party to choose whether to 'accept' the repudiation and terminate the contract, or to treat the contract as continuing. In the first case, he has to inform the other party that this is what he has chosen to do within a reasonable time. If he does nothing and continues to perform his obligations he will be taken to have treated the contract as continuing, as he will have 'affirmed' the contract. How long a reasonable time is will depend on the circumstances. Factors include how clear the breach was to the innocent party, how urgently the party in breach needs to know its position, and whether the delay might mislead the party in breach into thinking the risk of termination had passed.

Each case will be decided on its facts and whether any breach is of itself sufficiently serious to be a repudiatory breach should be considered with legal advisers before any action is taken.

The consequences of a repudiatory breach are severe. If there has been a breach of contract the other party is entitled to damages as compensation for the loss suffered as a result of a breach. Where there is repudiatory breach the other party is also entitled to all its losses flowing from the termination. In the case of a repudiatory breach by a professional, this would mean that in addition to the damages payable as a result of, say, negligent design, the client will be entitled to the costs of the delay to his project while he finds a new professional to take over, any additional monies he has to pay to that professional over and above those that would have been paid to the original professional and any additional monies payable to the other consultants, contractors and/or sub-contractors.

The ordinary rule of contract is that late payment or under-payment of an account during continuing work is not enough to bring the contract to an end unless there is an express right to terminate. However, the cases show that if the implication to be drawn from the non-payment by the client was such that a reasonable person would conclude from the behaviour of the party in default that he no longer intended to be bound by the contract provisions, the breach would be repudiatory.

The cumulative effect of the failure to pay has to be sufficiently serious or severe to give rise to the right to repudiate.

If the professional exercises the right to treat a breach by the client as repudiatory, but it is later determined by the courts that the breach was not repudiatory, the professional may have been in repudiatory breach and thus become liable to the client for all the losses described in this section. However, as happened in a recent case, sometimes a client may not be able to demonstrate he has incurred additional costs because the new professional was in fact cheaper than the original professional.

There is also the possibility that conduct by the client could amount to 'anticipatory' breach which again could entitle the professional to treat the contract as being at an end. If the client informs a professional that he cannot or will not pay in accordance with the terms of the contract, this could be 'anticipatory breach', giving rise to the same consequences and remedies as for repudiatory breach.

A recent case (*Mayhaven* v. *DABI (2009)*) considered whether a contractor's wrongful suspension of its works would as a matter of principle amount to a repudiatory breach of contract. There is no general right at common law for a contractor to suspend its works for non-payment. (Nor is there for a professional – see Section 8.7.) In this case, however, the contractor had a right to suspend for non-payment and did so because of a genuine but mistaken belief that it had not been paid sums by its employer pursuant to an adjudicator's decision. The court did not accept Mayhaven's submission that a wrongful suspension which gave rise to a failure to proceed regularly and diligently under an express term of the contract would necessarily also amount to a repudiatory breach. It would be a matter of considering the seriousness of the breach and the facts and circumstances of the case.

8.4. Professional's right to terminate

A professional's right to terminate should mirror the client's rights. In any event, there should be rights to terminate if the client has been in breach or becomes insolvent or approaches insolvency. Whether the professional also needs a right to terminate at will or on reasonable or specified notice will depend on the nature of the commission. The RIBA standard forms before the 2007 versions permitted the architect to terminate on reasonable notice. This reflected the fact that the architect's appointment can be a very personal one and if relations break down the architect will need to bring the commission to an end. It is also the case that the standards forms were used by small firms whose circumstances could change substantially during a project, leaving the architect unable to continue, for example, because he no longer had adequate resources or was suffering ill health. RIBA 2010 has reinstated this right.

Often, however, there are no rights in non-standard conditions of engagement for a professional to terminate, not even if there has been non-payment of fees after a certain length of time. This leaves the professional with the difficult task of deciding how long the professional should continue to work and whether the non-payment amounts to a repudiatory breach, as described in the preceding section.

In some appointments it may be necessary for the professional to have a right to terminate in certain circumstances – for example, if the client gives an instruction which goes against the professional's professional judgement or could result in something that is a risk to health and safety. The ACE agreements provide such a right – see Section 8.9.

8.5. Obligations after termination

Most obligations come to an end on termination. Some non-standard conditions of engagement, however, set out in detail what obligations the professional has in the event of termination, and what obligations continue; for example, the confidentiality obligations, the obligations to continue to cooperate or to assist in any hand-over, to give collateral warranties or third party rights, or to continue to provide a licence to use the professional's documents.

The question arises as to whether these are reasonable, especially if the reason for the termination has been the client's breach or insolvency where that breach has not been repudiatory. If the client has not paid the professional some of his fee or if the client has terminated at will early on in a project, would it still be reasonable for the professional to give collateral warranties in respect of the work he has done to date to third parties, especially if another professional is going to take over and do the bulk of the work? Why should the professional continue to give a licence in respect of his documents? As will be seen in Chapter 10, the standard conditions of engagement generally include a right to revoke the licence if the professional has not been paid. Non-standard conditions of engagement generally provide that the licence should be irrevocable.

8.6. Delivery up of documents on termination/suspension

Professionals are more frequently being required in their appointments to deliver up their drawings and documents to the client on termination and/or on suspension, for whatever cause.

Such clauses need to be considered carefully as they could give the client rights to documents which they would not have unless they brought a claim, or to documents they would not normally be entitled to. because they are privileged.

An obligation to hand over documents to a client or indeed a third party in the appointment will be treated as any other contractual obligation – it will be construed in accordance with its terms, and if the professional does not comply with the obligation he has undertaken he will be in breach of contract. It is also possible that the courts will enforce such an obligation, for example, by ordering specific performance; that is, making an order that the professional hand over the documents he has agreed to give to the client. Third parties will not usually be able to enforce such a term as they are not a party to the contract unless they have been given specific rights to do so, either under the Contracts (Rights of Third Parties) Act 1999 or through collateral warranties.

Professionals therefore need to consider both the extent of the obligation and the circumstances in which it is to arise.

A typical clause is as follows:

> Following any termination of the Consultant's engagement howsoever arising the Consultant shall deliver to the Employer or any person nominated by him copies, in such form as he may reasonably require, of all Documents, and other documents including (inter alia) negatives and CAD disks (whether in the course of preparation or completed), which the Consultant holds in relation to the Development together with the originals.

The definition of 'Documents' would need to be checked but if often covers everything produced by the professional, including internal memoranda, calculations, draft documents, etc.

So if the obligation is to deliver copies of every single document produced by the professional, as it is under this clause, that is what the professional will have to hand over, regardless of whether any of the documents already would have been delivered to the client or his consultants/contractors in order for the project to be carried out.

It will be immaterial whether the client needs them, or would only have been entitled to them if there was a claim, or if they were otherwise privileged. Everything, therefore, depends on the description of the documents to be handed over or on the definition of 'documents' or 'materials'. This must therefore be checked carefully and restricted to those documents, etc. which are delivered to the client or others as required by the performance of the professional's services and to enable others to have the information they need regarding the professional's services and to perform their own.

If the originals are required the professional must ensure he is not prohibited from taking copies or he will not have the documents he needs for his own records and in the event of any claim.

If the object of the clause is to make sure the client has what he needs for the health and safety file, then the obligation should be confined to this.

If there is some uncertainty about the extent of the clause or the definition, the professional should in the first instance only provide the originals or copies of those documents he has already delivered as a result of the performance of the services and then seek legal advice as to whether he can resist providing any further documents, and particularly whether he has grounds for resisting handing over documents that would otherwise be privileged and thus potentially seriously prejudice his position.

Whether or not calculations should be released or a licence given to use them is considered in Section 10.3.

An obligation to hand over documents and/or disclose information may be justified in certain types of contract, where, as in *John Youngs Insurance Services* v. *Aviva Services Limited (2011)*, a construction company was engaged to provide claims-handling and building-repair services to an insurer. Such a contract is likely to involve delegated authority to the contractor to settle claims and manage contracts for repairs and the insurer may want to be able to audit particular claims and contracts from time to time.

The question that arose in this particular case was whether the right to obtain information and documents survived the termination of the contract. The court held that the insurers could obtain such information in relation to the performance of fiduciary duties, which included the validation of claims on behalf of the insurers. There is a duty to account arising from such duties which survives termination of the contract. However, in relation to non-fiduciary duties, which included assessing the scope of and carrying out repairs, the contractual duties to provide information did not survive termination, save in relation to unpaid invoices or matters of an ongoing nature.

The court said clear words were needed for such an obligation to continue following termination.

ACE Agreement 1 says nothing as to what documents, etc. are to be handed over on any suspension or termination.

RIBA 2010 provides:

> 8.2.3 On termination of performance of the Services and/or other obligations, a copy of the Material not previously provided to the Client shall be delivered on demand to the Client by the Architect, subject to the terms of the licence under clause 6.3 and the payment of any outstanding fees and other amounts due plus the Architect's reasonable copying charges.

The 'Material' is the drawings, documents, software, etc. produced by or on behalf of the professional in performing the services, and so it extends the right to documents beyond those that would have been handed to the client or others during the project.

RICS 2011 says that on termination the professional:

> 15.10(b) provided it can lawfully do so, and only following receipt of a written request to do so, hands over to the Client copies of any Documents and other information relating to the Project in its possession or prepared by it (but not its own internal notes or memoranda or any Document which is subject to a confidentiality restriction in favour of a third party) subject to the terms of the licence under Clause 13.2 and to the payment of the Consultant's reasonable copying charges and the amounts due under Clause 15.9 [payment on termination].

'Documents' are widely defined but this clause does try to preserve the professional's rights in respect of internal and confidential documents.

PSC3 provides in clause 91.1 that on termination the client 'may complete the services and may use any material to which he has title' – it is assumed this is meant to cover both where he has copyright and where he has a continuing licence. 'After the final payment has been made, the *Consultant* gives to the *Employer* information resulting from work carried out to date and information the *Consultant* has obtained which he has a responsibility to provide under the contract'. Thus under PSC3 the professional's obligations seem to be limited to giving the client the documents and any other information he should have provided in the course of the services up to that date but it could be interpreted more widely than that.

Requests for a professional's documents can also be made based solely on the licence to use the professional's documents but this is based on a misconception of what licences are – see Section 10.3.

As described there, the purpose of a licence is to allow the client (or others) to copy the professional's documents (or adapt or otherwise exploit them). Without such a licence the client will be infringing the professional's copyright and could lay himself open to an action for damages or other remedies. However, the professional does not generally grant a completely open licence so that the client can copy and use the documents for any purpose whatsoever. Care is usually taken to define and limit the scope of the copying and therefore the use to which the professional's documents can be put.

Licences do not give the client a right to have the documents described. What the client actually receives by the way of deliverables is either set out in the services or implied by the services to be performed or, more unusually, covered by an express clause such as the one described above.

A demand from a client for all the documents described in a licence should be resisted – a licence does not give the client such a right.

8.7. Suspension and payment on suspension

The right of the client to suspend in appointments is usually tied in with the right to terminate, and generally the client reserves the right to suspend at any time. This may be essential because of the nature of the project. However, the professional needs to consider how many times the client can do this and how long the period of suspension should be before the fee and/or the services are to be reassessed. Non-standard conditions of engagement often do not recognise the disruption that can be caused to a professional's work by many or lengthy suspensions, and payments to cover the costs of demobilisation and start-up are almost never included. It is prudent, therefore, to consider whether there should be an aggregate period of suspension, after which the professional should be given the right to terminate the appointment. This at least would provide an opportunity to renegotiate the fee if the client wished the professional to continue. Otherwise, the fee would remain the same however many times the client exercised the right to suspend.

The entitlement to payment of fees on suspension needs to be checked in the same way as for termination. Will the payment to the professional include payment up to date or just the last instalment due? Will the costs incurred by the professional as a result of the suspension be paid? What if the fee payable is to be assessed on a fair basis? In the case of *Fitzroy Robinson Limited* v. *Mentmore Towers Limited (2009)* the architects' appointment was suspended and the basis of payment on suspension (as well as termination) was that 'the Consultant will be entitled to a fair proportion of the Fee for any of the Services properly performed up to and including the date of termination having regard to the instalment schedule set out in Schedule [2] and the payments already made to the Consultant under this Agreement'.

The architects had agreed a fixed fee and instalments. The architects argued (i) the whole of the fee should be recalculated on a time basis and credit given for fees already paid – the court held this was not encompassed within the relevant clause or the lump sum nature of the agreement between the parties; (ii) a calculation based on the lump sums in the contract with a reduction to reflect the value of these works not completed by the architects – the court found this 'slightly ambiguous'. The court held the proper assessment of the proportion of the fee due was by reference to the RIBA Work

Stages completed by the architects prior to this suspension. Where a work stage was incomplete at the date of suspension a fair proportion of the services completed could be assessed. This, the judge thought, was 'by far the fairest and most satisfactory'.

There is no common law right to suspend for non-payment of fees. Sometimes, although rarely, the professional is given a right in his appointment to suspend for a certain period, rather than terminate, in such circumstances, or if the professional has been prevented from carrying out services. Care needs to be taken in exercising a right to suspend and in complying with any notice provisions. If the professional were to suspend incorrectly (e.g., because the payment was not due) or where no right to suspend had been given, the suspension would be unlawful and could amount to a repudiatory breach with all the cost consequences that would entail – see Section 8.3.

8.8. Suspension under the Construction Act 1996 and as amended

The Construction Act 1996 gave a statutory right to suspend performance if there is non-payment of any sum due under a 'construction contract'. This provision still applies in respect of contracts entered into prior to 1 October 2011.

Section 112 provided the following.

(1) Where a sum due under a construction contract is not paid in full by the final date for payment and no effective notice to withhold payment has been given, the person to whom the sum is due has the right (without prejudice to any other right or remedy) to suspend performance of his obligations under the contract to the party by whom payment ought to have been made ('the party in default').

(2) The right may not be exercised without first giving to the party in default at least seven days' notice of intention to suspend performance, stating the ground or grounds on which it is intended to suspend performance.

(3) The right to suspend performance ceases when the party in default makes payment in full of the amount due.

It should be noted that the right was 'to suspend performance of his obligations under the contract', not simply to suspend the carrying out of the construction works. This would indicate that such matters as the contractors' obligations regarding insurance and security of the site could also be suspended.

The right could not be exercised unless a notice of intention to suspend had been given and this notice had to be of at least seven days 'stating the ground or grounds on which it is intended to suspend'. The only ground given in the Act was a failure to pay 'a sum due under the contract by the final date without giving the proper notice to withhold'. It is difficult to see, therefore, what other grounds there could be, but perhaps the legislators were anticipating multiple non-payments.

The right to suspend arose in respect of 'a sum due'. There could be an argument about whether a payment was due because, for example, the payment was in respect of work that is alleged to be defective, and advice should be taken before this right is exercised or the suspension could be unlawful and a repudiatory breach.

Once the payee was paid in full, he must recommence work. As to the consequences of suspension, the Construction Act provided that:

> Any period during which performance is suspended in pursuance of the right conferred by this section shall be disregarded in computing for the purposes of any contractual time limit the time taken, by the party exercising the right or by a third party, to complete any work directly or indirectly affected by the exercise of the right. Where the contractual time limit is set by reference to a date rather than a period, the date shall be adjusted accordingly.

As has been said, there is no right at common law to stop work if a payee has not been paid, although in some circumstances the failure to pay may amount to a repudiatory breach of contract – see Section 8.3. Some standard form appointments included the right to suspend if there was non-payment, but these provisions have now for the most part been superseded by the Construction Act provisions.

It needs to be remembered, however, that the Construction Act right to suspend only arises if the payer has not paid in full and has failed to serve an effective notice of withholding in the proper form and at the correct time – see Section 7.16. If the payer does serve an effective notice of withholding, the professional has to continue to work despite the non-payment. Any dispute over whether the grounds in the notice of withholding were sufficient or justified would then have to be referred to adjudication or to the courts or arbitration.

The Construction Act helps to make it clear when a payee is entitled to suspend performance; that is, when the final date has passed without the proper notice of withholding payment having been served at the correct time. This relieves the payee of having to pay the client the costs and losses the client incurs as a result of the professional's suspension (in the absence of a contractual right to do so) so long as the payee in turn has given the requisite notice.

The effect of the Construction Act is that any period of suspension is disregarded in calculating the date for completion, with the result that an 'automatic' extension of time is obtained. The extension does not extend, however, to the time taken for demobilisation and remobilisation but only to the strict period of suspension.

There is no right under the Construction Act to recover any loss or expense incurred during the suspension. If this is to be recovered it is thought that the appointment must specifically provide for it, although it could be argued that such loss and expense flows from the client's breach of his payment obligations.

It is not clear whether the parties to a construction contract may be able to contract out of the right to suspend. They might be unwise to do so, however, until this point has been tested in the courts. A client could, however, make it difficult to exercise the right. The notice of intention to suspend has to be at least seven days before the right of suspension is to be exercised. This can be extended by an express term in an appointment and so should be checked. A notice period of, say, 28 days or more could result in the professional incurring considerable further costs before he could suspend, with little hope of recovering those further costs.

Some non-standard conditions of engagement include suspension provisions in accordance with the Construction Act, but sometimes it is not clear how the statutory provisions will interact with other provisions in the appointment concerning suspension.

It should be noted that there is no need expressly to include the Construction Act provisions concerning suspension within appointments. Equivalent provisions to the suspension provisions in the Construction Act will be implied automatically into an appointment which is a ‘construction contract’ by operation of the Act itself.

The Local Democracy, Economic Regeneration and Construction Act 2008 (the amended Construction Act) has amended the Construction Act provisions concerning suspension – see Section 7.19. ACE Agreement 1 includes the amended Construction Act right to suspend and also gives the professional other rights to suspend. In the case of the amended Construction Act provisions there is a slight lack of clarity as to which sum it is the client has to have failed to pay by the final date for payment for the right to suspend to arise, and it is not absolutely clear that if the client has paid less than the amount in the professional’s invoice he must have served his pay less notice no later than the date specified in clause F5.10 (i.e. seven days before the final date for payment of the relevant sum) in order to prevent the right to suspend from arising. However, because these provisions are in a contract, this again should not matter in practice. A professional can choose to suspend under contractual provisions and/or rely on the amended Construction Act if his contract is a construction contract. Again, it is not thought essential to amend these provisions.

RIBA 2010 has both contractual and the amended Construction Act rights to suspend. In the case of the latter, the provisions in the amended Construction Act are not fully set out. If the wording of the Act had been followed clause 8.1.2 would refer to ‘obligations’ only and not ‘Services’ and to stating the ground or grounds on which it was intended to suspend performance and not ‘reasons’. Having defined ‘the notified sum’ earlier (in clause 8.1.2(a)) there could have been a reference to ‘the notified sum’ in this clause rather than to ‘any fees or other amounts’. Also there has to be a failure to pay the amount if any in the notice of intention to pay less and/or the failure to give an effective notice before the right to suspend arises.

Clause 8.6 seems to deal with only part of Section 112(4) of the amended Construction Act. Where there is a Construction Act suspension, time is to be disregarded in computing for the purposes of any contractual time limit the time taken to complete work directly or indirectly affected by the exercise of the right. This is not quite the same as the provision in clause 8.1.6 and could lead to some confusion.

The right to the reimbursement of the reasonable costs on resumption under clause 5.18 is not in the same terms as that under the amended Construction Act but the statutory provision would apply where the contract is a construction contract.

RICS 2011 includes the Construction Act's provisions concerning suspension, but does not give the professional a contractual right to suspend. PSC3 has no express rights to suspend, although it does have the equivalent of a force majeure clause. Option Y(UK)2 (including the Amendments of September 2011) brings in some of the Construction Act provisions and amended Construction Act provisions, but not specifically the right to suspend the professional's exercise of his right to suspend under the Construction Act or the amended Construction Act (presumably where his contract is a 'construction contract') is, however, a compensation event.

The statutory right to suspend will apply in any event if nothing is said if the appointment is a 'construction contract'.

8.9. Provisions for termination and suspension in standard conditions of engagement

ACE Agreement 1 gives a client a right to terminate on four weeks notice and to suspend a professional's performance of all or any part of the services at any time. It provides for the services to be ceased in an orderly manner and for an aggregate period of suspension of 39 weeks before the professional is entitled to treat the project or works as abandoned and the appointment terminated. Either party may terminate for breach (with a two-week period in which the breach can be repaired) or in the event of insolvency. In addition, the professional can terminate on not less than two weeks notice if the client does not pay a sum due by the final date for payment. The professional is given a right on notice to suspend performance if prevented or significantly impeded in the performance of services for a period up to 26 weeks. The professional is entitled to terminate on two weeks notice if circumstances arise for which he is not responsible and which the professional considers make it irresponsible for the professional to perform all or any part of the services. This is to protect the professional should the client, for example, ask him to do something that is against the professional's advice or professional's judgement, or could have adverse health and safety consequences. The new clause F9.10 in the Second Edition gives the professional the right to suspend under the amended Construction Act – see previous section.

The fee to be paid is different in different circumstances. Where the client determines by notice at any time, or suspends, or where the professional suspends or determines, any instalments of the fee due to date must be paid together with a fair and reasonable proportion of the next instalment due commensurate with the services performed to date (not just the amount due to date), together with outstanding expenses and a sum for loss and costs of disruption, calculated on the basis of the loss to the professional and costs to which the professional is committed in respect of planned future work. Where the client has determined the appointment because of the professional's breach or where the termination is as a result of insolvency of either party, the amount to be paid is a fair and reasonable amount on account of the fees due commensurate with the services performed to the date of termination and outstanding expenses. Where there is suspension under clause F9.10 (the amended Construction Act right to suspend) the professional is entitled to the costs and expenses as set out in the amended Act – see preceding section.

There is the usual provision in ACE Agreement concerning accrued rights.

RIBA 2010 allows either party to terminate by giving reasonable notice 'stating the reasons for doing so and the Services and obligations affected'.

It could be difficult to determine how long a notice might be 'reasonable'. In a short project or one that had just completed a particular stage, such as the application for planning, a relatively short notice period could well be reasonable. If the professional was a sole practitioner and had become seriously ill, an immediate notice could be reasonable. Where the project was in the construction stage, a month or longer could be required if this was necessary to effect a proper hand-over. This uncertainty would be removed if the parties agreed on a notice period of a certain number of days.

One wonders why, if there is a right to terminate, any reasons have to be stated. Do they have to be good or reasonable reasons?

Either party can terminate immediately by notice if the other becomes insolvent or in other circumstances which precede insolvency or if the professional becomes unable to perform the services through death or incapacity.

RIBA 2010 also allows the client to suspend performance of any or all of the services at any time on not less than seven days' notice.

The professional may suspend on not less than seven days' notice if the client has failed to pay by the final date for payment under the Construction Act and has not 'given effective notice under clause 5.15.2' (the reference should be to clause 5.15) of the intention to withhold payment (the reference should be to the notice of intention to pay less – see comments in the preceding section. The professional can also suspend if the client is in material or persistent breach, or if the professional is prevented from or impeded in performing the services for reasons beyond the professional's control or in the event of *force majeure*, which is not defined.

If the reason for a notice of suspension arising from a default (presumably by the client under the foregoing provisions) is remedied, the professional has to resume performance within a reasonable time. No time limit is given to the client for remedying the default. This leaves the professional in an uncertain position. How long does he wait? If it is not remedied (again, no time limit) the other party has a right to treat the performance of the services or other obligations affected as terminated on giving reasonable written notice. Nothing is said as to what is to happen if there has been no default and the professional has suspended, for example, in the case of *force majeure*. Where the client suspends and the services are not resumed within six months, the professional has the right to treat the performance of the services and other obligations as terminated on giving at least seven days' further written notice to the client.

Under clause 8.1.6, any period of suspension arising from a valid notice to suspend is added to the latest timetable for completion of the relevant services – see preceding section.

As to payment on suspension or termination, the professional has to issue an account on the expiry of the notice or as soon as reasonably practicable. However, the provisions that follow do not make any distinction between where all the services have been suspended and only part of them. It simply says that the professional is entitled in both cases to 'payment of any part of the fee or any other amounts properly due on the date of suspension or termination'. This is, it is assumed, meant to provide an entitlement to the fees earned since the last instalment was due, for example, if the suspension or termination takes effect half-way through a month, but it could be argued that what was 'due' at such a date was the last instalment only. The 2007 conditions and SFA/99 dealt with this by providing what the professional was entitled to for 'Services not completed'. This has been omitted in RIBA 2010. The provisions for payment when a tender is not made or accepted in clause 5.11 could help if this was the reason for the suspension or termination.

The professional is also entitled to payment of any licence fee due under clause 6 and the reimbursement of 'any loss or expense caused to the Architect by reason of the suspension or determination save where the Client suspends or terminates by reason of the material or persistent breach of the Agreement by the Architect'.

This is different from the 2007 Conditions and SFA/99. In the 2007 Conditions the professional was entitled to 'any direct costs properly and necessarily incurred' and in SFA/99 to 'any loss and/or damages'. It is thought that this should entitle the professional to damages for loss of profit or other damages should the appointment be terminated early, as he was so entitled under SFA/99. The unamended RIBA 2010 gave a right only to 'any loss or expense properly and necessarily incurred' which probably did not include loss of profit or other damages.

There is also a right to reimbursement of the reasonable costs of resumption of the services and other obligations if 'the reason for the suspension is remedied', which is a curious way of putting it. The client has been given a right for no reason at all to suspend the performance of any or all of the services on seven days' notice. Is this 'remedied' if he stops the suspension? If this is a reference to the professional's right to suspend because of the client's failure to pay fees, etc., why should it only be in those circumstances that the professional gets the costs of start up? The entitlement to reasonable costs where there is suspension under the Construction Act is not included – see preceding section.

There is no express provision that any termination is without prejudice to either party's rights and remedies.

RICS 2011 provides that the client can terminate at any time on 28 days' notice and either party can terminate immediately if the project is cancelled or cannot proceed for reasons outside the client's reasonable control. If either party is in material breach and does not remedy it within 14 days of a notice to do so, or becomes insolvent, the other party can terminate immediately on written notice.

The client can suspend the performance of all or part of the services at any time, but if the client does not instruct the professional to resume within a period to be specified in the appendix either can terminate on written notice.

The only other ground for suspension is if the client fails to pay the notified sum or, if a valid notice of intention to pay less is given, the sum in that notice by the final date for payment reflecting the amended Construction Act provisions, in which event the client has to pay the costs stipulated by the amended Construction Act as well as any other sums due – see Section 7.20 and previous section.

The only other ground for suspension is if the client does not serve an effective notice of withholding, reflecting the Construction Act provisions – see Section 7.15).

On termination or suspension the client pays the fee and any additional services fees, and any other sums that have accrued due, and a fair and reasonable proportion of the next instalment due (whether of the fee or additional services fee) commensurate with the services performed, less amounts previously paid, together with, except where the professional is in breach, all reasonable costs, disbursements and expenses properly and necessarily incurred, including those incurred in suspending and resuming performance of the services. These costs have to have been incurred as a direct result of the termination or suspension.

Termination is said not to affect the rights or remedies of either party in relation to any breach prior to termination.

PSC3 gives either party the right to terminate if the other approaches insolvency (the acts, etc. which give rise to this right are spelled out) and the professional can terminate if the client has not paid within eight weeks of the issue of a notice that payment is overdue. The client may terminate if he no longer requires the services or if the professional has substantially failed to comply with his obligations and not put the default right within four weeks of a notification to do so. In addition, the client can terminate if an event stops the professional from completing the services or from doing so by the date shown on the accepted programme, and is forecast to delay completion by more than 13 weeks, if it was one neither party could have prevented and 'an experienced consultant would have judged at the Contract Date to have such a small chance of occurring that it would have been unreasonable for him to have allowed for it'.

This last criterion could be difficult to satisfy. Would strikes, acts of terrorism, catastrophic fire on a neighbouring site, interruption of power supplies, all be things that have a 'small chance of occurring'? An experienced professional would probably contemplate that any of these sorts of things could occur, but how does he judge whether there is such a small chance of it occurring that he need not allow for it and how does he demonstrate this? Why is it the client who terminates in these circumstances, when it is what the professional 'judged' that determines whether the right has arisen?

As to payment, a final payment is made as soon as possible after termination and is to include 'an amount due assessed as for normal payments and other costs reasonably incurred by the *Consultant* in expectation of completing the whole of the services and to which the *Consultant* is committed'. This does not appear to entitle the professional to loss of profit if the appointment is terminated early, but only those costs he has already incurred in order to provide the services and which would otherwise have been reimbursed through later fee payments.

However, if the reason for termination is the professional's insolvency (and not the acts, etc. earlier referred to), or substantial failure by the professional to comply with his obligations, the amount due is to 'include a deduction of the forecast of the additional cost to the *Employer* of completing the whole of the services'.

This gives the client the opportunity to make a calculation of what it might cost him to get someone else to complete the services and, having given credit for the balance of the fee he would have had to pay the professional, to deduct that from any fees due. There is nothing to say this forecast has to be reasonable or substantiated and therefore there is very little the professional could do to challenge it. Such a calculation, however, does not allow the client to deduct any other costs he might incur as a result of the professional's breach: for example, the additional costs of delays to his project, additional payment to other consultants or to the contractor. It is not clear from the drafting whether this deduction is the only amount that can be claimed or whether the client would still have the right to claim these other sorts of losses or damages as a result of the professional's breach. In the absence of clear wording to the effect that that is all the client can recover, his right to other loss or damage is probably preserved.

Option X11 under PSC3 allows the client to terminate 'the *Consultant*'s obligation to Provide the Services for a reason not stated in this contract by notifying the *Consultant*'. It is thought this is the equivalent of giving the client the right to terminate at will.

There are no suspension provisions in PSC3.

8.10. Notice provisions

As has been said earlier in this chapter, it is important that any termination or suspension provisions are operated in accordance with their terms, or the termination or suspension could be unlawful and could amount to a repudiatory breach. Most such provisions require some form of notice to be given and these should also be complied with in accordance with the provisions in the appointment itself. The notice may have to be in a certain form – for example, in writing – and/or served or delivered in a certain way – for example, delivered in person, sent by first class letter, recorded delivery or fax only

– to a certain person and/or to a certain address, which may or may not be the usual address of the client. Often email is expressly excluded as a means of giving notice – as in ACE Agreement 1. The notice may also be of a certain number of days or weeks, or even months. If so, it will be important to check whether the appointment defines what this means. 'Days', for example, could be working days only, and bank holidays and/or other days may be excluded from the calculation. Failure on any head could render the notice ineffective, although the courts may take the view, if the error was procedural or clerical only, that there was no intention not to be bound by the contract and thus this did not amount to a repudiatory breach.

The same principles apply to any other notice that has to be given under the appointment.

Professional Services Agreements
ISBN 978-0-7277-4158-5

http://dx.doi.org/10.1680/psa.41585.145

Chapter 9
The Professional Services Agreement – PSC3

9.1. Introduction

The New Engineering Contracts (NECs) were developed by the Institution of Civil Engineers and are now in their third edition. It needs to be remembered that they are a 'family' of contracts and each is consistent with the others. As a result, the third edition of the consultant's Professional Services Contract (PSC3) is consistent with the Engineering Construction Contract (ECC) which is a contract between an employer and a contractor. PSC3 is therefore very different from the other standard forms of contract produced by the ACE, RIBA and RICS and much more like a contract with the contractor: there is a detailed programme which has to be accepted, a completion date and an option for sectional completion. If the professional is to obtain a change to the completion date he has to have a ground for doing so – a compensation event – and there is a procedure for ascertaining the extension to be given. There is an option whereby the professional is obliged to pay delay damages if the services are completed late. If the professional is to receive any addition to his fee the same basis applies. The professional has to comply with the same procedures concerning the risk register, key dates and early warning notices as does a contractor under the ECC. There can be key performance indicators. A defects liability period is to be stipulated with an obligation to put right defects within that period. The fee basis can be Option A: 'Priced contract with activity schedule', Option C: 'Target contract', Option E: 'Time based contract' or Option G: 'Term contract'. (There are no Options B, D or F. These are confined to ECC contracts.)

PSC3 is divided into Core clauses which set out the clauses that will apply to all PSC contracts and then there are the Main Option Clauses which set out the four options for the fee to be paid, two dispute resolution options known as W1 and W2 and then Secondary Option Clauses comprising 20 X options, two Y options, the first of which brings in the Construction Act payment provisions, and which was amended by the Amendments of September 2011, and the other the option to give a third party the right to enforce any term(s) under the Contracts (Rights of Third Parties) Act. Finally there is provision for additional Z clauses.

The NEC contracts each come with guidance – lengthy documents with many flow charts illustrating each contract's processes. They provide a commentary on all the provisions in the contracts and so it is advisable to refer to the guidance first when clarification of the underlying contract is needed, although they do not always provide an answer.

The PSC3 contract is created by the client completing the data in contract data part one and the professional the data in contract data part two, although the guidance accompanying PSC3 also contemplates a letter of offer from the professional, a letter of acceptance from the client and possibly a form of agreement.

Together, the two parts of the contract data provide the specifics for the contract and state which options are to apply. It is the client who designates the services and 'the Scope' and it is 'the Scope' which specifies and describes the services or states any constraint on how the professional is to provide the services. There is no definition of 'services' as such. This can give rise to some confusion, as the services are often no more than a brief description of the work – unlike other forms of appointment where there can be extensive services schedules – while the Scope contains everything else.

There is also the option for the client to add its own specific additional clauses (the Z clauses referred to above). These can be used to alter significantly the basis of the contract and so need special consideration.

PSC3 is an entire contract and no change to the contract has effect unless it has been agreed, confirmed in writing and signed by the parties unless provided otherwise.

The main feature of the PSC and other NEC contracts is that they are heavily procedural. Notices have to be given within certain timescales and likewise responses, meetings have to be held and decisions made, all within certain timescales. Acting promptly is a condition precedent to many of the rights that the parties enjoy under the NEC family of contracts. Professionals should not under-estimate the resources that will be needed simply to comply with these procedures.

The objective is to make the parties deal with delays, compensation events, risks and other matters that may affect the provision of the services or the project at an early stage and to resolve these. This is different from other contractor contracts where contractors are inclined to leave claims for additional time and/or money to the end.

In the author's view, these contracts work best where the work to be performed has been fully detailed before the works or services start, where only relatively few 'compensation events' should arise. The guidance says in relation to the clause 23.2 provisions concerning the holding of meetings: 'It is important that planning and programming of the work of the various consultants and Others are carried out before the start of the contract.'

Where projects have started before the works have been fully detailed, so that virtually every piece of design is a change to the original scope, the parties can struggle to keep up with and operate the machinery of the contract.

The contract also envisages cooperation between the parties – there is an express obligation to act in a spirit of mutual trust and cooperation – clause 10.1.

NEC contracts are different from the standard forms produced by the ACE and RIBA and the contractors' contracts produced by the JCT and the ICE in that they use the present tense (as does the RICS). There are therefore no obligations expressed as such in the traditional sense. It can as a result sometimes be difficult to work out when something is to be done. At the time of writing, there have been no court cases which have considered the language of the NEC contracts (although one case in relation to the notice provisions suggested the language was mandatory but found a way round it (*Anglian Water Services* v. *Laing O'Rourke (2010)*). So it is not yet known how the courts will interpret NEC language. For the purposes of this chapter, however, the provisions in PSC3 will be referred to as if they are obligations.

Finally, before turning to the detail, it may be helpful to comment on the role of the designer under ECC contracts. The designer has no role. The ECC only refers to the employer, the contractor, the supervisor and the project manager. The 'designer' does not appear – many professionals do not appreciate that they have no role other than that of producing a design for the client. If professionals do get involved in ECC contracts then it is not as a designer or even as 'engineer', but only as 'Supervisor' and/or 'Project Manager'. It should be noted that if professionals take on *both* of the last two roles then Chinese walls will have to be set up. PSC3 may also be used in both cases. PSC3 is not solely confined to design work.

Some of the provisions in PSC3 have been considered elsewhere in this book. Complying with the client's instructions is at Section 4.6. Key personnel provisions are at Section 4.12. Client's obligations are at Section 6.1. What has to be given to the client on termination is at Section 8.6. (There are no rights to suspend in PSC3.) Termination provisions are at Section 8.9. Confidentiality is at Section 11.3. The authority of the employer's agent is in Section 13.10. As to the tribunal to decide any disputes, this is at Section 24.5 with adjudication procedures at Section 25.9.

9.2. The professional's services and other responsibilities

As has been said above, there is no definition of services as such. It is anticipated that 'the work which the *Consultant* is to do' will be described in the Scope. The professional's obligation is to 'Provide the Services in accordance with the Scope'. To 'Provide the Services' is 'to do the work necessary to complete the services in accordance with this contract and all incidental work, services and actions which this contract requires'. The definition of 'Scope' does not necessarily oblige the client to describe the services, although the guidance says in relation to the Scope on page 28 that the client should use it 'to specify and describe the services as comprehensively as possible, including what he is expecting to achieve as a result of the contract and a statement of any constraints which the Consultant is to abide by'. It will be remembered how, if the client were to impose an obligation on the professional to achieve certain results, that could amount to a warranty for fitness for purpose and raise the problems for the professional discussed in Section 4.2.

The 'Scope' is 'information which either specifies or describes the *services* and states any constraints on how the *Consultant* Provides the Services and is either in the documents in which the Contract Data states it is in or in an instruction given in accordance with this contract'. However, the contract data part one does provide for the services as well as the Scope to be identified.

There is also, at clause 25.1, 'The *Consultant* obtains approval from Others where necessary to Provide the Services'. 'Others' are 'people or organisations who are not the *Employer*, the *Consultant*, the *Adjudicator* or any employee, Subconsultant or supplier of the *Consultant*'. This could encompass building control or planning authorities, but also other professionals and contractors. What about funders, purchasers or tenants who have an interest in the project or the ultimate end-user? Nothing is said as to where the professional might find whose approval he is to seek – this is probably the Scope, as its definition says it is information which '. . . states any constraints on how the *Consultant* Provides the Services'.

There is an obligation on the professional to cooperate with Others in obtaining and providing information which they need in conjunction with the services (clause 23.1) – this is not necessarily the same as the cooperation requirements under the CDM Regulations 2007 as these are dealt with separately.

The professional has also to act 'in accordance with the health and safety requirements stated in the Scope' – clause 25.4.

The professional has to hold or attend meetings with 'Others where necessary to Provide the Services' and inform the client of these meetings so he can attend them – clause 23.2.

This wide definition of 'Others' makes it unclear as to how far these and other obligations concerning 'Others' extend.

The professional has to notify the client as soon as he considers the Scope requires him to do anything which is illegal or impossible. If the client agrees, 'he gives an instruction to change the Scope appropriately' – clause 17.1. This would also be a compensation event under clause 60.

It is not clear whether the client has to be reasonable when he agrees or disagrees with matters that have to be put to him, and the contract makes no provision for the professional to be able to challenge the decision if the client disagrees on unreasonable grounds. It seems unlikely that the contract really intends that the professional should continue to perform an illegal or impossible service in these circumstances.

The same principle and uncertainty applies to the similar provision in clause 16 where the client or the professional have to notify the other as soon as either becomes aware of an ambiguity or inconsistency in or between the documents which are part of the contract. 'The *Employer* gives an instruction resolving the ambiguity or inconsistency.' What if it is not resolved or is incapable of resolution? This is not a compensation event unless it results in an instruction to change the Scope.

9.3. Skill and care

Clause 21.2 states 'The *Consultant's* obligation is to use the skill and care normally used by professionals providing services similar to the *services*'. This is wording similar to that in the JCT contractors' contract for the design obligations of a contractor.

This begs the question as to what is 'normally used'. Is it the same as or different (whether higher or lower) from the common law test for reasonable skill and care: that of the ordinary skilled and competent practitioner in the relevant profession? Why was not the usual wording 'reasonable skill and care' used? The courts could well imply that something different was intended.

Clause 21.1 states that the professional 'Provides the Services in accordance with the Scope'. 'Provides the Services' is defined as 'do the work necessary to complete the services in accordance with this contract and all incidental work, services and actions which this contract requires'. This is a strict obligation, not a reasonable skill and care obligation. Likewise 'Provides the Services in accordance with the Scope' is a strict obligation.

It is also not clear whether clauses 21.1. and 21.2 deal with two different obligations or whether clause 21.2 qualifies clause 21.1 only or qualifies clause 21.1 and every other obligation undertaken by the professional.

The guidance says clause 21.1 states the professional's basic obligation – implying that this is a stand-alone obligation. This could cause problems for both professionals and professional indemnity insurers. It is very rare for an obligation to perform the services to be undertaken on anything other than a reasonable skill and care basis or on a higher skill and care basis.

The guidance continues in relation to clause 21.2: 'This subclause' [i.e. not clause 21.1] 'states the level of skill and care required by the *Consultant*' and does not clarify the uncertainty referred to above. Further, the guidance does not make it clear that the professional is only to be liable if he fails to meet that duty of care. What it says is 'It follows that a Defect may not necessarily be the liability of the *Consultant*'. The definition of a Defect is 'a part of the *services* not in accordance with the Scope or the applicable law'. However, as has been said, this is a strict test, it is not a negligence test and the guidance is therefore misleading.

9.4. The client's responsibilities

This has been discussed in Section 6.1. The client's only expressed obligation in this section of the PSC3 is to provide information 'and things which this contract requires him to provide in accordance with the Accepted Programme' – clause 20.1 – and to 'provide access to a person, place or thing to the *Consultant* as stated in the Contract Data on or before the later of its *access date* and the access date for it shown on the Accepted Programme' – clause 25.2.

Nothing is said about the client having to behave reasonably. In a number of areas the client is given an unfettered subjective discretion or a right simply to decide or assess matters. The professional has no sanction if the client behaves unreasonably, or indeed illegally, which does not seem correct. It is not an answer to say the professional can take the matter to adjudication, because the contract gives no basis to do this. The professional could only point to the obligation under clause 10.1 to act 'in a spirit of mutual trust and cooperation' but this does not really address the point. It would be of assistance if there were a clause to the effect that 'all decisions, assessments and instructions made by the *Employer* are to be fair and reasonable'. A challenge could then be made if a decision or assessment, etc. were unreasonable.

9.5. Communication

In line with the procedural nature of this contract, there are detailed provisions as to how the parties are to communicate – clause 13.

Everything, that is 'each instruction, certificate, submission, proposal, record, acceptance, notification, reply and other communication', is to be in a form that can be read, copied and recorded. Writing is to be in the language of the contract, which the client stipulates in the contract data part one.

A communication has 'effect' when it is received at the last address notified for receiving communications at the address the recipient stipulates in his part of the contract data.

Where a reply is required to any communication, the period in which the party is to reply ('the period for reply') is completed by the client in the contract data part one. It is anticipated that this will be a number of weeks. This will apply to both parties and to all replies. This could make the contract a bit rigid – in some circumstances a longer period might be appropriate and in others a much shorter period. The guidance acknowledges that different periods for reply may be necessary and if so suggests they be separated out and listed in the contract data.

Where the contract provides that a communication has to be submitted to the client for his acceptance, the client has to reply within the period for reply unless another period is stated. The PSC3 is unusual in that it requires the client to give reasons for non-acceptance and then the professional has to re-submit the communication within the period for reply, taking account of these reasons. Clause 13.4 states that 'A reason for withholding acceptance is that more information is needed to assess the *Consultant's* submission fully'. It is thought this is not the only ground on which the client can refuse acceptance because of the wording 'A reason', but this is one example of many instances in the PSC3 where reasons, etc. are to be given and grounds set out, and it is not clear if this is meant to be exhaustive.

In clause 13.8, it says the client may withhold acceptance of a submission by the professional. It is not clear whether this is the same as not accepting a communication. Withholding acceptance for a reason stated in the contract is not a compensation event. The reasons stipulated in PSC3 have been included so that the parties can determine if a refusal does or does not give rise to a compensation event. If a reason is one given in the contract there is no compensation event. If it is not, this is a compensation event. Clause 60(8) says that it is a compensation event if 'The *Employer* withholds an acceptance (other than an acceptance of a quotation for acceleration) for a reason not stated in this contract'.

As the only reason for non-acceptance given under clause 13.4 is the need for more information, it would appear that any other reason could entitle the professional to a compensation event.

Acceptance by the client of a communication from the professional or of his work does not change the professional's responsibility to 'Provide the Services' – clause 14.1. See the comment on these sorts of provisions in Section 6.3.

The client can extend the period for reply to a communication if the client and the professional agree to the extension before the reply is due – clause 13.5. This implies that it is only a reply to a client communication that can be so extended but it is not clear if that is correct. If both parties have to agree, why should not any period for reply be capable of extension? Further, the client then has to notify the professional of the extension which has been agreed. It is not clear why this is thought necessary and such a requirement may well be overlooked if many such communications are in train at the time.

There is also a separate further specific requirement concerning 'notifications': these have to be communicated separately from other communications – clause 13.7.

9.6. Early warning and risk register

The early warning procedure has been in the NEC contracts from the beginning – the objective being, as has been said, to deal with potential problems early.

Clause 15.1 requires either party to give an early warning notice to the other as soon as either becomes aware of any matter which could

- increase the total of the prices (i.e. the fees)
- delay completion (i.e. of the services and putting right any defects)
- change the accepted programme
- delay meeting a key date
- impair the usefulness of the services to the client
- affect the work of the client, a client's contractor or another professional.

The professional may, but presumably is not obliged to, give an early warning of any matters which could increase the total cost. These early warning matters are entered by the client in the risk register. The parties do not have to give early warning of a matter for which a compensation event has previously been notified.

It should be noted that failure to give an early warning notice can have an effect on the assessment of a compensation event.

Clause 61.5 says that if the client decides that the professional did not give an early warning of the event which an experienced professional could have given, he notifies his decision to the professional when he instructs him to submit quotations (for compensation events). Then clause 63.5 says that in such circumstances the event is assessed as if the professional had given early warning. It is thought that this means that if extra time and/or cost has been incurred by the professional because of the failure to give an early warning notice, then that extra time or cost will not be recoverable in any compensation event, even if the professional has complied with the requirement to notify a compensation event within eight weeks of becoming aware of the event.

PSC3 introduced the concept of the 'Risk Register'.

It appears from the definition in clause 11.2(10), page 4, that this is to be compiled at the beginning of the project. The definition says 'The Risk Register is a register of the risks which are listed in the Contract Data'. Both parties have sections in the contract data where the matters to be included in the risk register are to be listed – in two or four lines respectively. The definition continues '. . . and the risks which the *Employer* or the *Consultant* has notified as an early warning matter (i.e. all early warning matters are now risks to go on the risk register]. It [presumably the risk register] includes a description of the risk and a description of the actions which are to be taken to avoid or reduce the risk'.

Nothing is said in PSC3 about who sets up the risk register and records what is to be done in order to avoid or reduce the risks listed in the contract data. It is assumed that these are the client's duties, for reasons explained below.

Once there has been an early warning notice PSC3 provides that the new procedure for dealing with matters on the risk register takes the place of the procedure which under PSC2 followed the giving of early warning notices.

It seems to be intended that all matters on the risk register, whether originally or added to by later early warning notices, are to be dealt with by risk reduction meetings but these are not compulsory, nor is any timetable set for them. Clause 15.2 says either the client or the professional may instruct (presumably at any time) the other and other people (if the other party agrees) to attend a risk reduction meeting and then pursuant to clause 15.3 everyone attending cooperates in

- making and considering proposals for how the effect of the registered risk can be avoided or reduced
- seeking solutions that will bring advantage to all those affected
- deciding on the actions which will be taken and who 'in accordance with this contract' will take them
- deciding which risks have now been avoided or have passed and can be removed from the risk register.

It is unclear what the status is of the matters on the risk register and of the decisions made at the risk reduction meetings under clause 15.3. Do the parties have to take the action decided upon; that is, is it a contractual obligation? What does 'in accordance with this contract' add? There are no specific obligations relating to who has to do what in relation to the risk register – although the professional does have to obey instructions given by the client – clause 25.3 – but the client's right to give instructions is limited to changing the scope or a key date (and after completion an instruction is given only if it is necessary to provide the services) – clause 20.3. Some other provisions in the contract also allow the client to give instructions in certain circumstances but not in relation to the risk register. Also, nothing is said as to what is to happen if there is any failure to take the action decided upon. Additional costs incurred as a result of complying will only be recoverable if they are also compensation events.

After the meeting it is the client who revises the risk register to record the decisions made at such meetings and who issues the revised risk register to the professional. If a decision needs a change to the scope, the client is to instruct the change at the same time as he issues the revised risk register – clause 15.4. The parties then move into the compensation event procedure.

It is possible that some of the actions decided upon as being needed to reduce the risk could also qualify as compensation events if they were not complied with even if they do not require a change to the scope – for example, if it was essential to reduce a risk that the client provided relevant information to the professional by a certain time and then failed to do so because he had been unable to obtain it from a third party, that would qualify as a compensation event.

This is all the contract says about the risk register and so this gives no idea as to the sort of risks it is anticipated will go on the register. All the guidance on page 28 says in relation to the definition is that the risk register provides a means of identifying the risks and managing them by eliminating them or reducing them or otherwise deciding how they are to be dealt with, and that sub-clause 15.4 provides a procedure for revising the register following a risk reduction meeting. The first part echoes the language of the CDM Regulations but it is not thought the register can be intended to cover health and safety aspects of the professional's design and/or the matters that need to go in the construction phase plan for which the regulations already provide procedures. (CDM is not mentioned in PSC3 at all – in the author's view, quite rightly. The CDM obligations are imposed by statute and should not also be a matter of contractual obligation.)

It is unclear whether clients will include in the original risk register matters that are different from and more extensive than those that are stated in the PSC3 clause 15 to be the subject of early warning notices, which are fairly comprehensive. Perhaps they are simply to be specific risks for that particular project which the parties foresee at the outset.

Professionals will need to consider any risks in the risk register which the client provides or which the professional considers should be included with these points in mind and their interaction with compensation events: should there be included on the risk register all the matters which could give rise to additional cost/time if they are not dealt with or which could not be dealt with as originally anticipated, and should these all be included as compensation events? It seems possible that they should.

Some professionals have commented that the status of the risk register is so unclear as to what should be included that it is little used.

9.7. Prevention

There is a new provision in PSC3 at clause 18 headed 'Prevention'. This is essentially a *force majeure* provision, and it is intended to be so. The guidance comment on this clause at page 30 is 'This subclause is designed to deal with what is commonly referred to as "force majeure" events, that is the kind of events which neither Party can prevent or control'. It is not, however, drafted in the traditional manner.

Clause 18 says if an event occurs which stops the professional providing the service or from providing the service by the date shown on the accepted programme *and* which 'neither Party could prevent *and* (emphasis added) an experienced consultant would have judged at the Contract Date to have such a small chance of occurring that it would be unreasonable for him to have allowed for it', the client should give an instruction to the professional stating how the professional is to deal with the event. If he does so, this could be (but not necessarily will be) a compensation event (clause 60.1(11)). The test is different for a compensation event. In clause 60, the reference is to an event which stops the professional 'completing the Services', or stops the professional completing the services by the date shown on the accepted programme, whereas in clause 18, the reference is to stopping the professional 'Providing the Services' or stopping the professional providing the services by the date shown on the accepted programme. No explanation for this discrepancy is given. An event could of course have both effects. But it is also possible that an act of *force majeure* could temporarily stop the professional from providing the services but not ultimately from completing the services. It is not clear whether in this case he is to have no adjustment to his fee/time for completion – does the difference in wording mean the professional is to take the risk of an event if it does not stop him completing the services but only stops him from providing them temporarily?

What about the description of the *prevention* event? The wording in clauses 18 and 60.1(11) is the same. Not only could neither party prevent it, but the professional has to establish that the event was something which had such a small chance of occurring that it would have been unreasonable for him to allow for it. This could be difficult. Are strikes, acts of terrorism, catastrophic fire on a neighbouring site, interruption of power supplies all things that have a 'small chance of occurring'? It is also not clear whether, if an event of this nature is in the contract data, it would not fall within this clause. Presumably not, if it in fact happens when it had been predicted and allowed for as a risk.

If the professional is to take the risks of foreseeable but likely events, which are not stated as compensation events, then he will need to include some time for them in his programme and in agreeing the completion date for his services.

The usual wording for an objective test in a traditional *force majeure* clause is 'beyond the control of either party' and it might have been clearer if this wording had been used, as it is in the guidance.

As has been said, once there is an event that comes within clause 18 'the *Employer* gives an instruction to the *Consultant* stating how he is to deal with the event'. The professional also has to obey any instruction given under this clause – clause 25.3: 'The *Consultant* obeys an instruction which is in accordance with this contract and is given to him by the *Employer*.' There is no provision for him to challenge it if he disagrees with the solution.

Professionals will have to consider the risks of accepting this clause unamended for any particular project, particularly for projects outside the UK, where *force majeure* clauses are probably more important.

9.8. Starting, completion and key dates

As with a contractor, the professional does not start work until 'the starting date' which is to be specified by the client in the contract data part one – clause 30.1. It appears to be expected that this form of contract will be completed before any work commences. The use of the present tense – 'The starting date is . . . ' – seems to preclude putting in a date that has passed and there is no provision in PSC3 to back-date the contract so its terms apply from whenever the services were first commenced. The starting date also fixes the payment assessment date.

PSC3 introduced provisions for key dates.

A key date is defined in clause 11.2(6) as 'the date by which work is to meet the Condition stated. The Key Date is the *key date* stated in the Contract Data and the Condition is the *condition* stated in the Contract Data unless later changed in accordance with this contract'.

It is assumed therefore that all key dates will be identified before the contract is entered into and the client will put them in the contract data part one by setting out where provided the stated condition that is to be met and then the key date.

There is no provision for adding key dates at a later date.

Presumably at that point, if the professional thinks he cannot meet the key date and/or condition he must refuse to agree the contract.

Then there are the requirements concerning key dates in relation to the programme set out in clause 31.2: they have to be set out in relation to the overall dates in the programme and the professional has to include in his programme 'the dates when the *Consultant* plans to meet each Condition stated for the Key Dates'. This seems curious – surely he will plan to meet the condition by the key date? In some circumstances, however, it might be that he plans to meet it earlier than the key date.

Key dates are different from the dates by which sections of the work are to be completed.

So what sort of thing could constitute a key date?

Clause 31.2 already covers 'the order and timing of the operations which the *Consultant* plans to do in order to Provide the Services', and 'the order and timing of the work of the *Employer* and Others . . . '.

The guidance page 28 says that key dates are 'dates by which the *services* are required to reach a stated Condition'. (The definition of key date in PSC3 refers to 'work', not 'services', but PSC3 seems to use 'work' and 'services' interchangeably.) Are they to be milestones, for example, completion of various stages of the design – design brief and concept approval, detailed design approval, etc.? But this could be covered by some of the other matters in the programme. This could cover stages in the consultancy process such as third party interfaces. However, it is thought the reason for having key dates is because of the strict obligation to meet them and the remedies the client has if the professional misses them.

There is no strict obligation to meet the various other dates in the programme, although there is a strict obligation to meet 'the Completion Date' – see below. By introducing the key dates the client will be able to fix some interim dates which the professional is also to meet.

The obligation to meet a key date is in clause 30.3: 'The *Consultant* does the work so that the Condition stated for each Key Date is met by the Key Date.' It is therefore a strict obligation. The professional will, it is thought, be entitled to additional time and a resultant change in the key date if the event preventing him from achieving a key date is a stated compensation event, even if the change to the key date is the only change required. The compensation events also include the client changing a key date. The professional has to notify the event within the timetable prescribed.

An early warning notice also has to be given if either party becomes aware of any matter which could delay meeting a key date (clause 15.1). This could result in a change to the scope. The client is entitled to give an instruction to change the scope which would be a compensation event – clause 20.2 – and so could reduce the scope so that the original key date is met. If a change in scope makes the description of the condition for a key date incorrect the client can correct that description and that is to be taken into account in the assessment of the effect of the compensation event – clause 63.9.

The client also has the authority to change a key date (clause 20.2) and this is a compensation event (clause 60.1(4)).

The question for the professional will be – will the agreed compensation events for any project cover all the likely events that ought to relieve the professional of the strict obligation to meet the specified condition by the specified key date in circumstances where he has not been negligent or where he has been prevented for reasons beyond his control? Such events could be difficult to predict. If not, further events will need to be added. The ACE Agreements deal with this more generally by qualifying an obligation to meet the programme with 'Subject always to conditions beyond the Consultant's control . . .'

The effect of compensation events on key dates is to be brought into the quotations/assessments for the costs and time consequences of all compensation events, which is why it is thought a professional can use the compensation event procedure if a key date cannot be met as a result of a compensation event (clauses 62.2 and 63.3). The latter clause says the delay 'is assessed as the length of time that, due to the compensation event, the planned date when the Condition stated for a Key Date will be met is later than the date shown on the Accepted Programme'.

There are consequences if the professional does not meet the condition stated for a key date. Clause 23.3 says that:

> If the *Employer* decides that the work does not meet the Condition stated for a Key Date by the date stated and, as a result, the *Employer* incurs additional costs either
> - in carrying out work or
> - by paying an additional amount to Others in carrying out work
>
> on the same project, the additional cost the *Employer* has paid or will incur is paid by the *Consultant*. The *Employer* assesses the additional cost within four weeks of the date when the Condition stated for that Key Date is met. The *Employer*'s right to recover the additional cost which is his only right in these circumstances.

(It is thought the 'which' in the last sentence has been included in error.)

Taking various parts of this clause, first the client 'decides' if the condition for a key date has not been met by the date stated. What if the client's decision is wrong? There is nothing in PSC3 about the parties acting reasonably. Yes, they have to work in a spirit of mutual trust and cooperation, but this does not really cover it. As has been said, it is not thought that there is any basis on which the decision could be challenged. The same applies to the assessment the client is to make. Again, it is not thought the professional would be able to challenge this assessment – the client has been given a right to do this – there appears to be no objective test and therefore no basis on which to found a challenge. If a challenge is possible, then it would be through the relevant dispute resolution procedure.

There is also the question this clause raises of whether the client will in fact be able to make an accurate assessment within the four-week period.

Presumably if the client fails to make the assessment within the four weeks he loses his right to the additional cost, but again that is not clear.

What is the cost he is to assess? This is the additional cost he or others incur 'in carrying out work on the same project'.

There is some uncertainty about this. Is the 'work' that is to be carried out only that necessary to meet the condition? It is implied, but not stated, that only the minimum is necessary but it will depend on the nature of the condition itself. It is also implied that either the client or others will carry out *work*. Is it anticipated that others will complete the relevant part of the professional's design which, for example, he has failed to complete by a key date. Will 'work' always be what is required? Do 'costs' cover the consequential costs of carrying out other work (e.g., additional work) the client and others incur because of delay? Yes, probably, because the clause refers to 'costs', but the client clearly is not entitled to any recovery for any other types of losses.

What if the professional does the work but does it after the key date? This might be a better and cheaper solution than the client employing others to do so. What about any delay to the project as a result? Will the client be able to collect damages or LADs (liquidated and ascertained damages) if the professional fails to meet the completion date as a result? It is arguable that the client may not be able to do so – as clause 23.3 provides this is the client's only remedy. There could therefore be some advantages in terms of the loss/damages payable to the client if a key date is missed – the remedy could be more restricted than if it were treated as a straight breach of contract.

So a professional will need to consider in respect of the key date provisions whether he is prepared to agree that the client should assess any additional costs if the conditions for key dates are not met – this will depend on his view of how the client may act in these circumstances.

There is also, as has been said, another strict obligation: the professional has to do the work 'so that Completion is on or before the Completion Date' – clause 30.1. How he gets to the completion date is a matter for the professional. If he is to have any extension to the completion date then he will need to bring himself within one of the agreed compensation events. 'Completion' is defined as:

> when the *Consultant* has
> - done all the work the Scope states he is to do by the Completion Date and
> - corrected Defects which would have prevented the *Employer* from using the *services* and Others from doing their work.

(There is a very wide description of 'Others'.)

The definition continues:

> If the work which the *Consultant* is to do by the Completion Date is not stated in the Scope, Completion is when the *Consultant* has done all the work necessary for the *Employer* to use the services and for Others to do their work.

This is a definition which is not unlike the meaning of practical completion in a contractor's contract.

It would appear from this that it is anticipated that the scope will set out what is to be done by the completion date, although that is not set out in the definition of 'Scope'.

The completion date is defined as 'the *completion date* unless later changed in accordance with the contract' – that is, as a result of the compensation event procedure. The completion date can either be decided by the client and put in the contract data part one (it is unclear whether the client can do this without reference to the professional's programme which does not have to be submitted until after the contract date, although it may be attached by the professional to his part of the contract data part two) or it can be decided by the professional and put in the contract data part two. The guidance simply refers to the contract data being specified by the client in part one or tendered by the professional in part two. No further guidance is provided as to how the parties arrive at the completion date. In view of the consequences for not meeting the date, this could be of significant concern for professionals and they would be advised to supply a realistic completion date. For some projects this may not be possible but PSC3 does not allow for that.

The guidance says the completion date may be of minor importance or critical in 'co-ordinating the work of several consultants'.

Without a completion date, however, the time consequences of compensation events cannot be operated. This is confirmed in the guidance at paragraph 30.1.

The client decides when the professional has completed the work, or, as clause 30.2 says, 'decides the date of Completion and certifies it within one week of the date'. Again there is no mechanism for challenging the client's decision – what if he got this wrong? What if he certifies this late? Technically this would be a breach of contract and thus a compensation event and so if this caused the professional loss, for example, because of the delay in payment, he might be able to recover this.

9.9. The programme

Clause 31 deals with the programme and it needs to be noted that this is the programme for the professional's services and not for the project. Nothing is said in PSC3 about what happens to the professional's programme and how far, for example, it is put into and coordinated with the programme for the project, the other professionals' programmes and so forth. The obligations simply relate to the professional's own programme. It is thought that the client will include in the scope of one of his professionals – say the project manager – the obligation to produce an overall programme and coordinate all the programmes of the professionals.

The provisions concerning the professional's programme are detailed and very much more extensive than anything in the standard forms produced by the ACE, RIBA and RICS.

The guidance says the programme can either be agreed at the contract date or can be prepared and submitted at an early stage in the contract. Clause 31.1 says if the programme is not identified in the contract data, the professional is to submit a first programme to the client for acceptance within the period stated in the contract data. In the latter event, as with any submission of the programme, the client is required to accept or not accept it within two weeks (clause 31.3) and, if the reply is not acceptance, it has to be given with reasons and the professional has to submit a revised programme for acceptance within the period for reply.

The 'Accepted Programme' is defined as the programme identified in the contract data or is the latest programme accepted by the client.

The guidance says it is an important document for administering the contract and it enables the client and professional to monitor progress and to assess the effect of compensation events. It also identifies when particular actions are needed from the parties.

Clause 31.2 sets out the very detailed requirements for the matters to be shown on each programme which the professional is to submit. The revisions in PSC3 have added to these and include the order and timing of the work of others 'as last agreed', the dates on which the professional plans to meet each condition for the key dates and provisions for float, time risk allowances, health and safety requirements and the procedures set out in the contract.

Although the reference to float is ominous there are no further references to it or the other matters referred to above.

The basis for the assessment of extensions of time has not been changed in PSC3. Nothing is said in the guidance about these additional matters.

Previously the client could only reject the programme if it did not show the information which the contract requires. Now the client has further grounds for rejecting the programme: that the professional's plans are not practicable or are not represented realistically or it does not comply with the scope (clause 31.3). Clause 60(8) provides that if the client withholds an acceptance for a reason not stated in the contract this is a compensation event – unless this is an acceptance of a quotation for acceleration.

Revised programmes have to be submitted not only when the client instructs this or the professional chooses to, but 'at no longer interval than the interval stated in the Contract Data from the starting date until Completion' (clause 32.2).

There are also detailed requirements as to what is to be shown on each revised programme

- the actual progress achieved on each operation and its effect upon the timing of the remaining work

- the effects of implemented compensation events
- how the professional plans to deal with any delays and to correct notified defects; and
- any other changes which the professional proposes to make to the accepted programme.

The additional information to be included in each programme and the ability of the client to require regular revisions of the programme to be submitted could all be administratively burdensome for the professional.

The client may instruct the professional to stop or not to start any work and may later instruct him that he may re-start or start it (clause 33.1). Such an instruction constitutes a compensation event but if it arises from the fault of the professional neither the price nor the completion date is changed.

The client may also instruct the professional to submit a quote for acceleration, that is to bring the completion date forward (clause 34.1). The client has to set out what changes there will be to any key date which are to be included in the quotation. The basis of the professional's quotation is proposed changes to the price, a revised programme showing the earlier completion date and the changed key dates, together with the professional's details of his assessment. He has to do this or give his reasons for not doing so within the period for reply.

It is assumed the professional does not have to have any particular reasons for refusing to do so and that on receipt of his quotation the client can accept or reject it as he wishes.

It does not appear that the professional is to be paid the costs of preparing this assessment as it is not a compensation event. Would a reason therefore for not doing so be the costs of providing it?

The guidance says 'There is no remedy if [the assessment] is not produced (by the professional). Acceleration can only be achieved by agreement between the Parties. It cannot be imposed on the *Consultant* without his agreement.' That is correct as a matter of contract law unless of course the contract provides otherwise.

9.10. Quality and correcting defects

Unusually among the standard forms considered in this book, PSC3 provides as a contractual obligation that the professional is to operate a quality management system, the requirement for which will be stated in the Scope – clause 40.1. To that extent the client will be fixing the quality management system.

Further, the professional has to provide the client within a specified period with a quality policy statement and a quality plan for acceptance which comply with the requirements stated. If the professional fails to comply with the quality plan, he has to comply with an instruction from the client to correct that failure – clauses 40.2 and 40.3.

Again, unlike any of the standard forms, such as ACE and RIBA, PSC3 has an express provision requiring the professional to correct defects up to the 'defects date' – clause 41. This is an arbitrary period fixed by the client in the contract data part one, being a number of weeks after the completion of the whole of the services, similar to the defects liability period under a construction contract but generally not the same length of time as for a contractor. In this respect, as in others in this form of contract, the professional is being treated as a contractor. However, it is important to recognise that the defects being referred to are defects in the services and the scope of work that the professional has undertaken for the client. The defects referred to are not defects resulting from faulty construction work undertaken by a contractor. It is also important to realise that the length of time a professional is liable under PSC3 for other matters is determined on the usual principles.

First there are provisions concerning notification. Until 'the defects date' both the client and the professional (and not just the client) have to notify the other as soon as they find a defect. At completion, when the professional completes his services, the professional has to notify the client of any defect that has not been corrected. After completion and until the defects date, the professional (only) notifies the client of any defect as soon as he finds it. These are overlapping provisions. Why does one need the final provision if both parties have to notify the other until the 'defects date'? Was the first period meant to be only until completion?

After the defects date there is no obligation to notify.

There is a statement in clause 41.1 that the client's rights in respect of any defect which the client has not found or notified by the defects date are not affected. This, it is assumed, is a reference to latent defects which could appear after the defects date, but why does it only refer to those the client has not found? Surely this should also refer to those the professional has not found or just apply to defects generally? The guidance page 36 says 'The final sentence preserves the *Employer's* rights in relation to Defects which are not discovered until after the defects date'. This would have been a simpler way to put it.

A 'Defect' is defined as 'a part of the *services* which is not in accordance with the Scope or the applicable law'. The first part of this definition is understandable, if misguided in the author's view. The logic is clear, however: if the professional has not provided the services described then that is to be a breach of contract, unless the client has waived that requirement or changed the scope to match. The reference to 'the applicable law' is discussed below.

Clause 41.2 then provides that the professional corrects a defect – i.e. those that appear before the defects date – whether or not the client notifies him of it, within a period which minimises the adverse effect on the client and others, but if he does not so correct it within that timescale the client can assess the cost of having this defect corrected by others and the professional has to pay this amount (clause 41.2). This implies the client can get others to correct the defect but it is not stated expressly. The wording could encompass not just errors in design discovered before the relevant part of the project is built but also a defect in the professional's design which has now become incorporated in the construction – say in a retaining wall built early on – and could only be 'corrected' as such by pulling the relevant part down and rebuilding it to a corrected design. How can the professional correct this defect? He can correct the design, but the contractor would have to do the demolishing and rebuilding. Would this fall into the part of the clause which says he has not corrected the 'Defect'? But the defect is the negligent design, not building works.

The whole concept of treating defects in this manner is the wrong way round and does not reflect the proper basis on which the professional should be held liable; that is, for a failure to use reasonable skill and care, not on the strict test of whether the services were in accordance with the scope where compliance may have been outside the professional's control and/or where he has not been negligent. Also, the scope might have been expressed in such a way that it contains a warranty or warranties for fitness for purpose and, as will be appreciated, this could take the professional outside his professional indemnity insurance.

A professional is not necessarily required by law to remedy or correct defects for which he is responsible (damages may be an adequate remedy), nor is the timescale within which he has to do this necessarily that which 'minimises the adverse effect on the Employer'. He is certainly not required to remedy defects for which he is not responsible. PSC3 recognises this to some extent by providing for the professional to notify as a compensation event the correcting of any defect for which he is not liable. However, this reverses the burden of proof. This results in the professional having to prove he was not negligent when it is the client who should prove that he is.

The guidance, page 44, states that 'Not all Defects are the *Consultant's* liability . . . Those Defects that arise notwithstanding the fact that the *Consultant* has exercised the skill and care in 21.2 are the *Employer's* responsibility'. It is not thought, as has been said, that the provisions in clause 41 achieve this, as the definition of defect does not include a reference to skill and care.

There could also be a problem if the client is allowed to assess the cost of correcting any defects. He may not do so on the common law basis of damages. Does this provision apply equally where the professional rectified the defect but not within the timetable required? If so, why should the basis be the cost of others correcting the defect? What about the situation where it might no longer be possible to rectify a defect, or where it is technically a defect but the defect has no significant effect on the works? What if the assessment is simply wrong?

As to that part of the definition that says a 'Defect' is something that is not in accordance with the applicable law, there is no obligation in PSC3 (and indeed there should not be) that the design is to be in accordance with the applicable law. A change in the law which affects the design is not a compensation event (although 'a change in the law of the project' can be added as a compensation event by option X2.1 – if that is the same thing). A change in the applicable law could also have consequences which do give rise to a compensation event.

The professional could therefore notify as a compensation event under clause 60.1(12), a change in the law after preparing his design as 'a Defect for which he is not liable' if he has amended or 'corrected' his design as a result. It is unclear whether this means he can also notify any other failure to comply with the law which results in corrective work having to be carried out, whether or not he is responsible for it. If there is no obligation to comply with the applicable law, however, there is a question as to whether he is 'liable' if he does not.

9.11. Payment

9.11.1 Basis of payment

The PSC3 provides for four different bases of payment.

- Option A: Priced contract with activity schedule
- Option C: Target contract
- Option E: Time based contract
- Option G: Term contract.

Options B, D and F are not used – they are used in the contractor's contract, however.

Detailed consideration of these four bases is outside the scope of this book – there is extensive commentary on them in the guidance notes to PSC3. In summary, the basis of payment is divided into 'the Prices' and 'the Price for Services Provided to Date'.

For option A: 'the Prices' are lump sum prices for each of the activities described in the activity schedule. These should be clearly and comprehensively described. As to 'the Price for the Services provided to date', the professional is paid the lump sum prices for the activities completed at the assessment date: the date for submitting invoices.

For option C: 'the Prices' are again lump sum prices for each activity in the activity schedule and 'the Price for the Services Provided to Date' is payment for each task on a time charge for services carried out. 'The Prices', however, are the target prices and are compared with the actual time charge at the time of final payment to determine whether the professional is entitled to his share of the 'savings' or has to contribute to the 'excess'.

For option E: 'the Price' is on a time charge basis and paid as such for the 'Services Provided to Date'.

For option G: 'the Price' is divided into time charges for items described as such in the task schedule and lump sum prices for any other item in the task schedule. As 'the Price for the Services Provided to Date' the professional is paid for each task the time charge for time-based work completed and a proportion of the lump sums representing the proportion of the lump sum work completed.

There are detailed provisions in each of the option clauses for the professional to prepare and provide forecasts and information about how the activities relate to the operations on the programme. There are also detailed provisions concerning keeping accounts and records for the client's inspection and for revising the activity schedule, together with information as to how the compensation events for that particular option are to be assessed and what could constitute additional compensation events. The option C clauses also provide for the detail as to how the professional's share of the difference between the total 'Prices' and 'the Price for the Services Provided to Date' is to be calculated and who is to pay what and when. There are also provisions for where sub-consultants are to be employed. For option G there are provisions to allow the services to be broken down into tasks in respect of which task orders are to be instructed, together with details about the information to be included for each task order and for completion dates to be given for each task with its own delay damages for late completion.

The payment provisions in PSC3 itself do not bring in the Construction Act provisions – in order to do this the parties have to include option Y(UK)2 of the Housing Grants, Construction and Regeneration Act 1996 – however, if the contract is a 'construction contract' the provisions of the Act will apply in any event but it will be the scheme provisions that apply if Y(UK)2 is not used. The Amendments of September 2011 amended Option Y(UK)2 to bring in the provisions of the Local Democracy, Economic Development and Construction Act 2009 (the amended Construction Act).

Option Y(UK)2 provides for payment to become due seven days after the date of the professional's invoice, and the final date is 14 days thereafter unless a different period is stated in the contract data. The option also provides for the notice of payment and notice of withholding under the Construction Act and the payee payment notice and notice of intention to pay less, as described in Chapter 7.

If the professional exercises his right to suspend under the Construction Act or the Amended Construction Act (because the client has not paid in full by the final date for payment and has not served an effective notice to withhold or paid the amount in any notice of intention to pay less) this is a compensation event and so would entitle the professional to his costs and extra time.

The payment provisions in the body of PSC3 are in general terms because of the variety of options that could apply each with their own particular details.

They refer to the professional assessing the amount due and submitting invoices therefore at each assessment date (or invoice date). The first assessment date is decided by the professional to suit 'the procedures' of the parties but it cannot be later than the assessment interval after the starting date. The assessment interval is specified by the client in the contract data part one. The assessment dates occur at the end of each assessment interval until eight weeks after the defects date and at completion of the whole of the services – clause 50.1.

Invoices have to include the details stated in the scope to show how the amount has been assessed. The first invoice shows the amount due and thereafter the invoices have to be for the changes in the amount due since the previous invoice. The amount due is 'the Price for Services Provided to Date', any expenses and any other amounts due, including tax, less amounts to be paid by or retained from the professional.

Clause 51.1 says payment is made within three weeks of receiving the professional's invoice unless a different period is stated in the contract data. It is not clear how putting a different period in the contract data would work if option Y(UK)2 is adopted, which gives 21 days. Similarly, clause 51.3 requires the client to give a notice if he does not accept the professional's assessment of the amount due but there is no timetable for this and he has to pay the amount he assesses. The professional has either to correct the invoice or provide further information to justify the invoice.

Again this is inconsistent with option Y(UK)2.

Interest is payable on late payment or payments which were disputed on a daily basis at the 'interest rate' and compounded annually. The 'interest rate' is to be specified in the contract data part one and is to be not less than 2% above the specified rate of a specified bank. This might not be sufficient to satisfy the Late Payment Act provisions – see Section 7.11.

9.12. Compensation events

Clause 60 in PSC3 sets out the events that will give rise to extra time and/or money. The possibility of extra time is necessary because of the obligation on the professional to complete within a certain timescale and the events include 'a breach of contract by the *Employer*' without which the whole mechanism could be put at risk. This is because, unless so provided, a breach by the client puts time at large and the professional then has to complete within a reasonable time. There are 12 compensation events and they are fair to the professional. Professionals will need to be wary of accepting any reduction in these through the additional Z clauses, as can happen. In the event of a compensation event the professional is entitled to a change in the prices (i.e. his fee), a change to the completion date and a change to the key dates. That is all. As to the basis for these changes, this is dealt with through the quotation procedure described below.

The provisions concerning compensation events are lengthy and set out a detailed procedure for notices, timetables, assessments, etc. It is important that the parties adhere to these, as the effect of non-compliance with some aspects is also dealt with in clause 60. The main difficulty with these clauses is a practical one – the procedure is exhaustive and elaborate and if there are many compensation events happening within a short space of time it can break down. The parties should not under-estimate the resources that could be needed to deal with compensation events.

The first point to make is that the happening of a compensation event requires one party to notify the other, either at the time or within a certain timescale. The professional is also required to notify an event if the client should have done and has not. There is a penalty if the professional fails to notify within the eight weeks of 'becoming aware of the event'. He is deprived of his rights in relation to any compensation event: a change to the prices, the completion date and the key dates – clause 63.4.

There is a mechanism for the client to notify the professional if he decides that an event notified by the professional arises from a fault of the professional; has not happened and is not expected to happen; has no effect upon the professional's costs, completion or meeting a key date; or is not a compensation event under the contract. Nothing is said in PSC3 as to what remedy the professional has if he does not agree with the client's decision – it is difficult to see how he can dispute it if nothing is said as to the basis of that decision, that is that it has to be reasonable. However, the guidance anticipates that the matter will go to adjudication.

If the client accepts that that has been a valid notification of a compensation event then the client notifies the professional and instructs him to submit a quotation. If the client forgets or fails to give the notice within a certain timescale, the professional may notify the client and this failure also qualifies as a compensation event if the delay exceeds two weeks. If the client decides that the professional should have given an early warning 'of the event which an experienced consultant could have given' he has to tell the professional so when he instructs him to submits quotations. Again, there is no mechanism for challenging these decisions.

No compensation event can be notified after the defects date.

Quotations by the professional have to deal with the proposed changes to the prices and any delay to the completion date and key dates. Alterations to the accepted programme have also to be included in the quotation if the programme for the remaining works is altered. There is a mechanism for the client to instruct the professional to submit alternative quotations and for the client to decide if the effects of a compensation event are too uncertain to be forecast reasonably: if so he then sets out the assumptions on which the professional is to base his quotation and if this turns out to have been wrong the client has to notify a correction which is in turn a compensation event.

There is a timetable for submitting the quotation (two weeks from the instruction) and for the client to reply (a further two weeks).

There are four possible client replies.

- An instruction to submit a revised quotation.
- An acceptance of the quotation.
- A notification that the proposed instruction or proposed changed decision will not be made.
- A notification that the client will make his own assessment.

The client has to give the professional reasons as to why he requires a revised quotation but it is not stated what these should be. The revised quotation has to be submitted within three weeks.

Time for submitting a quotation and for the client to reply can be extended by agreement but this has to be reached *before* the submission or reply is due and the client has to notify the extension that has been agreed.

There is also a mechanism whereby the professional can get his quotation accepted if the client does not reply within the stipulated time, but that does not apply to proposed instructions or proposed changed decisions which presumably then do not go ahead.

Clause 63 sets out the basis on which changes to the prices are to be assessed and likewise the length of time for the delay to the completion date and any key date. If the client has notified the professional that he should have given an early warning notice and did not, the event is assessed as if the professional had given early warning – that is the client will be able to exclude any additional costs/time that were incurred as a result of that failure.

The assessment is to include risk allowances for cost and time for matters which have a significant chance of occurring and are at the professional's risk under the contract. Assessments for work not yet done are based on the assumption that the professional will react competently and promptly and that the accepted programme can be changed. Assessments for work already done include only cost and time which were reasonably incurred.

Clause 63 also makes specific provision for how to deal in assessments with resolving ambiguities or inconsistencies in the scope, correcting descriptions of the conditions for key dates, providing a rate for staff if none have previously been given and for the deduction of the cost of events which the professional has to insure and other costs payable by insurers to the professional.

Under clause 64 the client is entitled to make the assessment in certain circumstances and to notify this to the professional. If he does not do so within the time allowed the professional may notify the client to this effect and state which of his quotations (if he has submitted more than one) it is proposed be accepted. In the absence of a further reply within two weeks from the client, that quotation is treated as accepted. Again, it seems to be anticipated by the guidance that any dispute about the client's assessment will go to adjudication.

Finally, this section of the contract states when a compensation event is treated as implemented, and also provides that an assessment is not revised if a forecast upon which it is based is shown by later recorded information to have been wrong. Thus, the parties have to do the best they can when making the assessments on the facts then known to them and cannot later employ hindsight to alter either the change in price or to the completion date and/or key dates. However, if an assumption upon which an assessment was based turns out to be wrong, then the notification of the correction is a further compensation event, as set out above.

This basis for assessing compensation events is harsher than the common law remedies for a breach of contract by the client. Under PSC3 this is to be treated as a compensation event and these provisions effectively restrict what would otherwise be recovered on common law principles by the professional in the event of any such breach. The consequences of the breach may not be properly addressed by having a price to the price, completion date and key dates. Why should the professional not be entitled to damages on the proper basis, when the client is in breach, when the client is so entitled when the professional is in breach?

9.13. Intellectual property

PSC3 has a straightforward approach to intellectual property. Clause 70.1 provides that the client has the right to use material provided by the professional for the purposes stated in the scope. Thus, it is the scope that will determine the extent of the licence to be given for any particular project. It is also the scope that could determine whether the professional can only use his own material for the project – clause 70.4. This would be the equivalent of the professional giving an exclusive licence – see Section 10.3.

There is an indemnity against copyright infringement in clause 80.1, although this could be worded more clearly. The clause refers to the infringement 'of the rights of Others' without specifying what those rights are and so although 'infringement' could relate to copyright claims and other intellectual property infringements, the reference to 'Others' could extend to providing an indemnity to clients in respect of any third party claims, which would not be acceptable to some professionals and their insurers. The guidance does not assist but it is not thought that this clause is intended to have this wider meaning.

9.14. Insurance and liability

This part of PSC3 deals both with the professional's obligation to maintain certain insurances and the extent of his liability.

Clause 81.1 requires the professional to provide the insurances stated in the 'Insurance Table' unless the contract data say the client is to provide them. The periods for which the insurances are to be maintained are from the contract date until the end of the period specified in the contract data.

The first insurance in the table under clause 81.1 is for professional indemnity insurance – expressed as being against 'liability . . . for claims made against him arising out of his failure to use the skill and care normally used by professionals providing services similar to the services'. The minimum amount required is to be stated in the contract data. This requirement refers only to the professional's failure to use reasonable skill and care and thus it would be satisfied if the professional were to have a negligence-only policy – see Section 15.1.

However, there are other obligations which are not negligence-based in PSC3 – for example, the obligations in relation to completion, key dates, notices, etc., breach of which would be outside a negligence-only policy. Also, the obligations in respect of 'Defects' might not be covered by such a policy.

The guidance contemplates something wider in that it refers to insurance for sums for which the professional could become liable 'as a result of any neglect, error or omission' in carrying out his professional activities.

As to the cover expected in the contract data, this is for 'each claim, without limit to the number of claims'. This does not reflect many standard professional indemnity policies where cover for claims such as pollution and contamination and asbestos are on an aggregate basis and that for terrorism on a very limited basis or excluded altogether. The guidance appreciates this: 'In some circumstances, some insurers insist on a maximum sum for any one claim and in the aggregate in the period of insurance.' The professional who has these restrictions on his professional indemnity insurance will have to ask the client to amend these requirements.

The second policy to be provided by the professional is against 'Liability for death or a bodily injury to a person (not an employee of the *Consultant*) or loss of or damage to property resulting from an action or failure to take action by the *Consultant*'. Again the minimum amount of such insurance is to be stated in the contract data. The guidance makes it clear that this is intended to refer to the public liability insurance carried by the professional, but personal injury, death and damage to property are also covered by professional indemnity insurance policies to the extent that they have been caused by the negligence of the professional. This overlap between the two policies can sometimes be forgotten.

Again, the contract data expect this insurance to be on a per claim basis. It is also anticipated that it should be maintained for a period to be specified after completion of the services or termination. This is not correct for public liability insurance which is an occurrence-based insurance and so it only needs to be maintained until the services are completed or terminated.

The third insurance is the professional's client's liability insurance described as 'liability for death or bodily injury to employees of the *Consultant* arising out of and in the course of their employment in connection with this contract'. As this insurance is compulsory in the UK but not necessarily in other countries, the table provides that the amount of cover required is 'the greater of the amount required by the applicable law and the amount stated in the Contract Data for any one event'.

Again, a period beyond completion of the services or earlier termination is to be set out, but this is not necessary for the same reason as for public liability policies.

The contract data have provision for other insurances to be provided if so required by the client.

The professional (or the client if the client is to provide any insurances) has to provide certificates from his insurer or broker stating that the required insurances are in force if so requested.

As to a limit of liability, PSC3 provides for limits of liability for certain matters but unlimited liability for others – this is a change from PSC2 which had an overall limit.

In PSC2, it said the professional's liability to the client 'resulting from a failure to Provide the Services is limited to the amount stated in the Contract Data'.

Although there could be doubts as to whether this clause would cover all claims that could be made against the professional, particularly those brought in tort for negligence, because it is not specifically referred to – see Section 16.1 – at least it was clear. It could not in any event apply to claims for personal injury or death because of the Unfair Contract Terms Act 1977.

Clause 82.1 in PSC3 says 'the *Consultant*'s total liability to the *Employer* for all matters arising under or in connection with this contract, other than the excluded matters, is limited to the amount stated in the Contract Data and applies in contract, tort or delict and otherwise to the extent allowed under the law of the contract'.

First, this is an aggregate limit, which is good but can lead to problems in fixing the amount. The guidance page 51 says that 'this total amount will be the amount stated in the Contract Data'. This aggregate approach will match those elements of the professional's professional indemnity insurance which are on an aggregate basis, and if say the professional has £5 million professional indemnity insurance per claim for most types of claim, and £5 million in aggregate for other types of claim, and the amount to be stated as a total aggregate limit is £5 million, that would be fine. If the amount stated were higher, say £7 million, to reflect the fact that there may be more than one claim, and the first claim was for £6 million, this would be more than the £5 million professional indemnity cover he has per claim and would thus be beyond his professional indemnity insurance but within the aggregate cap in the contract. Generally, the parties do just put £5 million as the aggregate limit of liability if that is the professional's professional indemnity insurance limit with, if necessary, lesser aggregate caps where this is the cover provided by the relevant professional indemnity insurance.

Secondly, there is again also a doubt whether this limitation could cover claims in negligence because this is not specifically referred to.

What is the purpose of 'to the extent allowed under the law of the contract'? Does this mean that if the law of the contract was a law which allowed no limitation at all or reduced or different limitations of liability, that would apply despite anything inserted in the contract itself? Where would this leave the excluded items? Would not it be better for limitations of liability in contracts not governed by English law to be dealt with separately? If this wording applies to a contract under English law, its effect is not clear. Is it a reference to the Unfair Contract Terms Act 1977 and what is or is not allowed under that Act and the test of 'reasonableness'? If so, what does that add? The wording regrettably does not meet one of the criteria for limitation clauses – that they must be clearly expressed.

Of most concern, however, is that this clause provides that the total aggregate amount in clause 82.1 will only apply to non-excluded matters.

It will therefore not apply to the excluded matters which are

- delay damages if option X7 applies, that is LADs
- the professional's share if option C applies, that is the target contract basis
- an infringement by the professional of the rights of others – presumably this is a reference to the indemnity in clause 80.1 which appears to cover copyright infringement, but it could be construed more widely
- loss or damage to third party property; and
- death of or bodily injury to a person other than an employee of the professional.

The amounts payable in respect of the excluded matters will, as has been said, be unlimited but will also be payable in addition to any aggregate amount agreed under clause 82.1. As has also been said, the guidance says at page 51 that the amount in clause 82.1 will normally be the amount of the professional's professional indemnity insurance. If that is the intention, then there would be no professional indemnity insurance for any of the excluded matters. Further, this basis for limiting liability ignores the fact that for some claims the insurance is on an each and every claim basis and for others on an aggregate basis, that the amount and basis of the professional's professional indemnity insurance could be very different by the time the claim is made and that the same insurance may have to cover this liability and liability for the excluded matters.

Taking the exclusions:

Exclusion 1 – delay damages: this is the fixed amount of LADs per day that is to apply if Option X7 is chosen. Some professionals accept this as a commercial risk. However, most professionals' professional indemnity insurance excludes LADs entirely or the amount by which the LADs exceed the damages for delay assessed on the usual common law principles. Careful consideration will therefore be needed before the professional accepts Option X7, as his firm will be responsible for these amounts. However, if there is professional indemnity insurance cover for some or all of delay damages, this is not covered by the limit of liability in PSC3. Indeed, if there is a claim which, say, results in £3 million damages for additional costs and £3 million in LADs under Option X7 and there is a £5 million aggregate limit for clause 82.1, then the professional will find he is paying the £6 million regardless of his insurance limit and the aggregate limit in the appointment.

Exclusions 2 and 3 may not be of particular concern – the professional's share of any loss when a target contract is agreed is generally a commercial risk not underwritten by insurance. And an indemnity for copyright infringement (if that is what it is) is also something professionals might accept as a commercial risk when this is within their control.

Exclusion 5 – death or bodily injury: liability for this cannot be limited in any event – Unfair Contract Terms Act 1977.

Exclusion 4 – loss or damage to third party property – this is a concern. A claim for loss or damage to third party property can arise under a professional's professional indemnity insurance or his public liability insurance, depending on the cause. It is not clear whether this reference to loss or damage to third party property is meant to refer to the sort of damages covered by public liability insurance and not such losses caused by the professional when carrying out his services. So, as presently worded, there is a real risk that loss or damage to property claims covered by professional indemnity insurance and not public liability insurance will be excluded from the cap in clause 82.1 and such claims can be very substantial. No explanation is given in the guidance.

A new option X18 at page 39 has been added to PSC3 which allows the parties to agree some other limitations. These are stated as a limit of liability for indirect or consequential losses, and a limit of liability for defects not found until after the defects date. These are to be put in the contract data.

Again, it is not clear how these two additional limitations if used with clause 82.1 would fit into the overall limit in that clause. Option X18 does not deal with this. Are they to be calculated as part of the aggregate limit in clause 82.1 or are they to be paid in addition like the excluded matters? Indirect or consequential losses and liability for defects would (subject to a caveat about the way PSC3 describes defects) normally be covered by the professional's professional indemnity insurance and subject to the same limit of indemnity. Are the limitations to be attached to these types of losses meant to be in addition to anything paid as a result of a claim that comes within the aggregate cover under clause 82.1, or are these limitations to provide for a lesser limitation for those particular losses? Again, there is no comment on this in the guidance.

As is said in Section 16.4, consequential losses have been held to be other than 'normal' losses – that is the 'normal' loss which every claimant in a like situation would suffer – and 'consequential loss' is loss that is special to the circumstances of a particular claimant. Thus, the consequential losses would cover indirect damages and therefore might not cover loss of profits or loss of rent, for example. Loss of profits and wasted overheads have been found in one case to be direct and not indirect or consequential losses.

It can be difficult to say in advance what specific losses could be incurred by a client if there has been a breach of contract and thus whether there will be indirect/consequential losses. It follows from this that a professional when capping indirect and consequential losses may not know exactly what he has capped. And in any event, the aggregate cap without this option would include such losses. If it is clear what the indirect or consequential losses could be, then it could well make sense for a professional to seek a lower cap for those than the overall cap provided under clause 82.1.

As to fixing a cap for 'Defects that are not found until after the defects date' – the question arises as to what is the liability of the consultant in respect of 'Defects' as defined in the PSC3 at the defects date. As is explained earlier in this chapter, the professional may or may not be liable under common law principles for 'services that are not in accordance with the Scope'. He is made so liable under clause 41 until the defects date insofar as he has an obligation to correct them until that date. Nothing in PSC3 says that he is to be liable as if clause 41 applied after the defects date and thus would be obliged to correct defects or pay for others to do so, and a professional should be wary of any additional Z clauses which might seek to provide for this. If he is not liable in respect of defects after the defects date, then this limitation is meaningless. If it does mean something, then a cap on this liability should only be used to bring in a lesser cap from that in clause 82.1 because such a claim would be within the professional's professional indemnity insurance which the overall cap is meant to reflect. The client will not be without a remedy. The professional could still be liable after the defects date, for example, if he has failed 'to use the skill and care normally used by professionals providing services similar to the services' (clause 21.2) but this limitation would not apply in that case, although the general aggregate limit under clause 82.1 would.

The parties can also as a result of Option X18.3 fix a time limit for claims, by agreeing 'the *Consultant* is not liable to the *Employer* for a matter unless it is notified to the *Consultant* before the end of liability date', that is a date also to be put in the contract data by reference to a number of years 'after Completion of the whole of the Services'. This again might not put a time limit on claims in tort for negligence or claims other than for breach of contract as they have not been specifically referred to. And 'Completion of the whole of the Services' may never happen. There could be early termination, for example.

Clause 82.2 is meant to be a net contribution clause. This is clear from the guidance at page 51. However, what clause 82.2 says is that 'The *Consultant*'s liability to the *Employer* is limited to that proportion of the *Employer*'s losses for which the *Consultant* is responsible under this contract'.

The principle which applies when two or more parties are liable in respect of the same damage is that each is 100% liable to the client. Thus, the proportion for which the professional 'is responsible' would be 100%.

In order to be able to limit that to a lesser percentage when another party with whom the professional is jointly liable has gone bankrupt or is unable to pay his share in full (the only circumstances in which a net contribution clause is needed), you have to have the assumptions set out in the traditional type of net contribution clause such as in the ACE Agreement – see Section 16.9.

There is a real danger, therefore, that this clause is ineffective.

9.15. The optional clauses

PSC3 has many optional clauses. Option W1 and Option W2 deal with dispute resolution (i.e. adjudication under the Construction Act and outside the Construction Act respectively). Provisions for payment and other options, such as bonus for early completion, delay damages, collateral warranty agreements, transfer of copyright to the client, performance bonds and key performance indicators are in X clauses. There are two further options – Y(UK)2 and Y(UK)3. The first brings in the amendments necessitated by the Housing Grants, Construction and Regeneration Act 1996 and the Local Democracy, Economic Development and Construction Act 2009 to the payment provisions. Y(UK)3 allows the client to specify under the Contracts (Rights of Third Parties) Act 1999 whether any person or organisation can enforce any term of the contract. If so, this has to be stated in the contract data.

Finally, there is the ability to add additional clauses, that is the Z clauses which can amend or add to the basic contract. Much use of Z clauses has been made by the client side of the industry. The guidance says this option should be used where the client wishes to include additional conditions but 'only when absolutely necessary to accommodate particular needs, such as those particular to the country in which the work is to be done'. It goes on to say that 'the flexibility of the PSC main and secondary options minimises the need for additional conditions. Additional conditions should not be used to limit how the consultant is to do the work in the contract as this is part of the function of the Scope'.

Unfortunately, many Z clauses are used to do that or to impose different or more onerous obligations on the professional. They need to be checked with particular care. Often they are drafted in a manner that is inconsistent with the procedures and obligations in PSC3 and with its language.

Professional Services Agreements
ISBN 978-0-7277-4158-5

http://dx.doi.org/10.1680/psa.41585.177

Chapter 10
Intellectual property rights

10.1. Copyright

Intellectual property can cover anything which is the result of mental labour – drawings, specifications and other documents of any kind, books, manuals, software programs, articles, inventions, information, ideas. The main intellectual property rights are copyright, patents, design rights, trademarks, confidentiality and database rights.

Broadly speaking, the purpose of intellectual property rights is to confer on their owner the exclusive right to use the particular intellectual property concerned. These rights can then be commercially exploited by the owner, by being licensed on appropriate terms to others.

However, not all intellectual property can be protected by all these rights. It depends on the nature of the intellectual property.

For example, an idea, by itself, can only be protected by confidentiality; that is, by ensuring that, if it is imparted to another party, it is done so only on a confidential basis – see Chapter 11. The same applies to information on its own. However, information may be effectively protected by copyright if it is in a form which can be protected by copyright, for example, if it is in a document or a drawing.

Copyright protects literary and artistic works (among others). They do not have to be works of any quality – any document is a literary work and any drawing is an artistic work. A software program is a literary work.

In the absence of any special agreement, copyright in a literary or artistic work belongs to the author or artist of that work. However, if the author or artist is an employee producing the work in the course of his or her employment, copyright belongs to the employer, in the absence of any agreement to the contrary.

A work in which there is copyright may consist, for copyright purposes, of a number of different works, the copyright in which could belong to different owners. Thus, in the case of a book containing a compilation of writing and illustrations by different people, the various writers and illustrators may own the copyright in the parts that they have contributed, and the compiler of the book may own the copyright in the compilation. This is typically the case with a software program, which often consists of bits and pieces originating from various parties, some of which may have had to be licensed to the compiler of the program itself.

Likewise, a work in which copyright subsists can (with the copyright owner's permission) be adapted or modified by another person and the other person will then own the copyright in the adaptation or modification.

Infringement of copyright in a work by another party includes copying it (i.e., reproducing it or a substantial part of it in any material form) or making an adaptation of the work or selling an infringing copy. Copying includes storing the work in any medium by electronic means. The transient reproduction of any work on a computer screen will amount to copying. In the case of a drawing, copying can include constructing the subject matter of the drawings: that is, reproducing the drawing in three-dimensional form.

In the UK, no formalities are necessary to own and protect copyright. A copyright owner can enforce his rights against any infringer through the courts. However, in some countries (although not now in the USA), it is necessary for the work to bear the international copyright symbol © followed by the name of the author or artist and the year of first publication. There is no harm, therefore, in putting this symbol on UK copyright work, and this is frequently done.

Protecting copyright is, therefore, straightforward, as long as it is clear who the copyright owner is and that no licence (express or implied) has been granted to the infringer.

Having established in which documents the professional has copyright, the provisions concerning copyright in the appointment need to be checked. Is copyright to remain with the professional or is it to be vested in the client? The latter should be resisted if at all possible (although the Crown and other statutory bodies will insist on having copyright in all the professional's materials) – see Section 10.4.

ACE Agreement 1, RIBA 2010 and RICS 2011 all provide that copyright in drawings and/or documents remains with the professional. ACE Agreement 1 also deals with the other intellectual property rights that could vest in or remain vested in the professional including patents, trademarks, service marks, registered designs, copyrights, database rights, design rights, etc.

RIBA 2010 simply refers to 'copyright in the original work produced in the performance of the Services' and RICS 2011 refers to 'copyright in all Documents' and has a wide definition of 'Documents'.

For reports and documents produced in circumstances in which there is no formal agreement about copyright, it can be helpful to include a statement concerning copyright and the extent, if any, of the licence given – see Section 14.14.

Copyright belonging to a third party, for example, to a sub-consultant, can only be vested in the client by that third party. Therefore, if it is a condition of the professional's appointment that all copyright, including that belonging to third parties, be vested in the client, the professional must ensure each third party agrees in its contract either to vest copyright in the client direct or in the professional so the professional may in turn vest copyright in the client.

Indemnities in respect of any infringement by a professional of a third party's copyright are frequently requested by clients. Whether or not the professional agrees to give an indemnity will depend on factors such as those set out in Chapter 17.

10.2. Remedies for infringement of copyright

The main remedies for infringement of copyright include payment of damages or the granting of an injunction.

An award of damages is usually based on any lost profits or licence fee, and it is not always necessary to prove actual loss – nominal damages may be awarded if the infringement only affects the claimant's reputation. No damages will be payable if the defendant did not know that there was copyright in the work at the time of the infringement. The court may, however, award additional damages depending on the circumstances of the case and the flagrancy of the breach.

The court may also consider any benefits obtained by the defendant through the infringement, and the claimant may claim any profits made by the defendant as an alternative to damages.

Injunctions are a discretionary remedy, and the court may not issue an injunction where it appears that the claimant is only interested in financial compensation. The grounds for granting an injunction to protect copyright have been set out by the courts (*American Cyanamid Co* v. *Ethicon (1975)*) – for example, an injunction will not be issued where the harm to the defendant is disproportionate to the benefit to the claimant.

Other remedies for breach of copyright include delivery up and seizure. The owner of copyright can apply to the court for an order that infringing articles or copies of copyright works be delivered up to the owner or to a third party, and any infringing copy of a work which is put on sale in public premises may be seized by the copyright owner. There is no need for a court order before exercising the remedy of seizure, but the copyright owner must follow detailed rules before seizing the infringing article, such as notifying the local police station of what is intended.

10.3. Licences

A licence in copyright works sets out the terms on which someone other than the copyright owner (the licensee) can copy or use the copyright works. Without such a licence the third party will be infringing the owner's copyright and lay himself open to an action for damages or other remedies. The licence agreement will, for example, deal with the right to copy, which includes building the subject matter of a drawing, adapting documents and, perhaps most importantly, the purpose for which the documents can be used. A professional will not generally grant a completely open licence so that the client can copy and use the professional's documents for any purpose whatsoever. Care needs to be taken to ensure that the purpose is correctly identified and does not, except in appropriate circumstances, extend beyond the particular project or purpose for which the information was provided.

If nothing is agreed in writing, when the copyright owner releases copyright work a licence may well be implied, but the terms of that licence may be unclear – for example, is it restricted to use by the client and project team or does it extend to purchasers and tenants? It is better, therefore, to have express licence terms included in the appointment documents. Where the documents are to be used in connection with a project the licence will be quite wide. They can, however, be much more limited and specific. For example, a licence to copy and use a report may be restricted to the client and for a particular purpose only.

That is all a licence is doing: defining the extent that a document can be copied and used without infringing the professional's copyright. It confers no rights to the documents themselves. It could perhaps be inferred from the terms of the licence (but more readily from the services) that documents or information will be released to the client and others during the project. However, what the client and/or others can do in relation to copying or using the information or documents is determined by the licence – see also Section 8.6.

It should be noted that a grant of a licence does not automatically carry the right to grant sub-licences.

The provisions in ACE Agreement 1, concerning the licence to be given to the client are as follows:

> F6.1 ... but the Client shall have a licence to use the Consultant's Intellectual Property Rights for any purpose related to the Project. Such licence shall enable the Client to use the Consultant's Intellectual Property Rights for the extension of the Project but such use shall not include a licence to reproduce the designs contained therein for any extension of the Project. In the event of the Client being in default of payment of any fees or other amounts due under this Agreement the Consultant may revoke the licence granted herein on giving seven days' Notice. Save as above, the Client shall not make copies of any of the Consultant's drawings or other documents or information nor shall the Client use any of the Consultant's Intellectual Property Rights in connection with any other projects or works without the prior written approval of the Consultant. The Consultant shall not be liable for the use by any person of any of the Consultant's Intellectual Property Rights for any purpose other than that for which the same were prepared by or on behalf of the Consultant.

A licence should not be qualified by such words as 'irrevocable' or 'exclusive', 'freely assignable', all of which alter the nature of the licence given and need therefore to be assessed individually. An 'irrevocable' licence would mean that whatever the circumstances, such as a breach of its obligations by the client, the licence will not cease. The professional is thus giving away any rights to argue that the licence has come to an end. It is rare to see an 'exclusive' licence, but this is tantamount to handing over copyright because it would confer on the licensee sole rights to copy and use the documents, etc., and this would prevent the other professionals, contractor, sub-contractor, etc., having this facility. It would also prohibit the professional from exploiting the professional's copyright work freely. If a licence is 'freely assignable' or the licensee is permitted to grant further sub-licences, the professional could lose control of the number of people who have the right to copy the professional's documents, and the uses to which they can be put if such use is not limited in the original licence. An assignee or sub-licensee can only have a licence in the same terms as the original licence.

As a general policy, the release of calculations or a licence to use calculations should be refused, as one purpose, whether stated by the client or not, can be to use them in litigation against the professional. Many calculations, if meticulously checked, could be found to contain an arithmetic error. This could have had no effect on the overall design nor amount to negligence, but the finding of such an error could give a client ammunition in a claim. If the client can supply a valid reason for needing the calculations, it may be considered commercially acceptable to release them. In such circumstances, the following would need to apply.

- The calculations should be the final calculations and not working drafts.
- The calculations should be in a satisfactory state for release.
- The costs of copying, etc., should, if possible, be reimbursed by the client.
- The calculations should be accompanied by an appropriate licence agreement.

The use in ACE Agreement 1 is 'for any purpose related to the Project'. It is essential to define the use to which the documents can be put. It would not be advisable for the client to have the right to use them for 'any purposes' because the documents may not be appropriate for any purpose and the professional may have no opportunity to review and advise concerning their use for other purposes. Further, the fee will not generally cover the preparation of documents of general application, say, to several projects. The phrase in the ACE licence clause assists in emphasising to clients that documents prepared for one project are not necessarily suitable for another.

The ACE Agreement 1 licence allows the documents, etc. to be used 'for the extension of the Project' but does not extend to reproducing the designs contained therein for any extension of the Project. A distinction is drawn between a right to use them for any extension (for example, to see where services are, what the structure of the original was) but not to reproduce the design – that is, build something identical to the original. This is particularly important to architects (see comments on the licence in RIBA 2010 below).

The ACE Agreement 1 licence provides that 'In the event of the Client being in default of payment of any fees or other amounts due . . . the Consultant may revoke the licence granted herein on giving seven days' Notice'. It can often be very difficult, in the absence of an express provision, to determine whether a licence to use drawings or similar documents is revocable if the client is in arrears with the fees due to a professional. There is some authority to the effect that if the client is in repudiatory breach – see Section 8.3 – a licence is revoked. However, it is much clearer if this is dealt with in the appointment. It is also clearer to express it as the revocation of licence, rather than making the licence 'subject to the payment of fees'. This is because drawings and documents are often released before payment is due and there will be no doubt that, at the time of their release, the client was given an implied licence, if not an express one, to copy and use them. It will then be a question of whether that licence could be said to be 'subject to the payment of fees' and thus revoked if subsequent fees are not paid.

The ACE Agreement 1 licence continues 'Save as above, the Client shall not make copies of any of the Consultant's Intellectual Property Rights in connection with other projects or works without the prior written approval of the Consultant'. This is again a useful reminder that any extension or variation to the licence expressly given must be agreed with the professional.

Finally, the ACE Agreement 1 licence says 'The Consultant shall not be liable for the use by any person of any of the Consultant's Intellectual Property Rights for any purpose other than that for which the same were prepared by or on behalf of the Consultant'. Although the professional may well not be liable in such circumstances, this statement is again a useful reminder of this fact to clients and, perhaps more importantly, to anyone to whom a further licence may be given.

The licence in RIBA 2010 is as follows:

> 6.3 The Client shall have a licence to copy and use and allow Other Persons providing services to the Project to copy and use drawings, documents, bespoke software and all other such work produced by or on behalf of the Architect in performing the Services hereinafter called the 'Material'.
> The Material may be used for the construction of the Project and for the operation, maintenance, repair, reinstatement, alteration, promotion, leasing and/or sale of the Project. The Material may not be used for reproduction of the design of any part of any extension of the Project, and/or for any other project except on payment of a licence fee specified in this Agreement or subsequently agreed.
> The Architect shall not be liable if the Material is modified other than by or with the consent of the Architect or used for any purpose other than the purposes for which it was prepared.

Some professionals might wish to put some restrictions on the right to use their documents to alter or modify the project. There also seems to be an implied right to modify the documents themselves – the latter part of this clause only deals with the liability which could result. Again, professionals may not be happy for this right to be given.

The licence does not appear to extend to contractors or sub-contractors who do not 'provide services' if this wording is meant to restrict use to those contractors or sub-contractors who provide services similar to those provided by professionals, such as design duties. However, contractors and sub-contractors may well need to copy or use the professional's drawings to build the project.

There are some provisos to the licence:

> Provided that
> 6.3.1 if it is intended to make any permitted use after the date of the last Service performed under this Agreement:
> (*a*) the Architect, following a request from the Client, shall confirm the degree of completion of the Material; and
> (*b*) the Client shall pay to the Architect any specified licence fee or a reasonable licence fee;
> 6.3.2 if at any time the Client is in default of payment of any fees or other amounts properly due, the Architect may suspend further use of the licence on giving 7 days' notice of the intention of doing so. Use of the licence may be resumed on receipt of such outstanding amounts;
> 6.3.3 the Client obtains or ensures that any third party obtains any necessary licence and pays any fees arising for access to any software used to produce any of the Material.

What the proviso to clause 6.3.1 does not make clear is that this should only apply if the termination is before the completion of the project.

Clause 6.3.2 gives the professional a comparable right to that under ACE Agreement 1 to suspend the licence, in the event of non-payment, on seven days' notice.

The licence in RICS 2011 is:

> 13.2 The Consultant grants to the Client a royalty-free, irrevocable (but subject to the right to suspend set out in this Clause), non-exclusive licence to use and reproduce the Documents and the designs contained in them for any purpose relating to the Project including the construction, completion, maintenance, operation, letting, sale, promotion, advertisement, reinstatement, mortgaging, refurbishment and repair of the whole or any part of the Project. Provided that the Consultant may suspend such licence if and for so long as the Client is in breach of its payment obligations under this Appointment by giving 7 days' notice of intention to do so. Such licence:
> (*a*) enables the client to use and reproduce the Documents for any extension of the Project, but such use does not include a licence to reproduce any designs contained in them for any such extension;

(*b*) continues notwithstanding termination of the Consultant's engagement under this Appointment;
(*c*) confers the right to grant sub-licences; and
(*d*) is transferable to third parties.

13.3 The Consultant is not liable for any use of the Documents for any purpose other than that for which they were originally prepared.

13.4 The Consultant warrants to the Client that, to the best of its knowledge, the Documents do not infringe the intellectual property rights of any third party.

This licence can also be suspended in the event of non-payment and the licence itself is very widely drawn. There is also an express warranty that the professional's documents 'to the best of his knowledge' do not infringe the intellectual property rights of any third party, which is not in the ACE and RIBA provisions.

10.4. Vesting copyright in the client or another party

Some appointments will provide that copyright is to be vested in or assigned to the client or another party. If copyright is vested in the client, all the rights belonging to the author also pass to the client who can then copy and as a consequence use the documents for any purpose and not necessarily for the project for which they were prepared. The client will also be entitled to exploit any of the rights in the professional's materials, for example, by licensing them to others for a fee. Assigning copyright will not necessarily mean that the client is protected from any disputes as to ownership of copyright or possible claims for infringement of copyright during the project, as these claims will be brought against the client in place of the professional. The reason usually advanced for requiring copyright is that the client does not want the professional to be able to halt work on a project by being able to revoke or suspend the licence given to the client to use his documents, for example, if there has been non-payment by the client or if there is a dispute about payment or if the appointment is terminated early.

Some non-standard conditions of engagement provide for copyright to be vested in the client for the duration of the project and then reassigned to the professional at the end, which reinforces this point.

It is not necessary, however, for the client to have copyright to achieve this. All he needs is an irrevocable licence which lasts for the duration of the project and not the duration of the professional's appointment if the appointment provides that this could be terminated early.

Having vested copyright in the client, the professional will be infringing the client's copyright and thus lay himself open to an action for damages or other remedies if the professional then copies his own documents, either for the purposes of carrying out his services or for providing copies to other members of the construction team. He will also be prevented from exploiting them himself, for example, by licensing them to others for a fee, or for using them in whole or in part for any other project or purpose. This could represent a very severe restriction on the professional's rights to exploit the essential know-how of his firm.

Whether the consequences of assigning copyright to the client are acceptable to the professional will be a commercial decision for him and may depend to some extent on his view as to whether the client is a 'good payer' and/or whether the appointment may be terminated early.

It is essential, if copyright is assigned to the client, that a licence back to the professional is included in the appointment and/or certain material is excluded from the copyright transfer such as that which is the professional's know-how. If something specific has been developed for the client, the professional may agree that copyright in this should vest in the client so that the client has the right to exploit it in the future, but the fee paid to the professional should reflect this.

The professional may well need a completely open licence in respect of some or all of his documents so that he can copy and use them for any purpose whatsoever. If a more limited licence is given, care will be needed to ensure that the rights given and the purpose for which the licence is given in respect of any of the documents is correctly identified and if necessary extends beyond the particular project or purpose for which the information was provided to the client.

If nothing is agreed in writing and the copyright owner (now the client) releases the professional's copyright work back to the professional so that he can perform his services, a licence may well be implied, but the terms of that licence may be unclear – for example, is it restricted to use by the project team only for that project only or does it extend to other uses the professional may wish to make of the material? It is better, therefore, to have express licence terms included in the appointment.

It should also be noted that a grant of a licence does not automatically carry the right to alter or modify the material, nor to grant sub-licences, which may be necessary.

10.5. Design right

A 'design right' is created by and defined in the Copyright, Designs and Patents Act 1988 (the 1988 Act) as a property right which subsists in certain original designs. Design in this context means 'the design of any aspect of the shape or configuration (whether internal or external) of the whole or part of an article', and there is no requirement for aesthetic quality. An original design will automatically qualify for protection under the Act as soon as it is recorded in a design document or an article is made to such a design. Design right, however, does not subsist in a method or principle of construction, nor does it subsist in features of shape or configuration of an article which enables it to be connected to another article so that either article can perform its function, nor to features of shape or configuration of an article which are dependent on the appearance of another article of which the article is intended to form an integral part. In addition, a design is not original if it is commonplace in the design field at the time of its creation. Commonplace is not defined, but according to Mummery LJ in *Farmers Build Ltd* v. *Carier Ltd (1999)*, in order to see if a design is commonplace, it is necessary to ascertain how similar that design is to the design of similar articles in the same field of design made by persons other than the parties. If a number of designers working independently of one another in the same field produce very similar designs by coincidence, the most likely explanation of the similarities is that there is only one way of designing that article. In those circumstances, the design in question can be fairly and reasonably described as 'commonplace'.

Unregistered design right subsists for 15 years from the end of the calendar year in which a design was first recorded in a design document or articles made from the design (whichever happened first). However, if articles made to the design were made available for sale or hire within the first five years in the 15-year term, the right will subsist for ten years from the end of the calendar year in which that first occurred. The owner of the unregistered design right has the exclusive right to reproduce the design for commercial purposes by making articles to the design or making a design document for the purpose of enabling articles to be made.

Design rights will therefore, in the main, subsist in designs of functional articles, such as chairs or spare parts for cars, which it is intended should be manufactured. There may be some overlap between these rights and the copyright protection previously referred to.

Design rights prevent the copying of any original three-dimensional design. As has been said, in order to qualify for protection, the original design must either be recorded in a design document or an article must have been made to the design.

However, in contrast to copyright, design rights vest in the person who commissioned the work (the client), not the designer. Since consulting engineers in particular might create such rights during the course of a commission, ACE Agreement 1 included these in the definition of the professional's intellectual property rights and they become vested in the professional rather than the commissioner by the words 'vest in' in clause F6.1.

Design rights created by an employee vest in his employer, as for copyright.

10.6. Registered designs

A registered design is a legal right which protects the overall visual appearance of a product in a particular geographical area. The visual features that form the design include such things as the lines, contours, colours, shape, texture, materials and ornamentation of the product which, when applied to the product, give it a unique appearance. The designer can also register a design showing the ornamentation alone, for example, a pattern to go on a product or a stylised logo.

A registered design can be a valuable asset, allowing the designer to stop others from creating designs which are too similar to the professional's within the same geographical area – registration gives a designer protection for the visual appearance of the product but not for what it is made from or how it works.

To be capable of registration, a design must be new and have individual character.

A design is new if no identical (or similar) design has been published or publicly disclosed in the UK or the European Economic Area (EEA). For example, a design would not be considered new if it has been ‘published’ on the internet on a site viewable in the EEA before the date it was filed.

Individual character means that the appearance of the design (known as the overall impression) is different from the appearance of other already known designs.

Registering a design right gives the designer a 25-year monopoly right over the appearance of the whole or part of the design. This means that the holder has the exclusive right to use and recreate any item incorporating the design, and to take legal action against anyone infringing the right, whether this is done intentionally or not.

RIBA 2010 prohibits the client from registering any part of the design without the consent of the professional. ACE Agreement 1 includes registered designs in the definition of the professional’s intellectual property rights.

10.7. Moral rights

The Copyright, Designs and Patents Act 1988 (the 1988 Act) also created several ‘moral rights’, including in Section 77 the right to be identified as the author of (or person who created) a copyright work.

For a professional, this would mean the right to be identified as the author of the drawings for a building, for example. Section 78 of the 1988 Act requires that such a right must be positively and formally asserted either on the copyright work or by an instrument in writing, for example the appointment document, before it can be enforced.

As far as architects are concerned, there are special provisions giving architects the right to be identified whenever copies of a graphic work of their building or of a model for a building are issued to the public. Also, where the building the architect designed is being constructed, the architect has the right to be identified on the completed building.

RIBA 2010 includes in clause 6.1 that the professional ‘generally asserts the Architect’s moral rights to be identified as the author of such work’, that is the original work produced in the performance of the services.

The right is infringed if certain acts are done without the author’s consent, for example, if the work is published commercially.

The 1988 Act also gives a right to object to a derogatory treatment, which includes in Section 80 the adaptation of copyright work and the right not to suffer false attribution of a copyright work.

Any infringement can give rise to an action for breach of statutory duty owed to the person entitled to the right.

These rights can cause administrative problems for a client and many non-standard conditions of engagement include a provision that all the professionals agree not to assert the right to be identified as the author by expressly waiving this right or by agreeing not to assert the right and usually do this by reference to the relevant sections of the 1988 Act, for example:

> The Consultant agrees not to assert against the Client or any other person who, with the permission of the Client, publishes commercially, exhibits in public, films, broadcasts, includes in a cable programme service, photographs or otherwise copies or deals with any image of the Project or any of the [copyright work] any right which the Consultant may have to be identified as author of the Project or any part of it or the [copyright work] pursuant to Section 77 of the Copyright, Designs and Patents Act 1988 ('the 1988 Act') or any other legislation which may supplement the 1988 Act. The Consultant also waives any rights it may have pursuant to Section 80 of the 1988 Act or any legislation which may supplement the 1988 Act.

Whether such a waiver can be accepted by a professional will depend on the nature of each particular appointment.

Other intellectual property rights may also be applicable, such as those concerning patents and trademarks, but these are beyond the scope of this book.

10.8. Building information modelling (BIM)

Building information modelling (BIM) is a relatively new way of providing information for a project. A summary of what it is and the issues it raises can be found at Appendix 1. These include the possible increase of the designer's liability and who has design responsibility for changes. As to copyright, the author of each drawing or part of a drawing or other information will still be the copyright owner, unless the appointments provide otherwise, and a licence probably implied to use it for the project utilising this particular form of resource. Whether such a licence is to be revoked in any circumstances, for example for non-payment of fees, or whether the author has given a right to others to alter the drawing or other information provided, will depend on the usual principles as to the extent of any implied licence or the terms of any express licence agreed by the parties.

Protocols are being developed to assist in regulating the way BIM is to be used on any project and to resolve the copyright/licence issues, as well as the respective responsibilities of the contributors, but these have not at the time of writing resulted in industry standards or norms.

Some professional indemnity insurers have become concerned about the liability implications of using BIM – will it be possible to identify who has been responsible for which piece of information? – and have excluded projects using BIM from professional indemnity insurance cover. Project insurance might provide a solution. Consideration of project insurance is outside the scope of this book.

10.9. Physical ownership of documents

The physical ownership of the piece of paper or electronic data containing copyright material is again something that can be dealt with in appointments. Generally, if the professional passes a copy of a drawing or electronic data to a client, the physical ownership of that particular copy drawing or data will pass to the client, but that is all that is passed. What rights the client have in relation to copying and using the piece of paper or data are governed by the express or implied licence from the professional and so physical ownership does not add anything significant to the client's rights.

Sometimes physical ownership does not pass outright, for example, where a professional has a lien on documents which he can seek to exercise if he is not paid. However, a lien is not as effective a remedy for a professional as the right to revoke the licence to copy and use his documents in the event of non-payment – see Section 18.2.

Professional Services Agreements
ISBN 978-0-7277-4158-5

http://dx.doi.org/10.1680/psa.41585.191

Chapter 11
Confidentiality

11.1. The duty of confidentiality

It is a common requirement of professional appointments for the professional to be asked to agree not to disclose to others confidential information provided by a client unless authorised to do so. It is not necessary, however, to impose an express duty of confidentiality on a professional – a duty of confidentiality will arise automatically under the common law where a client imparts confidential information to a professional in circumstances which give rise to an obligation of confidentiality and the client will have a remedy if the professional is in breach of his duty of confidentiality.

The court (*Coco* v. *A N Clark (Engineers) Limited (1969)*) held that the doctrine of breach of confidence required three separate elements as follows.

(*a*) The information must have the 'necessary quality of confidence'.
(*b*) The information must have been disclosed in circumstances which gave rise to an obligation of confidence.
(*c*) There must be an actual or anticipated unauthorised use or disclosure of the information.

11.2. What constitutes confidential information and when is the duty of confidentiality breached

The information can be of a technical, commercial or personal nature without distinction and the 'necessary quality of confidence' test is satisfied if the information is of limited public availability and is distinguishable from any generally available information. It must therefore be possible to determine the limits of the confidentiality and separate out the confidential information from other public information. The information must be sufficiently important and involve some original thought. The basic principle is that the law aims to protect an individual's valuable information until he or she chooses to put the information into the public domain or the information enters the public domain through an independent source.

The court (*Thomas Marshall (Exports)* v. *Guinle (1976)*) also held that, for the information to be confidential, the owner must believe that disclosure of the information would either harm the owner or benefit the owner's competitors. The owner must also believe that the information is not generally available, and this belief must be reasonable.

The information must also be disclosed in circumstances imposing an obligation of confidence. The generally accepted judicial test for when an obligation of confidence arises is 'if the circumstances are such that any reasonable man standing in the shoes of the recipient of the information would have realised that upon reasonable grounds the information was being given to him in confidence then this should suffice to impose upon him the equitable obligation of confidence'. This is an objective test and, once the obligation is assumed, it may be broken by conduct that is neither ill-motivated nor deliberate. Sensitive information about the client's proposals for a development, for example, or about the client's financial affairs, is likely to qualify as confidential information. If the relationship between the parties is one of a fiduciary nature, the duty of confidentiality will easily be implied, but the duty must exist at the time when the information is disclosed. An employee, for example, will always owe a duty of confidentiality to his employer, as will doctors, solicitors and bankers to their patients or clients.

For there to be a breach of the duty, there must have been some unauthorised use or disclosure of the information outside the limited purpose for which the information was revealed. The courts have established that innocent use can give rise to liability, even though the remedies might be restricted in this situation.

As the law of confidentiality is unclear in certain areas, it is advisable for clients and professionals to have specific ways of protecting information, for example, by marking documents as confidential when it is sensible to do so and by recording how information was received.

11.3. Confidentiality provisions in conditions of engagement

As to clients, the question of confidentiality is often so important that they do not wish to rely on this general duty and thus most standard and non-standard conditions of engagement contain an express obligation not to disclose confidential information.

The obligation of confidentiality should be there to protect the client's interest, but not to be exploited for commercial gain. The professional should not be required to keep his own work on the project confidential, save to the extent that this can be justified. To be justified, such a requirement would have to be very project specific. If a professional develops ideas or concepts in working on a project, he should not usually be prevented by a confidentiality agreement from using those ideas or concepts or developing them further on other projects, or from doing so without the client's permission, or indeed from using and building upon his general experience on the project.

An obligation of confidentiality should not apply to matters that are already in the public domain or to matters that have no quality of confidence. Nor should it prohibit the professional from pursuing his own legitimate interests under the contract, for example, pursuing a claim for fees, defending itself against a claim, or otherwise furthering or protecting his interests in any dispute. In such cases, the professional must be able to disclose or make use of any information in his possession, for the purpose in question.

ACE Agreement 1 provides:

F6.2 ... Neither party shall disclose to any other person any private or confidential information unless so authorised by the other party save in the proper course of that party's duties or to either party's professional advisers or insurers or as required or permitted by law.

These are the standard exceptions to allow the professional to use the information as necessary when performing his duties. Also, there are some circumstances where a person can be required to disclose confidential information, for example, in a court case or to HMRC, to a listing authority or stock exchange or under the Freedom of Information Act 2000 (assuming an exemption does not apply).

The reason why the professional's insurers and professional advisers should be excepted from any prohibition on divulging confidential information is because the professional must be able to seek advice from them without having first to seek the client's consent to the release of information. Insurers and other advisers will have a similar duty of confidentiality to the professional.

Once the information is in the public domain, however, through no act of the professional, the duty of confidentiality in respect of that information would no longer apply.

Sometimes there can be an express exclusion of information disclosed otherwise than by breach of a confidentiality agreement, but this should not be necessary if the confidentiality agreement is restricted to confidential information.

Frequently, such agreements can extend the obligation beyond confidential information to information that is not strictly confidential or to all information, which can cause practical problems and requires further consideration as to what the exceptions should be.

RIBA 2010 has an express obligation on both the professional and the client:

2.9 The Architect shall not disclose to any other person Confidential Information unless:
- 2.9.1 disclosure is necessary for the proper performance of the Services, or in order to take professional advice in relation to this Agreement or the Services, or in order to obtain/maintain insurance cover as required by this Agreement;
- 2.9.2 it is in the public domain other than due to wrongful use or disclosure; or
- 2.9.3 disclosure is required by law or because of disputes arising out of or in connection with this Agreement.

3.12 The Client shall not disclose to any Other Person Confidential Information unless:
- 3.12.1 disclosure is necessary to take professional advice in relation to this Agreement or the Services;

3.12.2 it is in the public domain other than due to wrongful use or disclosure; or
3.12.3 disclosure is required by law or because of disputes arising out of or in connection with this Agreement.

'Confidential Information' is defined as 'all information relating to the Client's or Architect's business and affairs which either party directly or indirectly receives or acquires from the other party or any representative of the other party either in writing or verbally'. It is not known whether the reference to 'business or affairs' is intended to restrict this to confidential information, but it is thought this description can be construed quite widely even if it does not quite extend to 'all information'.

RICS 2011 restricts the use and disclosure of any information (not just confidential information) relating to the project to specified recipients except in specified circumstances where the information can be disclosed. Their provisions are as follows:

13.6 The Consultant does not:
(*a*) release or disclose to any third party, other than the Client, Client Group Companies, other members of the Professional Team, the Contractor, sub-contractors, parties to Client Contracts, and persons entitled to receive Collateral Warranties or the benefit of Third Party Rights any information relating to the Project . . .
13.7 Notwithstanding Clause 13.6, disclosure is permitted if:
(*a*) the Consultant obtains the prior written consent of the Client;
(*b*) the information or documents are already in the public domain (other than through any fault of the Consultant);
(*c*) such disclosure is required by law; or
(*d*) the disclosure is to the Consultant's professional advisers.

PSC3 has a similarly wide obligation:

> The Parties do not disclose information obtained in connection with the *services* except when necessary to carry out their duties under this contract.

Non-standard conditions of engagement tend either to make all information 'confidential information' or define 'confidential information' in terms that are wider than the common law definition. Care needs to be taken in such circumstances to see that the professional is permitted to give the information he needs to give to others if his duties require it and can disclose it in the other circumstances set out above.

The provisions in non-standard conditions of engagement can also impose specific obligations in relation to the handling and storage of such information and contain restrictions as to whom of the professional's employees it can be released. The main questions for the professional will be whether the provisions are practicable, whether they require the professional to ensure compliance by third parties who are not under their control (e.g., sub-consultants) and whether they put unreasonable restrictions on the professional's personnel and working procedures. If so, these provisions will need to be amended.

Confidentiality provisions should not extend to matters such as exclusivity clauses, restrictions on competition clauses and/or those interfering with intellectual property rights.

The time during which information is to be kept confidential should be reasonable and clear and sufficient for the necessary protection of the information only. In the absence of an express stipulation the obligation under the common law would continue until the information no longer retains its confidential nature, for example, because it enters the public domain.

11.4. Confidentiality where there is novation

It will be seen from the preceding sections that a professional will owe a duty of confidentiality to his client. This could cause difficulties where a professional is novated from this client to another client or to a contractor. Once novated, he will have a duty of confidentiality to the new client, but his duty of confidentiality to the first client remains. As a professional he is also required to inform his client of anything that is material to the interests of that client. This could include revealing the confidential information he has received from the first client. This may not matter where the second client is a company in the same group as the first client, but could be of considerable concern where the second client is unrelated or is a contractor. It will be important therefore in this situation for the parties to agree expressly whether confidential information is to be given to the second client, and if so that the professional is released from his duty of confidentiality. Equally, if such information is not to be given, the second client should release the professional from his obligation to give him all information that is material to his interests where this comprises the confidential information belonging to the first client. Another approach, such as is adopted by the CIC novation agreement, is that the duty of confidentiality to the original client continues, save to the extent that this duty does not conflict with the duties owed to the contractor post-novation – see Section 20.11.

11.5. Separate confidentiality agreements

The same principles as set out above apply if the professional is asked to sign a confidentiality agreement separate from the appointment. This could happen, for example, where a client asks a professional to sign a confidentiality agreement before access is allowed to tender documentation. There should be clarity as to the information that is to be treated as confidential and as to the uses and disclosure that can be made of it. The standard exceptions also need to be incorporated and any restrictions on how the information is used and by whom should be practical and capable of being complied with. The length of time for which the confidential obligations are to continue should also be clear.

11.6. When professionals may need to protect confidential information

Confidentiality is just as likely to be an issue for the professional, although one seldom finds, in non-standard conditions of engagement, an equivalent obligation of confidentiality in favour of the professional. There is no reason, however, why one should not be inserted.

The particular skills and techniques that a professional applies on a project may, to some extent, be unique to it, and may have been developed in his own organisation in some secrecy. Copyright in the professional's drawings and other documents will – or should – be protected anyway by the appropriate licence. However, this may not be sufficient to protect the ideas that the professional is putting into practice.

The client therefore should also be required to treat with confidentiality any such matters as the professional identifies as confidential.

Confidentiality agreements may be required by professionals in other contexts, not merely as part of a wider agreement to perform services on a project, where it may be difficult in practice to protect ideas that are being put to practical use on the project. In a situation where the detail of a project is being considered and discussed without any commitment on the part of the client even to proceed with the project, let alone engage the professional on it, a confidentiality agreement may be essential for the professional to protect his ideas or information that he discloses to the client that he would not wish to be used without his participation (and without the terms of that participation having been agreed).

Another example is where a professional prepares an alternative tender for submission in a bidding process, incorporating a different design to the one on which tenders have been invited. It would be galling for the professional to find his alternative design being adopted without the professional himself being involved. The professional should at the least make it clear that his alternative design is confidential, and its use conditional upon terms being agreed for the professional's involvement in its implementation. An obligation of confidentiality cannot be imposed unilaterally but, if the alternative design is considered, the obligation of confidentiality will usually be held to have been accepted.

Other types of information may also be confidential, such as information about staff and financial information, or information relating to other clients, all of which can be requested from the professional when tendering. Again, the client should accept that this information is confidential or it should be marked as such.

If the professional employs sub-consultants he will want to ensure that his own confidentiality obligations are passed on to the sub-consultants through the sub-consultancy agreement.

Often, though, a properly drafted confidentiality agreement is the most sensible way of protecting an idea before it is disclosed. It need not be particularly long or complicated, but it could be very important and legal advice should usually be taken.

Confidentiality agreements are the only way of protecting ideas communicated orally. If the ideas are expressed in a drawing, in writing or in software, copyright can protect the professional – see Chapter 10.

11.7. Remedies for breach of the duty of confidentiality

The remedy if a professional is in breach of a duty of confidentiality is an injunction: whereby a person can seek to prevent the use or further disclosure of confidential information. However, injunctions are not automatically given for breaches of the duty of confidentiality and are at the discretion of the court. The courts may well not grant an injunction if

- the defendant was copying the information subconsciously or innocently
- the plaintiff's communication was gratuitous
- the defendant contributed to some extent to the information in question
- the information was economic or personal (i.e. if the information is economic, damages may be more appropriate as the loss may be easier to quantify)
- the information is mundane
- the information is already available to the public.

The granting of an injunction will depend on the facts of each case. While courts have a wide discretion in this area, it seems that an injunction will be considered most appropriate where confidential information is merely copied, rather than used as a 'springboard' for further independent development, or if a fraud or serious breach of duty has occurred.

Delivery up and destruction: an order for the delivery up and/or destruction of documents or other items obtained in breach of confidence may be made to give full effect to a prohibitory injunction. Such orders are again entirely within a court's discretion but, as a rule, delivery up is more likely to be ordered where the infringing articles belong to the plaintiff, and destruction where they belong to the defendant.

Damages: where unauthorised disclosure or use of confidential information has caused the owner to suffer loss, a claim for damages may be made. Where the breach of confidence is also a breach of contract, damages will be awarded according to normal contractual principles. The guiding principle is to put the plaintiff in the position he or she would have been in but for the breach of confidence. In general, the level of damages awarded will reflect the commercial value of the information.

If the liability only arises in equity (because there was no contractual obligation) then damages may be awarded in lieu of or in addition to an injunction. For example, damages could be awarded for damage already caused along with an injunction preventing future disclosure.

Account of profits: this is a discretionary equitable remedy. It may be combined in the same action with an injunction or an order for delivery up or destruction, but not with damages. However, ascertaining the amount of accountable profits is frequently difficult and damages are usually preferable.

Sometimes confidentiality obligations in non-standard conditions of engagement or in confidentiality agreements require the recipient of the information to acknowledge in advance that damages is not an adequate remedy for any breach and that the client is entitled to injunctive or other equitable relief. The objective is to make it easier for the client to obtain an injunction if there was any unauthorised use or disclosure of the information (or the threat thereof), as one of the criteria for an injunction is that damages are not an adequate remedy. Whether this will work in practice is uncertain but the professional would be unwise to agree to this sort of provision and thus deprive himself of any arguments that would otherwise be available to him as to the appropriate remedy if there were a breach. However, if the professional accepts that the information in question is so sensitive that it warrants such measures if there is a breach then such a provision might be reasonable. Equally, if the chance of the professional being in breach of the duty is very small, then there could be little risk attached to a provision of this sort.

Professional Services Agreements
ISBN 978-0-7277-4158-5

http://dx.doi.org/10.1680/psa.41585.199

Chapter 12
Liability for work carried out by others

12.1. Privity of contract

Under the doctrine of privity of contract, only parties to a contract (and not third parties) can acquire benefits under a contract or be subject to liabilities arising under the contract and only the parties to a contract can sue each other to enforce rights under the contract. Where there is an agency, the principal will be the party to the contract unless the agency is undisclosed and was not apparent. Where an employee contracts with a third party on behalf of his employer when he has authority to do so, the employer will be the contracting party. The Contracts (Rights of Third Parties) Act 1999, however, creates an exception to this and gives rights in certain circumstances to third parties to enforce terms of contracts that are for their benefit – see Section 2.3.

The effect of the rule is often avoided by the use of collateral contracts (or third party rights), where the main contract is accompanied by an ancillary contract with a third party relating to the same subject matter – see Chapter 21.

12.2. Responsibility for the work of others

The circumstances in which a professional can become responsible for the work or actions of others can vary. Such a responsibility can arise out of the appointment – for example, where a professional is instructed to check another professional's design – or arise through the sub-contracting of some of its work to another when the professional will remain fully responsible for the work of that third party sub-contractor. A professional can become liable for the acts of another party through joint liability, such as a joint venture.

It is possible for one party to be liable for another party's actions under the law of tort. In particular, a party is liable for torts which it has authorised or ratified, and it can also be liable for the actions of others which it has not authorised, such as the actions of employees, as long as they are within the course of the employee's duties. An employer is, however, not generally liable for the acts of an independent contractor, where, for example, a person employs an independent person or organisation to carry out cleaning or refurbishment work.

Some non-standard conditions of engagement may seek to make the professional responsible for (or 'guarantee' or 'warrant') the work or design of third parties with whom the professional has no contractual relationship and over whom therefore it has no control. This must be resisted.

It is essential to make clear in the appointment what the professional is, and is not, responsible for.

This can be usefully coupled with a statement of what the professional will do, especially where there may be a misapprehension on the part of a client as to the professional's duties, for example, in relation to coordination. For example, ACE Agreement 1 regarding the design of specialist contractors or sub-contractors – that is third parties who are not sub-consultants to the professional – says.

> F2.3 The Consultant may recommend to the Client that the detailed design of any part of the Works should be carried out by a Contractor or Sub-Contractor and the Client shall not unreasonably withhold consent to such recommendation. The Consultant shall examine such detailed design . . . and integrate it into the Services. The Consultant shall not be responsible for such detailed design or liable for defects in or omissions from it.

RIBA 2010 has a similar provision in relation to specialist services to be provided by third parties direct to the client, although it seems to be anticipated that these services could be part of the architect's services and so it is not clear why such a provision is needed in addition to the provision concerning sub-contracting – see below.

RIBA 2010 also includes a statement as to the responsibility of others. It provides:

> 3.8 . . . The Client acknowledges that the Architect does not warrant the competence, performance, work, services, products or solvency of any such Other Persons [widely defined].
>
> 3.9 The Client shall hold the contractor appointed to undertake construction works and not the Architect responsible for the management and operational methods necessary for the proper carrying out and completion of the construction works in compliance with the building contract or contracts.

12.3. Sub-contracting of services by a professional

A professional cannot assign or pass on his obligations under his contract to a third party without consent from the client – see Chapter 19. However, a professional may sub-contract or sub-let some of his obligations or services to another party without such consent unless his contract provides otherwise. In such circumstances the professional remains fully responsible to the client for all the acts and omissions of the sub-contractor, exactly as if the professional had performed the sub-contract services or work.

In some contracts a client may not be too concerned who undertakes the provision of professional services as long as those services are undertaken to at least the same standard and cost to the client and no more. Indeed, most construction contracts permit sub-contracting of building or civil engineering activities, although the consent of the client is usually required. However, there will be circumstances where a client has expressly chosen a professional and therefore specifically prohibits the professional from sub-contracting some or any of his services. The professional should be aware that this may be a serious requirement of a client, not to be ignored as a peripheral issue, when he considers whether he intends to sub-contract.

ACE Agreement 1 in relation to sub-contracting to sub-consultants states:

> F2.2 The Consultant may recommend to the Client that performance of part of the Services be sub-let to a specialist sub-consultant. The Client shall not unreasonably withhold consent to such recommendation and the Consultant shall integrate such sub-consultant's services into the Services. The Consultant shall be responsible for the performance and the payment of any such sub-consultant.

RIBA 2010 says:

> 4.2 With the consent of the Client, which consent shall not be unreasonably withheld or delayed, the Architect may appoint a sub-consultant or sub-consultants to perform part of the Services. Any such sub-contracting shall not relieve the Architect of responsibility for carrying out and completing the Services in accordance with this Agreement. Such consent shall not be required in respect of agency or self-employed staff.

RICS 2011 contains a similar provision to that of ACE Agreement 1:

> 14.5 The Consultant does not sub-contract or delegate the Services or any of them to any person without the prior written consent of the Client (which consent is not to be unreasonably withheld or delayed).
>
> 14.6 Any sub-contracting or delegation of the Services or any of them by the Consultant does not exclude or limit the Consultant's obligations and liabilities under this Appointment and the Consultant remains wholly responsible for the acts, omissions and defaults of its sub-consultants and delegates as if they are its own acts, omissions and defaults.

Most non-standard conditions of engagement will limit the professional's right to sub-contract so that consent is required. Other conditions can also be imposed, such as the obtaining of collateral warranties from the sub-consultant. It will be important therefore that these conditions are passed on to the sub-consultant in the professional's appointment of the sub-consultant.

If it is important to the professional to sub-contract some part of the services agreed, it would be prudent to agree this at the beginning with the client, and include in terms in the appointment that sub-contracting to a particular sub-consultant is permitted.

12.4. Implications of sub-contracting for the professional

There are two different types of sub-contracting. The first has been described above – where a professional sub-contracts some of his own services. The other is where one firm of professionals is appointed by the client to perform some or all of the professional services required for a project and then that professional appoints another professional firm or firms to perform the professional services which he cannot provide. So, for example, an architect as lead consultant may be appointed to provide all the architectural civil, structural and mechanical and electrical services required and he sub-contracts all but the architectural services.

There are some general principles affecting both forms of sub-contracting as follows.

- The professional will remain responsible for all the work of the sub-consultant and if the sub-consultant defaults or becomes insolvent during the project the professional will bear all the costs of having to put matters right, engage other sub-consultants and the additional fees that could ensue, make up any delays, etc. These could include the consequential costs for the client and the rest of the design team and contractors and sub-contractors. There would be no additional payment from the client for these unless this is expressly agreed. (There could be an argument that this should be the case where the professional is a lead consultant.) Also, the professional could find the new sub-consultant will not take on responsibility for what a previous sub-consultant has done, leaving the professional exposed in respect of claims arising out of the first sub-consultant's work.
- The professional will need to be satisfied as to the standing, assets and professional indemnity insurance of the sub-consultant so that he can take a view as to the likelihood of the sub-consultant being around and/or having sufficient professional indemnity insurance if a claim is made in the future. If there is a considerable disproportion between the size of the sub-consultant and its professional indemnity insurance, and the claims that could be made, the professional should consider either refusing to take on the sub-consultant or agreeing a different arrangement with the client as to what the responsibility the professional should have, and therefore what the liability of the professional should be in respect of that sub-consultant's work. Again this argument would have particular force in the case of a lead consultant where the client has chosen to whom the lead consultant is to sub-contract some of the services.
- The professional will need to check his professional indemnity insurance has no restrictions or conditions concerning sub-contracting, particularly where the sub-consultant is of a different discipline from the professional's. Additional premiums might be required. Further, because the professional will be liable for all claims against his sub-consultants, he should be confident his professional indemnity insurance is sufficient to cover all claims against the professional and sub-consultants.
- It is important that all the obligations and responsibilities of the professional in relation to the services that are to be sub-contracted are handed down to the sub-consultant; otherwise the professional could become liable to the client in respect of work done by the sub-consultant but not be able to recover the damages the professional has to pay the client from the sub-consultant, or may not be able to recover them to the same extent.

The way this is done is by a 'back to back' clause such as the following, which is used in the ACE Sub-Consultancy Agreement – Agreement 4 2009:

> The Sub-Consultant shall observe perform and comply with the obligations and provisions of the Main Agreement so far as they relate and apply to the Sub-Consultancy Services as if all such provisions were severally set out in this Agreement and save where the provisions of this Agreement otherwise require the Sub-Consultant shall so carry out the Sub-Consultancy Services and any further services pursuant to clause F2.3 so that no act or omission of the Sub-Consultant in relation thereto shall constitute or contribute to any breach by the Consultant of any of its obligations under the Main Agreement.

The professional will also need to consider whether there are any additional terms that will be needed in the sub-consultancy agreement to allow him to comply with his obligations to the client such as the following.

- Compliance with the programme for the performance of the sub-consultancy services or production of information.
- Provisions that allow the professional to manage and coordinate the work of the sub-consultant, as the professional will be responsible for this as well as his own services.
- Key personnel required by the client and/or the professional.
- Copyright licence: does the client require particular rights or even that copyright be vested in the client?
- Performance bonds or guarantees: are these required by the client?
- Indemnities: if the professional has had to give the indemnities then the sub-consultancy agreement should contain suitable indemnities so that the professional can recover monies payable under the indemnity to the client from the sub-consultant. These could be higher than those payable under the common law rules.
- Professional indemnity insurance: the sub-consultant should have equivalent obligations regarding maintaining professional indemnity insurance.
- Confidentiality: there may be specific requirements relating to sub-consultants in the professional's appointment with his client.
- Collateral warranties or third party rights: if these are required from the sub-consultant, whether just in favour of the client or to funders, purchasers or tenants in addition, the professional will need an obligation to provide them in the sub-consultancy agreement. Although having one consultant with one or more sub-consultants gives the client one-point responsibility – that is, one party to sue in the event of any claim – it is common for collateral warranties to be required from sub-consultants in case the professional becomes insolvent, leaving no direct contractual route from the client through to the sub-consultants. Such collateral warranties can also contain step-in rights so that the client has the option to take over the sub-consultant's appointment in such circumstances. All this rather militates against the rationale for having such an arrangement in the first place.
- Because the professional will not wish to be bound by the sub-consultancy

agreements if his own appointment is terminated, the sub-consultancy agreement should contain a provision allowing for termination in such circumstances, as well as for the sub-consultant's breach or insolvency, for example.

- If the professional's appointment is signed as a deed, the sub-consultancy agreement must be signed as a deed or the professional could find claims he wishes to bring against the sub-consultant are time-barred when the client's claim against him is not.

The definition of the services to be performed by the sub-consultant needs considerable care. The services should be expressed in the same terms as those in the professional's appointment so that there is no 'gap'. Also, it may be necessary to add further services to ensure cooperation with the professional and others and compliance with the professional's procedures.

As to communication, it is advisable that the sub-consultant should be prohibited from communicating direct with the client. If the professional is to manage his own work and that of the sub-consultant, he will need to have all communications directed through himself.

The provisions concerning payment of the sub-consultant also need careful thought. If the sub-consultant agreement is a 'construction contract', 'pay when paid' is outlawed – see Section 7.17. The professional is therefore under an obligation to pay the sub-consultant what he has agreed to pay him, regardless of whether or not he has received payment from his client. If he does not pay by the final date and does not serve an effective notice of withholding (or notice of intention to pay less if the amended Construction Act applies), the sub-consultant will have the right to suspend his services and the professional will bear the costs of that, while being obliged to continue performance of all the services under his appointment.

It would be sensible therefore for the professional to schedule his payment to the sub-consultant so that he will have received payment from the client before he has to pay the sub-consultant. Further, if the client serves a notice of withholding/notice of intention to pay less which could be as a result of defaults in the sub-consultant's work, the professional's timetable should be such that he can serve a notice of withholding/notice of intention to pay less on the sub-consultant before payment has to be made.

The professional is, however, permitted by the Construction Act to stop paying if the client becomes insolvent within the meaning of the Act – see Section 7.17. It would be prudent, therefore, to include an express provision allowing him to do this in such circumstances.

If the sub-consultant's appointment is not a 'construction contract', for example, because it is in respect of overseas work, 'pay when paid' is permitted.

As to additional payment, the provisions for this can be complicated. The professional and sub-consultant can agree what additional payment is to be made in what circumstances. Whether this is recoverable from the client will depend on the terms of the professional's appointment. It is usual for a sub-consultancy agreement, however, to make the payment of any additional monies conditional on an entitlement to such monies under the main agreement and the sub-consultant's right to that money would only arise when the professional or the client has determined it is due to the sub-consultant, coupled with an obligation on the professional to make reasonable efforts to obtain such payment.

Whether or not the sub-consultancy agreement should contain a limit of liability also needs consideration. If the professional has not succeeded in agreeing one in his appointment, there should be no limit in the sub-consultancy agreement.

If there is a limitation of liability in the professional's appointment, careful consideration is still needed before any limit of liability is included in the sub-consultancy agreement, including one in the same terms as in the professional's appointment. Generally, it should not be included. If the professional and his client have agreed a limit of liability, the sub-consultant will have the benefit of this without the need to repeat the limit in the sub-consultancy agreement as the professional will not be able to claim against the sub-consultant in respect of any client claims, any more than the client can claim against the professional. However, if, for example, the limit in the professional's appointment was found not to be 'reasonable' then the professional's liability to the client would be unlimited and the professional would not want the sub-consultant's liability to him to be limited in such circumstances. Also, if the limitation in the professional's appointment related to, say, the professional's fees or professional indemnity insurance, this might well not be effective as a basis for limitation in the sub-consultancy agreement. As to the professional's own claims against the sub-consultant, for example, in relation to defective work, additional costs, etc. which he cannot recover from the client, it is unlikely that he would want to agree a limit of liability for these except in very clear circumstances when these can be accurately assessed.

Further, a limit of liability in the professional's appointment does not protect the professional from claims made direct against the professional by third parties and which the professional will wish to pass on to the sub-consultant if his default caused the claim. (An indemnity from the client would be needed in respect of these if the professional were to be protected, but that would only be effective if the client has sufficient assets to meet any claim at the time.)

Claims from third parties therefore should be expressly excluded from any limit of liability in the sub-consultancy agreement.

The usual circumstances where the limit of liability will be included in the sub-consultancy agreement is where the sub-consultant takes on only a small part of the services and/or has less professional indemnity insurance than that of the professional. If so, it may be appropriate for the professional to agree that the sub-consultant's liability to the professional, including for third party claims, is the same as the professional's under the professional's appointment or on a different, lesser basis reflecting the sub-consultant's professional indemnity insurance and his input to the project.

It will be seen from the matters discussed in this section that it is important that negotiations and agreement on all sub-consultancy agreements keep step with those for the professional's appointment by the client and that revisions to that appointment are sent to sub-consultants as they arise. If not, there is a danger that that professional agrees something and then cannot persuade his sub-consultant to go 'back to back' with him. This can be a costly and time-consuming process.

12.5. Implications of sub-contracting for the sub-consultant

The considerations for the sub-consultant are the opposite to those for the professional employing him. He will need to decide what level of liability he can accept, whether he is prepared to go 'back to back' with the professional's obligations under his appointment (the professional may have accepted strict obligations or indemnities or liquidated damages), whether the 'back to back' clause should only apply if it is his negligent acts or omissions that give rise to the breach by the professional, whether he can accept any additional obligations imposed directly by the professional, whether he is prepared to give the collateral warranties and third party rights required, or whether the time for payment and provisions for additional payment are reasonable, etc.

It is reasonable for a sub-consultant to see the professional's appointment (except perhaps the provisions specifically concerning the amount of the lead consultant's fees). Then he can check the provisions which may affect his own services and fee, as well as the obligations with which he may be required to go back to back.

Professionals should be very cautious about accepting the obligations of contractors, for example, to carry out a site investigation, and carrying these out by way of a sub-contract to a contractor.

First, professionals are not contractors and would find it difficult or impossible to perform the contractor's obligations if the contractor failed to do so. Further, the professional's professional indemnity insurance would not cover any of the contractor's obligations, leaving the professional at much greater risk in the event of any non-performance or later claims. It is, however, something that can be requested by clients but it may be perfectly possible for the professional to arrange this for the client without taking on the contractor's obligations through the agency arrangements described in Section 13.1.

This is also not without risk to the client – he would have no contractual redress direct against the contractor if the professional were no longer around to meet any claim arising out of the work the contractor had done.

12.6. Taking over from another professional mid-project

In recent years, some clients have found that the professional originally appointed has not been able to complete a project, either because of a loss of qualified personnel or because of insolvency. In these circumstances, the client is often left with an incomplete design which may not yet have been constructed or is partially constructed. The client is therefore anxious to find a new professional who will take on responsibility for what has already been done and to complete the project, as there is unlikely to be any recourse against the first professional. This is a situation fraught with danger for the second professional. It is unlikely that, even with a thorough check of what has been done to date (assuming all the information is available, which it may not be), the second professional will ever feel he can take full responsibility for all that has been done. He was not privy to discussions with the client and others, so cannot know all that was discussed and agreed, why certain decisions were made, what warnings, etc. were given. It would be a very unwise, therefore, for a professional to accept full responsibility, even if this is acceptable to his professional indemnity insurers. (The professional needs to remember that it will be the policy in force at the time any claim comes in that responds to a claim, not the professional indemnity insurance he has when he enters into the arrangement.)

The better solution is for the professional to discuss and agree with the client the level of checking he is to carry out in respect of the earlier design and advise the client as to any matters that this may not reveal and therefore for which he cannot take responsibility. The obligation therefore should be in relation to that checking exercise only and the test should be whether the second professional has carried out the check with reasonable skill and care. The appointment should then go on to deal with the services that have yet to be performed and the second professional's obligations in relation to these. Any collateral warranties to be given by the second professional will then relate to the check he has done and the performance of his own services and will not extend to warranting that the first professional carried out his services with reasonable skill and care: a state of affairs that the second professional cannot know.

12.7. Joint and several liability and the Civil Liability (Contributions) Act 1978

Where a number of people join in making a promise, for example, to perform services that promise can be made by those people jointly, severally, or jointly and severally. The consequences, and thus liability, of those people depend on the nature of that promise.

Joint liability arises where two or more people, say A and B, jointly promise to do the same thing – for example, that together they will perform the services or pay a certain sum to the client. There is only one obligation and each of A and B is liable to perform all the services or pay the whole sum. If only one of them performs the services or pays the sum, that discharges the other.

Since only one obligation is created, all the promisors (A and B in the above example) must be joined as defendants to any legal action, and if only A is sued he can join B into proceedings as a co-defendant. Thus, although the client could choose to sue only one of the joint promisors, if that person does not have sufficient assets to meet the claim, the client cannot then sue any of the other parties who are jointly liable.

If one of the promisors dies, then the liability devolves on to the survivor but the liability of the last surviving promisor passes to his estate.

Several liability arises where two or more people make separate promises to another, for example, where A and B each promise to perform certain but different services for a client or that each will pay a certain but different sum to the client.

These promises are cumulative. The client may recover A's sum from A and B's sum from B but payment by B of the sum he owes does not discharge A's liability to pay his sum. There are therefore in effect two separate contracts – one between A and the client and one between B and the client. The client therefore would have to sue A under his contract if he failed to perform his services or pay the sum due, and similarly in respect of the contract with B.

Joint and several liability arises where two or more people join in the same contract in making a promise to the same person and at the same time each of them individually makes the same promise. For example, A and B jointly promise to perform services or pay a sum to a client, but both A and B also promise separately that the services will be performed by either A or B or that the sum will be paid by either A or B. These are not cumulative promises. As a result, performance of the services by one of A or B or payment of the sum by one of A or B discharges the others. Where joint and several liability occurs, all or any of the promisors (A and B) can be sued. The client has the option to decide whether he sues one or more in separate actions or all of them together. If one of the promisors dies, his estate, through his personal representatives, is liable jointly and severally with the other promisors. If the client sues one promisor but he is unable to meet the liability, the client can sue the other promisors.

A common example where joint liability can arise is through partnership or joint ventures. Here an innocent party can become liable for the negligence of another – a partner is liable for the negligence of another partner without having been involved in any way in the negligence concerned.

Whether the promise made by two or more people is several, joint, or joint and several is one of construction and will depend upon the intention of the parties as expressed in the contract. No particular words are necessary – they are construed according to the ordinary meaning of the terms used. Generally, some words of severance are required to create several liability. The courts will look for intention of severance or an indication that the parties intended that their potential liabilities were to be restricted to their own acts. It is best therefore if the parties state if they are jointly or jointly and severally responsible for the performance of the obligations undertaken.

Another situation where joint liability can arise is where two or more people are liable in contract or tort in respect of the same damage – whether jointly or otherwise. In such a case a person liable may obtain a contribution from any of the others as a result of the Civil Liability (Contribution) Act 1978. This liability arises from the damage itself. Before that the parties may have had no relationship with each other at all (e.g., different drivers both being responsible for a car accident), or may have been working on the same project but not in a relationship that would itself create joint liability. They do not have to have been in contract with the same party for such liability to arise. In such a situation neither party is innocent and both are 100% liable for the damage in question. In this case, therefore, being jointly liable is a potential benefit to each party, because it can claim a contribution from the other party. If the other party is insolvent, however, this right is ineffective.

The Civil Liability (Contributions) Act 1978 states that 'any person liable in respect of any damage suffered by another person may recover contribution from any other person liable in respect of the same damage (whether jointly or otherwise)'.

The Act goes on to provide that the amount of the contribution is to be determined by the courts, having regard to the extent of the person's responsibility for the damage. The court makes the assessment on equitable principles, with the result that the court may hold that the cost should be shared unequally, or that one party should pay nothing at all. There is uncertainty as to whether the courts should consider a party's overall blameworthiness or relative blameworthiness in relation to others, but the court cannot consider the fault of parties not involved in the proceedings.

Because of the interaction of all members of a construction team, it may often be difficult to avoid incurring joint liability to the client or others. It can therefore be important that the professional knows that the other professionals in the team have signed up on the same basis as he has (so that if he is found liable another professional does not escape liability because a different and lesser obligation was placed on him) and that the other professionals are of sufficient financial standing, have sufficient professional indemnity insurance, and are likely to be around when any claim comes in.

The inclusion of a 'net contribution' clause in the appointment can assist in these circumstances – see Sections 12.11 and 12.12.

12.8. Joint ventures

Professionals may sometimes be asked to join with other organisations to create a joint venture.

There are several different sorts of joint venture arrangements. The most usual is known as a 'horizontal joint venture', where the parties to the venture are relatively equal as regards their involvement in the enterprise (irrespective of their economic power) and the venture has been set up for a common and related purpose.

There is also the 'conglomerate joint venture', which describes an entirely new enterprise for all of the participants.

There are three forms of joint venture: a corporate joint venture, a partnership (unincorporated) joint venture set up through the medium of a partnership, and a consortium.

(*a*) A corporate joint venture is effected by setting up a company, most commonly a private, limited liability company. It is most used where the intended venture is significant, either in terms of the resources to be committed by the parties, the number of parties, or the likely duration of the enterprise. The principal reason for choosing a company structure is the fact that the company can be registered with limited liability. Thus, unless individual guarantees are given, the participants in the joint venture, as shareholders, will be protected from liability for the company's losses and therefore from the liabilities that can arise during the operation of the joint venture. The company can be used to hold assets and incur contractual commitments of its own, and it has a defined management structure. The relations between the parties are governed by a memorandum and articles of association, but the most important document is the shareholders' or joint venture agreement which regulates and determines the relationship and rights of the participants.

Some of the matters to be considered in relation to a corporate joint venture are:

(*i*) the costs of setting up the company and the shareholders' agreement and sub-contract arrangements – these can be considerable and may be irrecoverable if the tender is not successful

(*ii*) the costs of running the company – preparation of accounts, Companies Act requirements, separate taxation, etc. as well as the management time needed

(*iii*) the financing arrangements

(*iv*) the individual directors' responsibilities under the Companies Act, etc.

(*v*) whether a parent company or other guarantees are required

(*vi*) proper sub-contract arrangements with the joint venture shareholders to allocate the respective responsibilities and liabilities as between themselves. The arrangements should if necessary include indemnities (although indemnities are of course only worth having if there are sufficient assets to meet the liability)

(*vii*) a procedure for resolving differences of opinion, conflicts of interest and disagreements or disputes

(*viii*) proper arrangements for the execution and management of the joint venture's contract with the client

(*ix*) the basis of the arrangement for sharing any profits or contributing to any losses or costs.

(*b*) A partnership joint venture can be set up under the Partnership Act 1890, but whether a joint venture arrangement is actually a partnership will depend on a mixture of fact and law. The definition of a partnership is 'the relationship which subsists between persons carrying on business in common with a view of profit'. If it is a partnership, some provisions of the Act will apply to the joint venture, irrespective of any contrary intention or any agreement drawn up by the joint venture parties. Special care is needed to ensure that the joint venture is not inadvertently constituted as a partnership unless that is the specific intention. Under a partnership arrangement, each partner is deemed to be the agent of the other and each participant can therefore bind the enterprise and all the participants in respect of any act or transaction carried out by any of them in the ordinary or usual course of business.

The main difference between a corporate and partnership joint venture relates to liability. Under a partnership arrangement the participants are jointly liable to the full extent of their assets for all the debts and obligations of the joint venture and are liable jointly and severally for all the tortious acts and omissions of any of the participants in the joint venture. This means one of the participants could find itself liable to the full extent of its assets for the liabilities of another participant – see Section 12.7. It is for this reason that many professional indemnity insurers require professionals to tell them specifically if they are proposing to enter into a joint venture arrangement. Insurers will only usually provide cover for the consequences of acts of their insured. They will be concerned if a joint venture agreement does not include a provision that permits their insured to claim a contribution from the other liable partners in the joint venture. There may be express clauses which prohibit this. Another provision may be to impose 'joint-named' insurance. This results in the parties being co-insured and as an insurer normally cannot sue a party it is insuring, this has the same effect as removing a right to claim contribution.

(*c*) A consortium is the simplest structure for a joint venture, and it is based upon a specific contract drawn up between the participants. This contract is generally described as a consortium or cooperation agreement. Under this agreement the parties to the joint venture agree to form their association solely as independent contractors, as opposed to being partners under the Act or shareholders in a company registered under the relevant Companies Acts. The consortium agreement should define the rights and duties of the members and determine the rights and duties between the members of the consortium and the third parties who will deal with the consortium, for example, the client.

A consortium agreement is most frequently used to avoid having to set up the more formal arrangements necessary for a joint venture company or to avoid incurring joint and several liability under a partnership. The consortium contract usually provides the members with a direct interest in the assets and revenue of the venture. It is an arrangement commonly used by parties seeking to tender on a joint basis for a substantial contract or for a business enterprise which is of short duration or consists of a single undertaking or venture.

A consortium agreement should in summary

(*i*) set out the scope and duration of the joint venture
(*ii*) define the specific and detailed obligations of the individual members
(*iii*) devise the machinery for financing the venture
(*iv*) provide a mechanism for determining the sharing of profits or bearing of any losses which may result from the running of the enterprise.

The consortium agreement should also contain particular provisions to deal with any property to be used by the members and should determine, for example, which assets remain the exclusive property of the members and which assets are to be joint property.

Because the members are not partners, they are not subject to any joint responsibility for the liabilities and obligations of the other members of the consortium, nor are they liable for the other members' acts or omissions. Nevertheless, a member of a consortium can be exposed to claims (which may be unlimited) by third parties and for liabilities incurred by the consortium to third parties on the basis that a member has expressly assumed responsibility or is vicariously liable under the consortium agreement. This arises through the general presumption that each member of a consortium is the agent of the other. It is, therefore, important that the consortium agreement should define the extent to which each member will be responsible for the acts and omissions of the other members and also define the areas where they have no authority to bind the other members. Each member should indemnify the others against liabilities that member has incurred because it has acted outside any expressed authority. Such indemnities are, however, only of any value if they are, and will continue to be, backed up by sufficient assets.

Where a project is undertaken by all the members of a consortium, their liability for that contract will be joint. This will occur, for example, where the client enters into one contract to which all the consortium members are parties. Where an individual member of a consortium assumes sole responsibility for a liability undertaken on behalf of the joint venture, for example, where the client contracts with only one member of the consortium, the agreement should contain indemnities to protect that particular participant.

12.9. Joint venturing with a contractor

Professionals can be asked by clients to propose solutions which include a contracting element. Many professionals are, quite rightly, concerned as to the implications of their taking on such contractual responsibilities themselves (particularly from the point of view of liability, for which they may not be covered by professional indemnity insurance) and so may consider a joint venture with a contractor so that the contracting risks can be properly allocated and managed, but any such arrangement could still leave the professional at risk of having to undertake and/or become liable for matters that were or should have been carried out by the contractor. It is advisable in these circumstances for the professional to arrange for the contracting element to be carried out under a separate contract between the contractor and the client which the professional organises and manages as the client's agent – see Chapter 13.

It is particularly important if there is such a joint venture that the work of the professional and the contractor is properly described and separated.

The professional will not want to take on fitness for purpose warranties or indemnities which are commonly imposed on contractors. The agreement with the client should be drawn up on the basis that the professional and the contractor will only be liable for the work they perform and not the work of the other, but if this is not possible the professional should consider with his insurers the liability implications of such an arrangement as professional indemnity insurance may not be available in the event of a claim arising from a default of the contractor.

12.10. When joint or concurrent liability can arise

A professional undertaking checking, reviewing or inspecting obligations can find himself at risk of becoming jointly liable with the consultant or contractor whose work he was checking or reviewing or inspecting. If the professional performs that work negligently and fails to pick up an error or omission in the work of the other party, it is likely that both will be found liable to the client. If there has been no default on the part of the professional, joint liability cannot arise, although it may not be until a claim is well advanced that this is established – see Sections 5.3 and 5.6.

A professional can also assume responsibility for the work of others by, for example, commenting on the work being carried out by a contractor, by managing or directing the work of others, or by coordinating or integrating the work of others. The extent of that responsibility will depend on the circumstances of each case and the reliance placed by others on the professional's advice. This, however, illustrates the perils of being 'helpful' – if a professional steps in to try and resolve a problem, even though this is not required by his appointment and no fee has been paid for it, he can become liable in tort.

A professional could also accept responsibility for the work of another by expressly agreeing to do so in his appointment, although this would be unwise!

12.11. Net contribution clauses generally

It follows from the principles of joint and several liability that professionals and contractors who become 'liable in respect of the same damage' (Civil Liability (Contribution) Act 1978) are taking the risk of another member of the construction team becoming insolvent or having no (or insufficient) assets and/or professional indemnity insurance. This was perceived as unfair when liabilities to people other than the client were incurred through collateral warranties and it was in collateral warranties that 'net contribution' clauses first started to appear. Even where the client is concerned, there is the argument that the client should take this risk because it alone chooses the contractor and the professionals and can therefore choose reputable organisations with financial standing and proper professional indemnity arrangements, who are likely to be in business if and when a claim is made.

The objective of these clauses is to try to ensure that each party 'liable in respect of the same damage' only has to pay to the client (or holder of the collateral warranty, as the case may be) that proportion of the damages for which it is responsible. Thus, if a contractor is held 70% responsible for some damages and a professional 30% responsible for the same damages – see the example in Section 5.6 – each will pay their respective 70% and 30% only. If all the parties are around at the time of the claim and have sufficient assets, a net contribution clause is not needed as all will be able to pay their proportionate part. Such a clause is needed, however, when one or more of the other parties is no longer around and/or has insufficient assets. In a case such as the one described above, and if the contractor were insolvent, the professional would have to pay 100% of the damages as each party is 100% liable to the client. A net contribution clause could therefore ensure that the professional pays his 30% only and the client takes the risk of the contractor's insolvency.

The main elements of such clauses where included in an appointment would consist of the following.

(*a*) Limiting liability to a sum which it would be just and equitable for that party to pay, having regard to the extent of its responsibility for the damage in question (this echoes the words of the Civil Liability (Contribution) Act 1978).

(*b*) Identifying the other parties who could have contributed to the same damage. In the case of a professional, this could comprise the other professionals, the contractor and any sub-contractors responsible for detailed design or construction of the parts of the project to which the professional's services relate.

(*c*) Assuming those other parties to have undertaken contractual obligations to the client on terms no less onerous than those of the professional who has contracted with the client. If they were less onerous the other party could escape responsibility for 'the same damage' and therefore not be brought in to any assessment of contribution. As this is to be assumed it will not matter whether or not their contractual obligations are less onerous and the professional would not have to check them.

(*d*) Assuming there are no exclusions of or limitations of liability between the client and any of those other parties, and that those other parties are contractually liable to the client for the same loss or damage. In the absence of these assumptions, the other party's assumed liability for his proportionate part could be reduced or be assessed as nil which would have the effect of increasing the proportion of the party relying on the clause.

(*e*) Assuming there are no joint insurance or co-insurance provisions between the client and any of those other parties. This assumption was thought necessary as a result of the *Co-operative Retail Service* v. *Taylor Young Partnership (2002)* case described in the next section.

(*f*) Assuming all those other parties to have paid such proportion of the loss or damage to the client as it would be just and equitable for them to pay having regard to the extent of their responsibility for that loss or damage. Thus, again, it will not matter whether or not the other parties have actually been able to pay their respective proportions.

There could be a more fundamental problem, however the net contribution clause is worded, when the courts come to consider a net contribution clause in detail. How can the court determine whether another party was at fault, and the degree of that party's fault, if that party was not a party to the action (or even if he is a party to the action but, due, for example, to his insolvency, takes no part in the action). Say, for example, the court had to consider the respective liabilities of a professional and a contractor but the contractor was insolvent. How would the court decide the amount which it would have been just and equitable for the contractor to pay, if it has not heard any evidence concerning this put forward on behalf of the contractor? The court would hardly be willing to decide the matter solely on the evidence put forward on behalf of the professional. It would have to be for the claimant to argue the case on behalf of the contractor. A trial of the contractor's liability would, in effect, have to take place in his absence, with the claimant arguing the case for his defence. This might be thought to be impractical or, in any event, put an unreasonable burden on the claimant. It is not impossible that the tribunal would be driven to the conclusion that a net contribution clause was unworkable and, therefore, of no effect.

Net contribution clauses ought to work in simple situations, for example, where the other party (from whom the contribution is sought) is before the court and participates fully in the action and is found liable but is then unable to pay its contribution. In those circumstances, the amount which the first party (who has the benefit of the net contribution clause) would pay should be reduced on the footing that the insolvent party is deemed or assumed to have paid its contribution. Another situation where net contribution clauses should work is where all parties are agreed that the other party was at fault and the nature of its fault is well understood. For example, suppose the claimant's case is that there has been a basic error of workmanship on the part of the contractor, which the professional has failed to notice. In those circumstances, there may be no difficulty for the court in deciding what it would have been just and equitable for the contractor to pay, even though he is not a party to the action.

See Section 16.9 for a discussion on the net contribution clauses in the standard appointments and Section 9.14 for the PSC3 provisions concerning net contribution.

12.12. The Co-operative Retail Services case and its effect on net contribution clauses

In this case, the defendants were architects and engineers on a construction project. The works were badly damaged by fire and the defendants were sued by the developer, who alleged that the fire was caused by their negligence. The defendants denied negligence but they also said that, if they were to blame, others (the main contractor and the electrical sub-contractor) were to blame as well. The case concerned the defendants' attempt to join the contractor and the sub-contractor to the action in order to claim a contribution from them if necessary.

The contractor and the sub-contractor were co-insured with the developer under a joint names policy taken out to cover various losses, including damage by fire to the works. Also under the terms of the JCT and DOM/1 contracts which applied, the developer was not entitled to claim damages for consequential delay.

This is a situation which can arise quite often. Policies to cover damage to the works are frequently taken out in the joint names of employer and contractor, as is required under some forms of building contract.

The case went to the House of Lords who said the effect of the relevant clauses in the JCT contract between the developer and the main contractor, including the obligation to take out a joint names policy, was to exclude any liability on the part of the contractor to the developer for losses that would be covered by the joint names policy, including damage by fire. These same features of the main contract were reproduced in and therefore applied also to the sub-contract between the contractor and the sub-contractor, with the result that the sub-contractor could not be held liable to the developer under a warranty given by the sub-contractor to the developer. Therefore, as neither the main contractor nor the sub-contractor could be held liable to the developer for the damage that occurred, the architects and the engineers could not make a claim against them for contribution in the present proceedings.

It is for this reason that more recent versions of net contribution clauses include the assumption that none of the other parties are jointly insured or co-insured with the client so that the proportion for which the other party is responsible is assessed on the usual basis and not on the basis that he has no liability because of the terms of the insurance policy.

The judgement also contains some interesting comments on the date when liability (of the party from whom contribution is sought) needs to be established in order for the claim for contribution to succeed. The ruling in the judgement is that the relevant date is when contribution is being sought, not when the damage in question occurred. It follows that, if the architects and engineers had been able to persuade the court that the contractor had a contingent or potential liability under the main contract in respect of the fire damage (in the event that insurance cover failed or proved to be inadequate), the claim for contribution would still have failed because, at the time the contribution was sought, the fire damage has been covered by the joint names policy and made good in accordance with the contract. It could not therefore be said at that time that the contractor had any liability.

Professional Services Agreements
ISBN 978-0-7277-4158-5

http://dx.doi.org/10.1680/psa.41585.217

Chapter 13
Agency

13.1. Appointment of the professional as agent for the client to employ others, such as contractors

The situation in which a professional is appointed by an agent on behalf of a client is discussed in Section 2.2, as is the question of when the agent or the principal can be sued. The situation in which a client asks a professional to act as its agent (i.e., to create contractual or legal relations between the client and a third party) is considered in the first section of this chapter. A professional may also have authority to act on certain matters in relation to a third party and this is described in Section 13.10.

Particular considerations arose in the past where a professional acted as an agent for a client in respect of the client's duties under the CDM Regulations 1994 but this role was abolished under the CDM Regulations 2007.

The sort of circumstance where the type of agency being considered in this section arises is where a client asks the professional to engage a third party on his behalf, for example, a materials testing house or a contractor to do a site investigation, or another professional to provide a specialist report. The Services Schedules to ACE Agreement 1 contain examples of several services, such as site investigations, which the professional is to 'arrange as agent for the Client'.

It is clearly important to establish whether the intention of the client is for the professional to act as an agent for the client or as a principal. If there is any doubt at all, the professional should clarify this with the client – the question being: 'Is the client the person who is to be the employer [of the contractor or plant operator] and therefore responsible for their payment, or is it to be the professional?'

If the professional is to employ the third party direct, and thus become responsible to the client for the performance of the third party, the professional will need to consider the implications of this, both from a liability point of view and also whether his professional indemnity insurance will cover the acts and omissions of the third party. This is covered in Chapter 12. Frequently, the client will have no objection to employing the third party. What he may well be looking for is someone who will undertake the responsibility for arranging what the third party will do, the terms of his employment and so forth, all of which the professional may well be prepared to do. In such a situation an agency arrangement with the professional is the best solution, but the professional needs to understand the implications of acting as an agent.

13.2. The extent of the agency

An agent has to act within the parameters of the express or implied authority given to him by his principal – the client. This will extend to all acts which are necessary or ordinarily incidental to the exercise of his express authority. The professional should therefore first establish the extent of the authority he is being given in respect of the employment of the third party. Sometimes this can be as broadly expressed as a request to arrange the necessary site investigation with some discussion as to what that will entail and a budget for what it will cost. The client and the professional may well not discuss the detail of what is to be done, arrangements for access, whether the professional will attend site while the contractor is there, in what format the result of the site investigation will be delivered, and what the professional will advise the client as a result of the investigation. All of this, however, would probably necessarily be implied as a result of the original request and thus be within the scope of what the professional is authorised to do on behalf of the client.

However, the extent to which the professional can give directions or instructions to the third party, change the nature of the work, add to the scope of the work or do anything that might result in additional payment, could be unclear and generally the professional should obtain clarification from the client before proceeding further unless these matters have been previously discussed and agreed with the client. The client may wish to impose some limitations, for example, as to the extent to which the professional can commit the client to additional payment without authorisation.

Where a professional acts regularly for a client in selecting and arranging for the employment of third parties, such as contractors, it would be helpful if a suitable agency provision were included in the professional's appointment. This, for example, could authorise the professional to select the contractor and determine the terms of the contract on which the contractor is to be engaged, including what he is to be paid and the scope of the work to be carried out. The professional could also be authorised to enter into the contract on the client's behalf, administer it and give any necessary instructions, including changing the nature or scope of the work or other terms if this is necessary, and to determine the amount of any additional payment.

If the professional and the client discuss a provision of this nature it should become apparent as to what matters have to be referred back to the client, for example, any substantial changes to the scope, or if something unforeseen occurs on site which alters the nature of what was originally to be undertaken and/or any or all additional payments.

The agent needs to remember that any contract entered into as agent for the client must be in accordance with the instructions given by the client or he will have exceeded his authority and will become liable to the client as a result.

It would be prudent, in any event, if the professional confirms to the client what is going to be done, and that it is being done on the client's behalf as its agent, and gives brief details of the contract that is going to be organised and the arrangements to be made about payment.

In most cases, it is usual for payment to be made direct by the client to the third party. In some cases, however, the client may wish the professional to pay and then the client will reimburse the professional. This is perfectly consistent with the principles of agency. Professionals who enter into an arrangement on this basis, however, need to appreciate that if the client becomes insolvent or does not honour payment commitments in good time the professional will be taking the risk on payment, unless the third party knows that the professional is an agent for the client, as only the principal can be sued for payment if the agency is disclosed.

13.3. The contract with the third party

It is essential that the professional makes it clear to the third party that the contractual arrangements are being made by him 'on behalf of' the client so that the third party is clear that the professional is acting as an agent and that the principal (i.e. the client) is to be the employer. If the contract is formed by an exchange of letters between the professional and the other party, the professional should confirm the terms 'on behalf of' the client, and make it clear that the contract is direct with the client and that the client will be responsible for payment.

If there is a formal contract, the parties to that contract should be the client and the third party. Although the professional can sign the contract in his own name on behalf of the client if given authority to do so, it is better if it is stated after the signature that the agent is signing 'for and on behalf of [the client]'.

Copies of any contractual arrangements should be sent to the client immediately they have been entered into, with a statement that this has been done on the client's behalf. The client should also be reminded of such things as when payment is due and how much.

If this is not done the third party could be entitled to treat the professional as the principal as the agency has not been disclosed. If he is to be treated as the principal, then he becomes responsible for all the obligations of the client as far as the third party is concerned, although the professional may well be entitled to an indemnity from the client.

Once the agency is established the third party is entitled to rely on the ostensible authority of the agent and so is entitled to assume, unless told to the contrary, that everything that might reasonably be expected of a client in relation to the contract will be performed by the agent. The third party does not need to concern himself as to the detail of the arrangements between the client and the professional, nor as to the extent of the agency, in the absence of any notice of any particular restrictions.

13.4. Duties of an agent

An agent has to use reasonable skill and care in carrying out the tasks undertaken for the client and therefore will be liable to the client in damages if he is in breach.

As has been said, an agent should act only within the extent of the express authority given by the client, but this will extend to subordinate acts which are necessarily and ordinarily incidental to the exercise of that express authority, for example, giving information to a contractor which it may need, even though this has not specifically been referred to between the client and the professional. For such acts the agent will have implied authority.

An agent is expected to carry out the tasks agreed to be undertaken in person, or through the agent's organisation. If given a definite instruction by the client as to how to carry out the commission, that must be followed strictly, provided it is lawful; if the agent does not do so he could be liable. The agent has no discretion to disregard express instructions from a client, even if the agent believes that he is acting in good faith in the interests of that client in doing so. Further, the agent must not delay unreasonably in carrying out instructions. If the task is or becomes impossible, the agent must promptly inform the client.

Where an agent has expressly been given discretion as to how to carry things out, the agent must be guided by 'the honest exercise of its own judgement and the interests of the client' (*Chown* v. *Parrott (1863)*), or in accordance with the ordinary usages of that business.

In negotiating a contract on behalf of a client, an agent must take all reasonable precautions to protect the client. The form of the contract must be capable of being enforced by the client.

Once the contract negotiations are completed and signed, an agent is not entitled to rescind the contract or vary its terms unless it has express authority to do so.

An agent must disclose any conflicts of interest between itself and the client, and it cannot enter into any transaction that is likely to produce that result unless it has given full disclosure to the client and obtained his consent.

As the nature of the client–agent relationship is fiduciary, if an agent receives money for or from the client for a specific purpose, the agent becomes a trustee of that money and has to keep accounts, or account for the monies received, and take no secret commission or bribe. The agent should employ the materials and information obtained as a result of the agency solely for the purposes of the agency and not in any way in a manner that is inconsistent with good faith, for example divulging them to another party or by using them himself in unfair competition with his principal.

If, in the course of a transaction or contract, an agent receives notice, or acquires knowledge, of material facts (e.g., defects in work being carried out), it is the agent's duty to communicate this immediately to the client.

13.5. An agent's rights against a client

A professional acting as an agent has a right to be indemnified by the client for any liability incurred and any losses suffered which were in contemplation when the agency was undertaken, together with any expenses occasioned by the agent's employment. Thus, if the client does not reimburse the professional for any sums paid by the professional to the contractor, or for any expenses incurred, the professional can rely on this implied indemnity in order to obtain payment from the client. It is better, however, if an express indemnity from the client is included in the professional's agency agreement. Whether the indemnity is implied or expressly stated, money may not be recoverable if the client has become insolvent in the interim.

The agent will generally only be entitled to be remunerated for his services if that has been expressly agreed, but such a right could be implied from the circumstances of the case.

13.6. The relationship between the client and the third party

A client is responsible for all the acts of the agent which were authorised by the terms of the agency agreement between them. If, however, the professional did something that was not within the scope of its express or implied authority from the client, the client would not be bound. For example, if the professional instructed some extra work which had not been specifically authorised by the client, the client, although he would have to pay the contractor if the instruction was within the apparent scope of the agent's authority, would be able to recover the extra costs of that work from the agent.

Where a contract with a third party has been made by the professional on behalf of, and with the authority of, the client, this can be enforced by the third party against the client directly. For example, if payment has not been made, the contractor can go straight to the client. If the contractor were successfully to pursue the professional for payment, for example, because he had no knowledge of the agency and was therefore entitled to treat the professional as the principal, the professional can rely on the express or implied indemnity from the client to recover that payment from the client.

If the agent has acted outside the actual or ostensible authority of the principal, those acts cannot be enforced by or against the principal unless the principal ratifies the contract.

If a third party, say a contractor, contracts with the professional, honestly believing the professional to be the principal and not an agent, the real principal (that is, the client) cannot sue the contractor direct (e.g., if there were allegedly defective works and the client did not want to pay the amount due under the contract). The professional, however, could be sued by the client if he has acted outside the terms of its authority.

13.7. The agent's liability in tort

An agent can incur a direct liability in tort to a third party. If a client should give the professional express authority to do a particular act which is wrongful in itself, or necessarily results in a wrongful act, the client would be jointly and severally liable with the professional to a third party for any loss or damage. The professional should refuse to undertake anything that it thought might have this result.

Similarly, a professional has to be careful when making representations or statements on behalf of the client, because the professional could become liable if he makes a negligent misstatement or fraudulent misrepresentation. In this situation the client would be liable jointly and severally with the professional to the third party if the professional was acting with express authority.

If the act was not expressly authorised but was within the agent's implied or ostensible authority, again the client would be jointly and severally liable with the professional. It is only if the act falls entirely outside the scope of the authority that the agent will be solely responsible. If the agent makes a representation which he honestly believes to be true, but which the principal knows to be false, the principal will be liable if there was actual fraud or dishonesty on his part, as in the situation where he employs the agent in order that the misrepresentation might be made. The principal may be vicariously liable for the negligent misrepresentation of the agent.

13.8. Criminal liability of an agent

As a general rule no criminal liability can be imposed on a client as a result of any criminal act by an agent unless the principal takes parts in, authorises or connives in such an act. Otherwise professionals will be solely responsible for any criminal act they commit during the course of the agency.

13.9. Termination of the agency

Termination of an agency can happen either by agreement, by an act of the parties, or by the operation of law. Examples of the latter are where the professional has completed the performance of the undertaking (e.g., the organising of the particular contract), frustration (if it becomes impossible for that contract to be performed), or if either party becomes incapable through death, bankruptcy or being of unsound mind.

However, a client cannot revoke the previous authority given to the professional as its agent and thereby destroy the professional's right to an indemnity from the client, for example, to reimbursement of a sum that the client originally agreed the professional should pay on the client's behalf to a contractor.

A professional acting as an agent should advise the third party if a client terminates an agency arrangement, not least so that there is no confusion as to the status of the relationship if the professional and the third party continue to be involved on the same project, albeit under other contractual relationships.

13.10. Agents appointed on behalf of the client, the delegation of their duties and provisions in the standard conditions of engagement

Clients may appoint project managers or others to act on their behalf, for example, in relation to negotiating the terms of the other professionals' appointments; they may delegate their obligations to give instructions and information and make decisions during the project; and they may give authority to act on their behalf in relation to the project. In the first instances, this is a genuine agency as described above. In the second case, this will become part of the professional's duties. However, in either case, it behoves the client to make the extent of the authority given clear.

Where the delegation concerns the client's obligations under the appointment, the client needs to clarify where the project manager's remit ends and that of the architect or other professional starts. What is the project manager to do and what is the client to do? Are the other professionals only to communicate with the project manager or are there to be some matters that must be discussed or notified to the client? If the client is also entitled to give instructions or make decisions, what are the other professionals to do in the event of conflicting decisions? What if the client later countermands any decision of his project manager? Some non-standard professional services agreements now have provisions dealing with this but they may not fully cover all the situations that could arise.

ACE Agreement 1, although it deals with the people within the client's organisation who have the authority to give or receive notices or to discharge his obligations, does not, except in the cases in the Services Schedule referred to above, create any agency arrangements.

RIBA 2010 has a clause in the following terms:

> 2.5 The Architect shall act on behalf of the Client in the matters set out or necessarily implied in this Agreement or in project procedures agreed with the Client from time to time, but has no authority, without the Client's prior approval:
>
> 2.5.1 to enter into any contractual or other commitment on behalf of the Client;
>
> 2.5.2 to terminate the employment of Other Persons appointed by the Client;
>
> 2.5.3 make or cause to be made any material alteration to or addition to or omission from the Services or the approved design.
>
> In the event of an emergency, the Architect may issue instructions to a contractor to prevent danger to persons or material damage to the Project without the Client's prior approval, and shall confirm such action in writing to the Client without delay.

It is not entirely clear whether this establishes a genuine agency because of the prohibition on entering into a contractual commitment on behalf of the client without prior approval. However, if he were to give that approval the professional would be acting as an agent of the client.

The role of contractor administrator is not technically that of an agent but this clause deals with the authorisation given by the client to the professional to do certain acts on the client's behalf, although some of those acts are subject to the independent judgement of the professional and to the obligation to act fairly or impartially.

RIBA 2010 also deals with the authority of the architect's own representative as follows:

> 2.6 The Architect's Representative shall have full authority to act on behalf of the Architect for all purposes in connection with performance of the Services but not to vary the terms of the Agreement.

The latter part is a little curious – the professional usually acts through his representative and if that representative does not have power to vary the terms of the agreement it is difficult to see how the agreement can be varied in the absence of any other provisions in these conditions dealing with variations.

RICS 2011 also has provisions restricting the authority of the professional in relation to the matters he can perform on the client's behalf as follows:

11.1 Subject to Clause 11.2, but notwithstanding any other provision of this Appointment, the Consultant has no authority to do any of the following on the Client's behalf without the Client's prior written consent:

(*a*) vary the agreed design or specification of work or materials or their quality or quantity from that described in the Building Contract;

(*b*) subject to any greater limit of expenditure stated in the Appendix, issue any instruction or notice under the Building Contract or any Client Contract which either delays completion of the Project or increases the cost of the Project (per item or in the aggregate); or

(*c*) terminate the Building Contract or any Client Contract or the appointment of any other member of the Professional Team, agree any amendment of their contract terms or waive compliance by the Contractor or any party to any Client Contract or any other member of the Professional Team with their respective obligations.

11.2 In the event of an emergency, the Consultant has authority to issue any instruction reasonably required to prevent danger to persons on or adjacent to the Site or material damage to the Project. The Consultant promptly notifies the Client of any such instruction and of its cost implications and effect on the Programme.

This implies that there are matters that are within the professional's authority and this would have to be found in the specific services to be provided or necessarily implied from them. They are quite stringent and wide-ranging restrictions and a professional might find it helpful to discuss with the client at the beginning how far these are to be applied to the particular project concerned.

There is also a provision to nominate the person who can act on the client's behalf subject to any limitations to be set out in the appendix, and for that authority to be delegated, revoked or amended.

PSC3 in Option X10 allows the client to designate an employer's agent to act on his behalf but the authority of the employer's agent is to be set out in the contract data.

Professional Services Agreements
ISBN 978-0-7277-4158-5

http://dx.doi.org/10.1680/psa.41585.225

Chapter 14
Statements, certificates and reports

14.1. Legal principles

Professionals are at risk of claims arising out of the statements that they make in reports, studies, certificates and surveys, etc., not only from a client but also from third parties. Such claims can be without limit and could therefore exceed the professional's professional indemnity insurance cover.

The law on this subject is complex and the following is a brief, simplified summary. It is necessary to understand the principles so the professional can see to whom and in what circumstances liability may arise.

14.2. The duty owed to a client in respect of statements, etc.

First, as has been explained in Section 4.1, the contractual duty of care owed by the professional to his client should be specified in the appointment which commissions a report, or requires a certificate, etc., and should be to use reasonable skill and care only. If it is not so specified, the duty to use reasonable skill and care will be implied by common law and by statute; that is, that the professional will exercise reasonable skill and care.

The professional also owes the client a duty of care in tort so could be sued in negligence for negligent misstatements contained in his advice – see Section 14.4. There are other possible causes of action in tort, such as misrepresentation – see Section 3.8.

Professionals can owe concurrent duties to a client in contract and in tort – although they may not necessarily be of the same extent – and the client may bring a claim in either – see Section 4.1. This can give a client the longer limitation period that applies to an action in tort and hence bring a claim that would be statute-barred in contract – see Section 16.6. However, in the case of a professional engagement, the duty a professional owes to his client in tort will normally be the Hedley Byrne-type duty described in Section 14.4, which would be the same as the duty in contract.

14.3. The duty owed to third parties in respect of statements, etc.

In the absence of a contractual relationship, a third party may be able to bring a claim for negligence against a professional if it can establish that the professional owed it a duty to take reasonable care. The third party can in such cases generally only recover damages for personal injury or damage to third party property, such as chattels and building contents (*Murphy* v. *Brentwood District Council (1991)*). This is consistent with the general principle that economic loss is not normally recoverable in tort. Whether economic loss is recoverable in tort depends on the scope of the duty. In the case of Murphy the cost of rectifying a defect in a building and damage caused to the same building by that defect was economic loss – see also Section 4.1.

One of the important exceptions to that principle relates to negligent misstatements, which are one form of negligent advice only.

14.4. Negligent misstatements

Statements made by a professional in reports, surveys and similar advice, as well as in certificates, could be negligent misstatements if a duty of care was owed to the person to whom the statement was made and the statement was made without the proper care.

Various formulae have been devised by the courts over the years to help in deciding whether a duty was owed, involving concepts of proximity, fairness and assumption of responsibility, whether voluntary or deemed. None can be described as definitive or universally applicable. The incremental approach, however, is adopted by the courts and the following three factors were the ones that were applied in the case of *Hedley Byrne* v. *Heller (1963)*, the leading case on negligent misstatement. These set out what must be established in order for a duty of care to exist in connection with the making of a statement.

(*a*) There must be a special relationship between the maker of a statement and its recipient.
(*b*) There must have occurred a reasonable reliance by the recipient on the statement.
(*c*) It must be proved that the maker of the statement must have, or ought to have, known that the recipient would rely on the statement.

If these criteria are met the professional can become liable to the third party for economic loss in addition to being liable for damages for personal injury and damage to third party property.

14.5. Those to whom the professional can become liable

Clearly, a professional would be liable to his client. However, what constitutes a 'special relationship' (or sufficient 'proximity', as the cases often refer to it) to a third party and 'reliance'?

It has been established that a relationship of proximity cannot exist between the maker of a statement put into general circulation and a recipient who has relied on the statement for a purpose not anticipated by the statement maker. For a third party to succeed in a claim for breach of duty of care against the maker of a statement, it is likely that the following four conditions would have to be satisfied.

(*a*) The maker must know the statement made is being communicated to a person or class of a specific nature. If, for example, a client has agreed with a professional

that the professional's report can be shown to its funder, the client cannot use it in a prospectus for the sale of land, for example, and make the professional potentially liable to future purchasers.
(*b*) The communication must have occurred in connection with a specific transaction or transaction type. So, in the example just given, it can be relied on for the purpose of lending money but not for the purpose of a sale.
(*c*) The communication must have been made to a person or class likely to rely on it in determining if it or they should enter into a certain transaction without making a separate independent enquiry.
(*d*) The person or class that has relied on the statement must have suffered some loss as a result.

The Law Lords (*Caparo Industries plc* v. *Dickman (1990)*) have restricted the liability of professional advisers to those third parties with 'reasonably close proximity'. Otherwise, they thought this would '... confer on the world at large a quite unwarranted entitlement to appropriate for their own purposes the benefit of the expert knowledge or professional expertise attributed to the maker of the statement'.

If a statement or report is prepared by a professional for a client, for example, and is then altered unilaterally by that client and incorporated into another document and given to a third party without the professional's knowledge that this would happen, it is unlikely that any proximity or special relationship will occur between the professional and the third party who relied on the altered document. See also Section 14.14 concerning liability to third parties for reports.

The case of *ADT Limited* v. *BDO Binder Hamlyn (1995)* illustrated when a Hedley Byrne-type duty can arise and also illustrates the serious consequences there can be for a professional who makes a negligent misstatement. In that case, a firm audited a company's accounts. A prospective purchaser of that company asked the audit partner about those accounts and the partner stated that he stood by his firm's audit opinion. It was found that the audit partner owed a duty of care to the purchaser under the Hedley Byrne principle and had made a negligent misstatement. This led to a judgement against the firm of £65 million plus £40 million in interest and costs. The firm appealed and the case was settled before the hearing of the appeal.

14.6. Necessary expertise of the maker of a statement

The duty of care only arises where the statement maker holds himself or herself out as 'possessing the necessary skills/competence to give advice and the necessary diligence to give reliable advice'. A professional will fall within this category and will therefore be a person who has responsibility to make any statement with reasonable skill and care.

14.7. Circumstances in which there can be a breach of the duty of care

The courts have established that a professional, in writing a report, should display and apply reasonable care and a reasonable standard of professional competence. The exercise of professional judgement is almost bound to lead to different opinions.

There is a particular rule for valuations that the valuer will not be liable if his valuation falls within a certain range. Evidence of what is acceptable practice within the profession or what standards are laid down by its professional bodies can be crucial in determining the standard of care to be exercised by a professional.

The case of *Watts and Watts* v. *Morrow (1991)* provides a useful illustration. In this case the plaintiffs wished to purchase a second home and so arranged for a full and detailed survey to check that no unexpected major expenditure would be required. The report was generally reassuring, recommended only minor repairs, and contained no warnings that heavy immediate expenditure would be required. On the basis of the report the plaintiffs bought the house and subsequently discovered that there were a number of serious defects requiring some heavy expenditure. The plaintiffs therefore sued the surveyor for preparing his report negligently. It was held that the surveyor had not displayed and applied reasonable care and a reasonable standard of professional competence in preparing his survey report and was therefore liable for the loss in value of the house suffered by the plaintiffs. Against his professional body's practice, the surveyor had dictated his report while walking around the site which led to a lengthy and diffused report with a totally incorrect conclusion. The report failed to deal with its specific purpose, 'Will we have any unexpected major expenditure?', gave no warnings about this and in fact only dealt with the need for minor repairs.

14.8. Gratuitous advice or statements

Professionals can mistakenly believe that no liability can be incurred to a third party unless the professional has been paid by that third party – in other words, that there is a contract in place with that third party. The question of payment is immaterial, however, when considering whether liability can arise in tort to a third party or indeed to the client.

14.9. Disclaimers

Disclaimers are the means by which clients or third parties are notified that the professional does not accept liability to them and that they cannot rely on some matters in a report or in a statement. For example, a professional in a report may say that he has been unable to consider or give an opinion on a certain piece of information.

Where the professional is engaged under a contract with his client, disclaimers in reports, etc., like other limitations of liability, are subject to the test of 'reasonableness' under the Unfair Contract Terms Act 1977 and also the Unfair Terms in Consumer Contracts Regulations. These Regulations appear to apply to the provision of advisory services to consumers and so would seem to cover any report written, for example, for a domestic client. The Regulations require that terms in a consumer contract, including a disclaimer, must satisfy a 'requirement of good faith'. This is not properly defined and could cause confusion as to its relationship with the 'reasonableness' test under the Unfair Contract Terms Act 1977 – see Chapter 16.

Disclaimers can also be used, however, to try to avoid any liability arising in respect of a report or statement to a third party. The case of *McCullagh* v. *Lane Fox & Partners*

Limited (1996) concerned an estate agent who prepared particulars for a property, which stated that the exact size of the property was 0.91 of an acre. The particulars included the disclaimer that 'None of the statements contained in these particulars are to be relied upon as statements or representation of fact'.

The actual size of the property was 0.48 of an acre and when the purchaser discovered its true size he sued the estate agent. The court held that the estate agent knew or ought to have known that its representation would be relied upon by the purchaser. However, the receipt of the particulars containing the disclaimer had the effect of cancelling out the proximity between the purchaser and the estate agent and the assumption of responsibility on the part of the latter.

As has been said, disclaimers are subject to the 'reasonableness' test under the Unfair Contract Terms Act 1977, but in this case the disclaimer was held to be reasonable as the purchasers

(*a*) were sophisticated members of the public
(*b*) could have looked at other properties after receiving the disclaimer
(*c*) would have had legal advice before exchanging contracts for the purchase of the property.

In addition, the court took into account the fact that disclaimers are commonplace with estate agents' particulars.

In another case the courts (*Omega Trust Company Limited and Banque Finindus* v. *Wright Son & Pepper and Barker & Co (1998)*) had to consider a disclaimer in the case of a valuation of three leasehold properties that a firm of surveyors had been asked to send to a bank which was providing a loan to the owner. The valuation, which had originally been prepared for another lender, was readdressed to the bank and a nominal fee was paid, making that bank the surveyor's client. The valuation contained a disclaimer that:

> This report shall be for the private and confidential use of the clients for whom the report is undertaken and should not be reproduced in whole or in part or relied upon by third parties for any use whatsoever without the express written authority of the surveyors.

Unknown to the surveyors, another bank had been asked to participate in the loan and the valuation was passed to it. When the security was found to be insufficient, both banks sued the surveyors. However, the Court of Appeal held that no duty of care was owed to the undisclosed bank and that it was fair and reasonable for the surveyor to rely on the disclaimer. It drew a clear distinction between commercial transactions and those with domestic householders and considered that it would have been reasonably practicable for the bank to have obtained a separate valuation, or to have asked the surveyors directly whether they were prepared to assume responsibility to the bank.

Where it is not feasible to prevent a report or the equivalent prepared by the professional from being shown to other parties, a disclaimer along the following lines may help to prevent liability to any such party arising:

> This report has been prepared by [professional] on behalf of [client] in connection with [project] at [site] and takes into account their particular instructions and requirements. It is not intended for and should not be relied on by any third party and no responsibility is undertaken to any third party.

or

> [The professional] accepts no duty or responsibility (including in negligence) to any party other than [original client] and disclaims all liability of any nature whatsoever to any such party in respect of this report.

The disclaimer should be repeated to the client and if possible to the third party if the professional is or becomes aware that the report is being passed to any party.

14.10. Certificates

Certificates which have to be given by professionals under construction contracts (e.g., as to the value of the work carried out by the contractor) are outside the scope of this text. This chapter deals with certificates required by non-standard conditions of engagement in relation to the professional services, often to be given to third parties, describing what the services may or may not have achieved. Clients such as government departments often require such certificates. They are usually in a fixed form and may contain statements describing a level of performance different from, or beyond, what can reasonably be expected from the professional. Examples include:

> The [professional] will certify on completion of the works, that the works have been constructed in accordance with the drawings and specification. The [professional] will further certify that all calculations, designs, drawings and contract documents prepared by the [professional] are to the highest standard utilising staff well experienced in works of this type.
>
> The resulting structure, if constructed as designed, will be capable of meeting the Client's requirements of the scheme made known in writing.
>
> The said Design Data comply with the Construction Requirements, as amended . . . and comply in all other respects with the Agreement.
>
> The Temporary Works are satisfactory for the safe and proper discharge of the Client's obligations.
>
> The resulting engineering services, if installed as designed and correctly serviced, will be capable of meeting the Client's environmental criteria and policies which have been made known to [the professional] in writing and upon which the design has been based, and will provide proper access for safe operation and maintenance purposes.
>
> [The professional] has exercised reasonable skill, care and diligence in performing the services required by its commission in relation to inspection of the work to

> see that the Contractor has generally fulfilled its obligations with regard to the Building Regulations, Codes of Practice (including the Department of Health Firecode) and British or European standards specification or European technical approvals or common technical specifications, as appropriate, relevant in the design calculations and drawings, subject to any alterations or modifications made by [the professional] under its commission.

The latter two quotations have been taken from the Department of Health's Agreement for the Appointment of Architects, Surveyors and Engineers for Commissions in the National Health Service.

Claims by clients or third parties usually arise out of such certificates under what is known as the Hedley Byrne principle – see Section 14.4. Clients may also be able to bring claims for breach of contract. The claim arises when financial damage results to the person relying on the statement, if that statement is not correct but was acted upon. It will be seen from this that a professional is likely to face a claim if he makes a statement, whether it is in a prescribed form or his own words, which turns out to be untrue. Professionals are expected to consider carefully the words by which they describe a state of affairs or by which they give an opinion and to qualify them appropriately so that the recipient is given an accurate description and not one, for example, that is outside the professional's knowledge.

This is a particular problem where the client has stipulated in advance what the professional is to say by including the giving of a certificate in a certain form as a contractual obligation. If the professional knows at the outset that he will not be able to make the statement required, because, for example, the matters it covers will be outside his knowledge, he should refuse to agree to the form of certificate. Taking one of the examples above: how can the professional know that 'the works have been constructed in accordance with the drawings and specifications'? Even if he has visited the site regularly during construction, he cannot know whether every bit of the works has been so constructed. Also, there is no description of to which drawings and specifications the certificate refers so this could well be construed as all the relevant drawings and specifications for the works – this would therefore include the drawings and specifications of other consultants, contractors and sub-contractors. It would certainly not be possible for a professional to be able to make this statement about others' designs and specifications. The most he may be able to say in relation to his own drawings and specifications is that the works have been constructed generally in accordance with them. This reflects what is in the services schedules to the ACE Agreements: '... make other periodic visits to site ... to monitor that the Works are being executed generally in accordance with the contract documents and with good engineering practice'.

It will be remembered in this context that 'the Works' are those elements of the project in connection with which the professional is providing his services and therefore not necessarily the whole of the project.

It would be more appropriate if any contractual obligation to give a certificate was phrased as 'The professional shall give a certificate as to whether or not ...' because of course if the state of affairs is not as set out in the prescribed certificate the statement the professional makes will be untrue.

A further problem for professionals in relation to certificates can arise if the professional makes a statement about the fitness for purpose of the works or other matters and this turns out to be untrue. It is possible that the damages payable if the professional has been found to have made such a statement negligently could be those that would be payable in the event of an actual breach of a warranty for fitness for purpose. If so, professional indemnity insurance may not provide an indemnity to the professional in respect of those damages. (See Sections 4.1 and 4.2 on the principles concerning the duty to use reasonable skill and care and warranties for fitness for purpose.)

Certificates such as the ones described above often blur the distinction between professionals' and contractors' duties. For example, in one of the foregoing examples the words 'Construction Requirements' were used. 'Construction Requirements' will generally include matters with which the contractor is to comply. It should therefore be the contractor's responsibility if there is non-compliance. Likewise, whether the temporary works have been safely constructed is primarily a matter for the contractor. If the professional makes a statement such as 'the said Design Data comply with the Construction Requirements', and if, in fact, the professional's design does not achieve the criteria set down in the 'Construction Requirements', the professional would have made a statement that is untrue even though he may well have exercised 'reasonable skill and care' in the preparation of the design data. The professional could then be liable for a negligent misstatement. Worse, if the 'Construction Requirements' contained warranties for fitness for purpose, he could be taken as having made a statement that amounts to a warranty for fitness for purpose with the consequences set out above.

On the other hand, the statements in the NHS certificates are expressly qualified so that the test is whether the professional has exercised reasonable skill and care in performing his original duties. This is very different from making a statement describing the condition of the works, for example. The statements in the NHS certificates include:

> [The professional] has exercised reasonable skill care and diligence in performing the Services to see that . . .

or similar wording, and there is an opening paragraph in each NHS certificate which provides:

> Notwithstanding anything to the contrary contained in this certificate this Firm is obliged to exercise reasonable skill and care in the performance of the services required by that commission and this Firm shall not be liable except to the extent that it has failed to exercise reasonable skill care and diligence and this certificate shall be read and construed accordingly.

It is thus clear that the intention is that the scope of the certificate is limited to whether or not the professional has exercised reasonable skill and care in the performance of his services, which makes these types of certificate much closer to the giving of a collateral warranty – see Chapter 21.

The reason why such certificates are sought from professionals is that it may well be easier to make a claim against a professional which is based on negligent misstatement – *prima facie*, all the client has to do is show that the statement is untrue and that the professional did not make it carefully – than to make a claim based on negligence where he would have to show the professional had failed to use reasonable skill and care in the performance of his services. Also, recovery can be easier in the event of a claim against a professional who has professional indemnity insurance than in a claim against a contractor who may have insufficient assets or no longer be around.

There is no need for a professional to give any certificates of this nature. If the professional has been negligent in the performance of his services or in breach of any contractual obligation the client will have a remedy and be entitled to the damages prescribed by law.

14.11. To whom the certificate is addressed

As has been said, a certificate is a statement and a professional is obliged to make such statements carefully. Generally, a certificate will be requested by the client so that he is reassured as to the condition of the works or certain aspects of them.

Sometimes, however, the certificate can be required by a third party, such as a future occupier of the building, or to the ultimate client for the project when the professional has been employed by the contractor. If the certificate is given without the proper care and is untrue, the third party may well be able to bring a claim including a claim for economic loses on the Hedley Byrne principle described above. The professional needs to consider in such circumstances whether he is prepared to do this as he could, as a result, incur a liability to a third party when otherwise there would be none. An agreement that a certificate given to one recipient can be passed on or relied upon by one or more other parties could have the same effect.

It could be better in such circumstances to give a collateral warranty for the reasons set out in Section 14.13.

14.12. Council of Mortgage Lenders' Certificate

The Council of Mortgage Lenders' (CML) Certificate was produced by the CML in 2007 at a time when lenders were finding that some new-build residential property was not covered by schemes such as the NHBC's (National House Building Council's) warranty/insurance or Zurich guarantee schemes. This was because some contractors could not afford to register under such schemes, had become insolvent, or the residential part of a project was only incidental to a much larger commercial development. Lenders on residential properties had generally required the property to have the benefit of such schemes so that there was a remedy if the property was later found to have defects.

It was also felt that purchasers of these properties should not be required to have their own surveys carried out when the developer or contractor should be able to certify the structural integrity of the property.

The purpose of the certificate therefore was to provide lenders with some reassurance that the purchaser to whom they were lending would be able to recover the costs of remedying any defect from the certifier. CML said that it would accept a certificate in it's prescribed form signed by a professional affiliated to a member of prescribed bodies including the RIBA, RICS and CIAT (Chartered Institute of Technologists). In fact, consulting engineers as well as architects and surveyors have been signing these certificates.

It is worth considering this certificate in some detail as it provides a further example of the liabilities that can be incurred in respect of certificates referred to earlier in this chapter.

Although it is envisaged that the professional who gives the certificate will be employed by the house builder, developer or the contractor, the certificate is to be given to 'The Applicant'. This is the first purchaser of the property. Thus, the professional will become liable to him as a third party as well as to his original client. Without such a certificate there may not be any liability to the purchaser as it could be difficult to prove the necessary duty of care arose. However, there can be circumstances in which a designer can become liable to future occupiers for economic loss – that is for physical damage caused, for example, by flooding due to undersized drainage.

The certificate also states that in addition to the first purchaser the professional is 'aware this certificate is being relied upon . . . by [the lender] when making a mortgage advance to that purchaser secured on this property' and that the professional 'will remain liable for a period of six years from the date of this certificate. Such liability shall be to the first purchasers and their lenders and upon each sale of the property the liability for the remaining period shall be transferred to the subsequent purchasers and their lenders'.

Although the wording is not entirely clear, it seems the purpose of this is to show that the professional has accepted that future purchasers and lenders will be relying on the certificate and thus have a right to claim if the statements in the certificate are untrue, however many of them there are, until six years have expired.

Without such wording the professional would only have been liable to the original purchaser and their lenders for up to six years.

The CML certificate requires the professional to certify that the professional visited the property at appropriate periods from the commencement of construction to the current stage to check generally progress and conformity with drawings approved under the building regulations and drawings and instructions issued under the building contract. The professional has to state at what stage he last inspected the property. This stage should be at least practical completion or it is difficult to see how the professional can make the further statements set out later in the certificate. The notes to the certificate say this allows the professional to say if there were any limitations of any inspection, for example, that certain items were not complete at practical completion and what is being done about it. However, the further statements in the certificate allow for no qualifications to be inserted to take account of these 'limitations'.

The professional then has to certify that 'so far as could be determined by each periodic inspection' the property has been generally constructed (a) to a satisfactory standard; and (b) in general compliance with the drawings approved under the building regulations. The certificate covers the whole property. It is not restricted to the elements which the professional designed or inspected and thus an architect could be certifying in relation to the designs of a mechanical and electrical engineer or structural engineer and vice versa.

As has been explained, the statements in the certificate constitute a representation as to the facts stated, and if the statement is not correct an action will lie against the professional for negligent misstatement if that statement was made without the proper care. The courts therefore will expect the professional not only to have made the inspections but to have carried them out. If the professional did not carry out the inspections he cannot give the certificate, or he should qualify the statements so that they express what he can say. There are some notes to that part of the certificate which state 'I have visited the site at appropriate periods from the commencement of construction to the current stage to check generally . . .' These notes say that 'In the absence of inheriting "normal/reasonable skill and care" as the appropriate standard from the initial appointment one could add it to this clause by inserting "and exercised reasonable skill and care" after the word "stage" ', that is so that it reads 'I have visited the site at appropriate periods from the commencement of construction to the current stage and exercised reasonable skill and care to check generally that . . .'.

By this it is supposed that the CML was accepting that the certificate did not relate back to the skill and care with which the original services were carried out, which is correct, and that this could be rectified by adding this wording. This is probably right because that part of the statement would then only be incorrect if the professional had failed to exercise reasonable skill and care when checking the matters set out in the certificate.

The courts will also expect the professional to have had all the information, drawings, etc. necessary to make the statements in the certificate. Again, if he did not, he cannot give the certificate.

His liability in the absence of the amendment set out above will have nothing to do with whether the inspections on which the professional based his opinion or his original services were carried out with reasonable skill and care.

It can be seen from this that despite the qualification of 'generally', a consultant is going to find it difficult to defend a claim when the claimant can demonstrate that the construction of the property was in breach of any building regulations and/or was not in some respect of satisfactory quality, even when the defects arose out of matters for which other consultants or the contractor were responsible.

The claimant will have a claim against the builder, as the person primarily responsible for the defect, and possibly the other professionals, but if the builder becomes insolvent or has insufficient assets when the defects appear, it is going to be easier for the claimant to bring a claim against the certifying professional, who is likely to have professional indemnity insurance.

The CML certificate offers no opportunity to the professional to limit his liability, and thus liability under such certificates is unlimited. Professionals are, at the time of writing, finding that claims are coming through for hundreds of thousands of pounds. The fees for these inspections and giving the certificate are often in the hundreds of pounds. The risks therefore being taken on by the professional signing these certificates can be wholly disproportionate to their fees.

The professional has to confirm in the certificate that he will maintain professional indemnity insurance at a certain level, but this of course does not limit his liability to that amount. It is also possible that professional indemnity insurers could take the view that that statements in the certificates amount to warranties for fitness for purpose, in which case the insurance may not cover them at all.

The certificate also requires the professional to state who he was retained by and the basis of his appointment. The purpose of this is not clear because the notes acknowledge that 'more detailed reference making this certificate back to back with that appointment is difficult to achieve simply' but say the professional could try adding 'The duties and liabilities under this certificate do not extend those of the original appointment contract'. It is thought this addition is aimed at making the professional liable only if he was in breach of his original appointment, but the wording could be clearer. It might also be an attempt to allow the professional to rely on any limitation of liability in the original appointment, but it could be argued that clearer wording would be needed to achieve this.

Finally, the professional is required to confirm he has the appropriate experience in design and/or monitoring of the construction or conversion of residential buildings. This is not only to remind the professional of the CML's requirement as to the professionals who are to sign such certificates, but also that a professional should not sign a certificate unless he has the requisite experience. One would hope this was self-evident. The notes carry a warning about making positive claims about experience 'sometimes considered risky' and say that 'it can also be limiting in that one is not claiming expertise beyond "residential buildings" '.

14.13. The basis for giving certificates

If certificates must be given, the professional should try to ensure the following.

(*a*) The terms of appointment do not compel the giving of a certificate in a certain form to a certain person, regardless of the actual circumstances at the time.

Otherwise, the professional will be in breach of contract if the professional then refuses to give or tries to qualify the certificate, even though the professional may have good reasons for wishing to do this. This will give rise to a claim for damages. It might be possible, if the reasons for not complying with this obligation are fully explained to the client, for the client and the professional to agree to vary the obligation in relation to the certificate's form.

(*b*) If the certificate is to be given a third party, consider the unlimited liability that could result and whether that liability could be limited, for example, through the use of a collateral warranty. If no third party is to have the benefit of the certificate, consider including a disclaimer in respect of any reliance on the certificate by third parties.

(*c*) The wording of the certificate should be such that it accurately records only what the professional has or has not done and makes any statement as to the conditions of the works, etc. with appropriate caveats and qualifications and does not amount to a statement of what others have done or what they or the works will achieve. If it is a choice between being in breach of contract through not giving a certificate in an agreed form and amending the certificate so that it says what the professional can properly say, the professional should opt for the latter.

(*d*) Express wording to limit the scope of the certificate to a statement that the professional has used reasonable skill and care in the performance of the relevant services should be agreed, if possible.

(*e*) Wording similar to that in the NHS certificates could be used (although it has not been tested in court and thus it is not known for certain how effective it is) if statements have to be made as to the condition of the works and what the works might achieve, as this may mean the professional will only be liable if he has been negligent.

14.14. Particular contract considerations for reports, surveys, etc.

All the principles concerning professional services agreements discussed in this text apply equally to contracts to undertake reports, surveys and so on. However, in relation to these, the professional should consider the following in particular.

(*a*) Who is entering into the contract on the professional's behalf and the nature of the report. For a very risky project it might be wished to establish a separate limited liability company and use what is known as 'ring fencing'; that is, confine all the liabilities and risk to that particular company. There are practical considerations about this which would need to be looked at carefully.

(*b*) Defining the brief: have the terms of reference been fully understood and discussed and any limitations on the brief or on the enquiries to be undertaken discussed, agreed and recorded, preferably in, or at least referred to specifically in, the contract?

(*c*) Defining the services, that is, whether they set out what the professional will be doing in detail and the extent to which, for example, the professional may be inspecting or monitoring any contractors' work or other third parties and, if so, for what purpose. It might also be helpful to be clear as to what will not be included or covered.

(*d*) Whether the contract includes all that the professional needs and whether, for example, the client is going to provide it or whether the client will be arranging for something to be done by someone else. For example: is there any information which is needed from the client or third parties, will it be sufficient, is more needed, will the client obtain it, is there anything else the client should undertake to do?

(*e*) Defining in detail what it is the professional is going to provide (e.g., the report).

Other matters which may arise in non-standard conditions of engagement relating to reports and surveys, etc., could include the following, which also need careful consideration.

(*a*) The client requiring the professional to give guarantees or specific assurances directly or indirectly to third parties. For example:
  (*i*) The remedial work carried out will ensure that the site will be free from all contamination.
  (*ii*) The site investigation will ensure that all potential sources of contamination are identified.
  (*iii*) The work will be carried out so that no pollution of the aquifer takes place.

  This could result in the professional becoming liable to the third party and could also amount to warranties for fitness for purpose with the consequences for professional indemnity insurance, referred to in Section 15.6.

(*b*) Statements of future uses of the premises or site, for example, by third parties; whether third parties are to be entitled to copies of the report and/or there is a statement that the professional must have 'due regard' to the interests of third parties, such as future owners/occupiers, in preparing the report. This should alert the professional to the possibility that third parties could be given, and therefore be relying on, the report. Whether this is the case should be clarified. Remember that if the professional knows that a statement is being communicated to a person (or class) in connection with a specific transaction it could incur liability to that person (or class). If a professional includes such a statement in the appointment, it will be considered to have such knowledge.

  If the professional agrees that the third party should have a copy, the professional should also be told for what purpose, so that the report can be limited to that purpose. Some appointments could set out the purposes very widely, for example, letting, sub-letting, sale, purchase, mortgage, charging, etc. The professional should not agree to include a provision which would allow anticipated uses or purposes to be changed, for example, to things notified to the professional at a later date. It might then be impossible to amend or qualify the report to take account of this.

(*c*) Statements that a client or third parties will 'rely' on the professional exercising reasonable skill and care in the preparation of the report. These should be deleted. They are designed to pre-empt any arguments that the professional may have as to whether a client or third party did rely on the report – they could have taken separate advice on the same matters from another professional and relied on that.

It will be remembered that for third parties to succeed in suing a professional they must show not only that they relied on the statement but that the professional knew or ought reasonably to have known that they would.

(*d*) Provisions which allow the client to disregard any matters in other reports or advice and to rely instead totally on the professional's report where someone else is being asked to advise on aspects which overlap with or complement the professional's terms of reference. This could inhibit any arguments the professional may have about the extent to which the client or third parties should have relied on the professional's report.

If third parties should obtain their own advice (e.g., obtain a survey) this should be stated clearly either in the appointment or in the report.

(*e*) The professional's right to the copyright in the report and the ability to restrict or prohibit licences should be used to protect the professional. The professional may not always know what use the client is going to make of the report or to whom it will be given. This should be discussed and agreed with the professional. If not, the professional should consider prohibiting any licence to copy the report, or make the copying of the report subject to the express consent of the professional, so that it is known to whom it is going and whether anything in particular needs to be said to those persons about the nature and extent of the report and the reliance that can be placed on it.

A limitation of liability ought to be considered in appropriate cases, for example, where the risk is out of all proportion to the fee, such as where the fee for the report is, say, £1000 but if the professional gets it wrong the claim could be for millions – see Chapter 16. Limiting or excluding liability may be particularly necessary where environmental site investigations are being carried out and there is the possibility of pollution and contamination claims. Liability for such claims could be excluded or limited to clean-up costs or to a monetary amount.

14.15. Client requests to pass on reports

In recent years, professionals have been asked more frequently if they will 'assign' their reports to third parties or allow them to be passed to them. It is probably worth expanding on this topic because of the liability implications for the professional.

Taking assignment first: it is not possible for the report itself to be 'assigned'. What the client may do is assign to a third party the benefit of the client's contract with the professional under which he produced a report for the client unless assignment is expressly prohibited in the contract. Any such assignment would usually take place when the client sells or transfers its interest in the project to a third party, but it could take place some time later. If the contract requires the professional's consent to the assignment, that must be sought and given for the assignment to be effective.

Provided that notice of the assignment has been given to the professional, the third party would then be entitled to sue the professional in the assignee's own name for any breach of the contract committed by the professional – for example, that he had failed to exercise reasonable skill and care in preparing the report.

As to the effect of the assignment, see Section 19.1.

The professional may be asked directly to allow a third party to rely on the professional's report or for consent to his report being passed on to a third party after it has been prepared. For example, prior to committing to the purchase of a site for development, the purchaser may wish to rely on the professional's site investigation report, condition survey, or such other report which had been prepared for the client. Some common scenarios are

- the client asks the professional to 're-address' its report to a third party
- the client asks the professional to 'confirm to the third party that the professional owes a duty of care to a third party in respect of the report'
- a third party asks the professional to confirm that it can rely on the report.

Such requests might also come in the form of the professional being asked for a collateral warranty in respect of the report – this is dealt with in Chapter 21.

The professional does not of course have to agree to such a request but may be willing to do so for commercial reasons. However, the professional should bear in mind that for the reasons explained in this chapter it may incur liability to the third party for losses for which it would not have been liable to its client and that the liability may not be subject to any limitations contained in the appointment with its client.

14.16. The Technotrade case

Some of the issues explored in this section are usefully illustrated in the Technology and Construction Court case of *Technotrade Limited* v. *Larkstore Limited (2005)*.

The case concerned a site investigation carried out by a company, Technotrade Limited, on the stability of a site, prior to the site being developed. The company that engaged Technotrade later decided not to carry out the development and sold the site to another developer, with the benefit of planning permission that had been obtained with the assistance of Technotrade's report. The developer acquired the report, used it in connection with a further planning permission and proceeded with the development. During the development a landslip occurred.

The landslip caused damage to various houses uphill from the area of the slip and led to stabilisation works having to be carried out.

The developer wanted to make a claim against Technotrade. The original client agreed to assign to the developer the benefit of its contract with Technotrade. It was now more than five years since Technotrade had prepared its report and completed its services for its client, and nearly five years since that client had sold its interest in the site to the developer. In return for the assignment, the developer agreed to pay Technotrade's original client one half of any monies it recovered from Technotrade, after deduction of costs.

The judge made the following findings on preliminary issues.

(*a*) The assignment was valid, and so the benefit of any claim for breach of contract against Technotrade arising out of the latter's investigation and report was passed to and could be pursued by the developer.
(*b*) The argument that the original client had suffered no loss as a result of any such breach, having sold the site for full value, and that the developer only had a claim for nominal damages, was rejected by the court, on the authority of *Linden Gardens Trust Ltd* v. *Lenesta Sludge Disposals Ltd (1994)*. The judge's ruling on this point was upheld by the Court of Appeal.
(*c*) The landslip did not give rise to a claim in tort against Technotrade in favour of the original client, because it occurred after that client had sold the site to the developer. No such claim could therefore be assigned to the developer.
(*d*) Technotrade did not owe an independent duty of care in tort to the developer in respect of economic loss (i.e. the cost of stabilising the site, save for any part of that cost for which Technotrade may be liable under *(f)* below). Technotrade did not know at the time of its report that the development would be carried out by another party, to whom its report would be passed, or that its report would be 'recycled' in connection with a further application for planning permission.
(*e*) Technotrade owed a duty of care to the house owners in respect of the physical damage to the houses. This meant that the developer could pursue a claim against Technotrade under the Civil Liability (Contribution) Act for contribution towards its own liability to the house owners, as owner of the land from which the houses derived support.
(*f*) Technotrade owed a duty of care to the developer in respect of the physical damage to the houses.

The findings on liability were findings in principle on preliminary issues and subject to actual liability, for example, negligence and causation, being proven.

14.17. Guidance on report writing

The professional should make the report as self-contained as possible. If a claim arises against a professional, it is the report itself that will be scrutinised by the court, possibly many years later, when some of the details of what happened at the time have become vague, warnings or discussions with the client may not have been recorded, a covering letter forgotten or misplaced and so on. It is on the report itself that the professional might be judged and it is therefore vital that it is comprehensive. Any pressure from the client to put any matters in a separate letter or document should be resisted.

Some matters that should be included in a report are listed below. They may overlap and are not exhaustive because this will depend very much on circumstances.

(*a*) Identify the client for whom the report is prepared and its purpose, so there is no doubt to whom it is addressed and who should be using it.
(*b*) State the scope of the report as defined by the brief or terms of reference, including any limitations or restrictions on that brief. In particular, the specific

transaction or transaction type in connection with which the report is being prepared should be identified. So, make sure there is no doubt or ambiguity about the scope, or any likelihood that it could be interpreted more widely than was intended.

(*c*) State what was done, what was not done and what could only be done to a limited extent. If something further should be done to check and/or assess the risks, this should be set out.

(*d*) Describe the data and information received from the client or others, the nature of it, its adequacy or otherwise and say what effect it has had on the work undertaken and what reliance has been placed on it. Set out, if relevant, what information others are, or should be, providing.

(*e*) State what has not been received or what could not be obtained, and if and where this has been requested or promised.

(*e*) Record any restrictions imposed on the professional or the professional's enquiries by the client, outside agencies or circumstances.

(*f*) State the limitations on the information/advice contained in the report if, for example, it is preliminary or based on limited information. Spell out any risks inherent in the advice or in any reliance that is placed on it. Put in all necessary caveats and limitations. Expressly state any warnings.

(*g*) Detail any assumptions implicit in the report and/or its conclusions.

(*h*) Do not be generally reassuring. Be specific.

(*i*) Qualify any conclusions appropriately within the report, not only in any accompanying letters. Do not assume that any limitations, warnings or qualifications given to the client about the report will be remembered or recorded elsewhere.

(*j*) Make it clear that the report is not for the use of, or reliance on, by third parties – see Section 14.9 on disclaimers.

(*k*) Prohibit any copying of the report for third parties without express consent. Then, if the client says it is needed for a third party, the professional can discuss the terms on which a copy of the report is released and whether any further caveats or limitations are needed so that the third party does not use it for purposes for which it was not intended – see Section 14.5.

(*l*) Date it so that anyone reading it later can be alerted to the fact it may be out of date.

The object is to make the report's purpose, the limitations of any advice or conclusions and the nature of that advice or conclusions, so clear that no one reading it could possibly mistake its purpose, content or rely on it for things that it simply was not reasonable for them to rely on.

Professional Services Agreements
ISBN 978-0-7277-4158-5

http://dx.doi.org/10.1680/psa.41585.243

Chapter 15
Insurances and liability

15.1. Professional indemnity insurance generally

Professional indemnity insurance policies are a form of liability insurance taken out by professional people to protect themselves against liability in respect of claims arising out of the conduct of their professional business. Such policies may be written either on a 'negligence' basis where they cover only the insured's negligent acts or omissions, or on a broader 'civil liability' basis where they cover any civil liability of the insured. The latter have become more common and the Architects' Registration Board (ARB) has imposed mandatory professional indemnity insurance for architects on a civil liability basis. Professionals with negligence-only policies will therefore not have any professional indemnity insurance cover for claims that could arise out of strict obligations or indemnities in appointments whereas civil liability policies would *prima facie* cover such claims, subject to any exclusions. Cover may be on an 'each and every claim' basis or 'in aggregate'. The former provides cover for each claim up to an agreed limit of indemnity for each claim whereas the latter provides cover up to a maximum fixed sum in respect of all claims made during the policy period.

Most professional indemnity insurers of construction professionals now have a form of hybrid cover – indemnity is provided on an 'each and every' basis generally but with an aggregate sub-limit for particular types of claim, such as pollution and contamination claims and/or asbestos claims. Some insurers treat date recognition problems in the same way. A few introduced an 'exclusion and buyback' for such claims. Under these policies, cover is excluded but can be bought back if certain conditions are fulfilled.

Most professional indemnity policies will indemnify the insured against loss resulting from any claim which is made against the insured by a third party in respect of any negligent act, error or omission by the insured or an employee or partner of the insured which arises in the course of the insured's business. The policy will not usually provide cover in respect of any negligence of the insured's clients, or any losses suffered by the insured itself because of its own negligence.

The policy is an indemnity policy, so that the insured is not strictly able to recover anything under the policy or make a monetary claim against the insurer until the insured's liability has been ascertained by a judgement or settlement agreement. Proof of payment to a third party is not usually required unless this is stipulated in the policy. Insurers will usually of course take over conduct of the claim but this principle can have some significance in the context of claims under the Third Parties (Rights

Against Insurers) Act 1930 and also in cases where an insured has been forced to settle a claim itself and is then seeking an indemnity. Such settlements need to be very carefully drafted in order to ascertain the loss.

It is usual for the insurer to impose a limit on the amount of the indemnity recoverable and also provide that the insured will be responsible for bearing some of the loss. The usual requirement is for the insured to bear the losses up to a specified amount, called the 'excess' or 'deductible', in the case of each individual claim under the policy. The insurer therefore only has to bear the remainder of the losses up to the amount of the indemnity agreed.

There is, however, a difference between an excess and a deductible. In a policy for £1 million cover and a £250 000 excess, the £1 million cover comes into play once the excess is reached. If there is a deductible of £250 000, the maximum payable by insurers would be only £750 000. It is possible with some professional indemnity policies to agree an aggregate limit to the total number of excesses payable by the insured within certain specified time periods.

15.2. No general duty to minimise damage

The Court of Appeal (*Yorkshire Water* v. *Sun Alliance (1996)*) held that unless there was an express provision in the policy, the insured was not under a duty to take reasonable steps to avoid or mitigate a loss. If the insured did take steps to avoid such a loss and incurred expense as a result, the insurer would not be responsible for meeting that expense. The policy in question did, however, contain a clause requiring the insured to take steps to prevent losses 'at his own expense' and the court held that the insurer could not therefore be made to pay for preventative steps.

Where the insurance policy requires the insured to take reasonable steps to minimise a loss, it has been held that the insured would have to be at least reckless as to the action which is required before the policy becomes inoperative. The insured is not under a duty to take measures to avoid any liability which it did not foresee, nor take measures which a reasonable insured party would not take.

15.3. 'Claims made' basis

Professional indemnity insurance is on a 'claims made' basis; that is, the insurance policy that covers the claim is the one that is in operation when the claim is notified. The policy that is in force at the time an appointment is entered into, or when the breach occurs, will therefore be immaterial, unless the claim is made (notified) within the same period of insurance. This has to be borne in mind when considering whether certain risks can be accepted under an appointment – the risk may be covered by the professional's current professional indemnity insurance but will it be covered by the policy that is in place when a claim is made in the future? Also, limitations of liability may be agreed by reference to the terms of the professional's current professional indemnity insurance but that may have altered by the time a claim is made. Obligations to maintain professional indemnity insurance also need to be agreed on the basis that professional indemnity insurance terms may change over time – see Sections 15.15 and 15.16.

The difference between a claim and a circumstance is fundamental to understanding professional indemnity insurance and the disputes that can arise, between an insurer and an insured but also frequently between two insurers, in respect of two policy periods. Such policies are written on a claims made basis, as has been said, meaning they respond when a claim is made during a policy period against an insured. Some seek to add a further trigger, namely that the claim has to be made and also notified during the policy period. Claim will usually be defined in a policy but if not will be accorded its natural meaning, namely some communication by the client of discontent, which will result in some remedy expected from the insured. Unless defined, it does not have to be made in writing.

Professional indemnity policies also contain a deeming provision in respect of circumstances. Many policies define a circumstance as the discovery of some act or omission that is likely (i.e. greater than 50%) to give rise to a claim. Some define it as one that may give rise to a claim, a much lower threshold. Policies will contain a condition requiring all circumstances to be notified promptly. If a circumstance is notified during the policy period, then even if the claim is received several years later, it will be deemed to have been received during the policy period in which it was first notified.

The deeming provision is necessary because insureds are also under a duty to disclose material facts to insurers when renewing a policy. This will include any circumstance of which they are aware. A new insurer will exclude any circumstances disclosed or known about and without such a provision enabling a policyholder to notify his outgoing insurer the insured would have no cover when a claim eventually arrived.

Disputes frequently arise between insurers and insureds when claims or circumstances arrive at the end of a policy period and the insurer is changing. Further, where additional losses subsequently become apparent, arguments will ensue as to whether those losses fall within the scope of the claim as originally made or within the scope of the notified circumstance.

In *Kajima UK Engineering Ltd* v. *The Underwriter Insurance Company Ltd (2008)*, the insured notified its insurer of a circumstance in 2001 relating to a block of flats and movement that had caused roofing, balconies and walkways to distort. They changed insurers in 2002 and by 2005 it was clear the problems were much more serious and the block at risk of collapse. The insurers argued the wider problem was not within the scope of the notification and the insured settled the claim by purchasing the entire block and then sought an indemnity. The Court held the insurers were not liable as the notification was not sufficiently wide and had not been a 'can of worms' type notification. This is a good illustration of the care that needs to be taken when notifying insurers, and it is important to receive guidance from brokers and sometimes lawyers in advance of doing so.

15.4. Aggregation of claims

Where the policy provides for cover on an each and every claim basis, there will also be an aggregation clause under which claims in the same policy period may be aggregated together for the purposes of the limit of indemnity and the excess. This can occur, for example, where one negligent act in relation to an aspect of the professional's design gives rise to claims from more than one person, such as all the parties holding collateral warranties or where the same fault is repeated in a number of different buildings. In theory it has been held that different aggregation clauses can apply to both the excess and the limit of indemnity but in practice they are usually the same.

The types of clauses differ. Broadly speaking those allowing for all claims arising from 'one event or occurrence' to be aggregated allow the narrowest search for a unifying factor. At the other end of the spectrum clauses involving an 'originating source or cause' allow the least scope to aggregate claims. Clauses involving 'one act or omission' or a 'series of acts or omissions' are also common. Each will turn very much on the specific wording and all the facts giving rise to the claims.

Aggregation is important as it can quickly exhaust a limit of indemnity, meaning there is no cover left. Alternatively if claims do not aggregate and there is a large excess then all of the losses can fall to the insured.

15.5. Losing the right to indemnity

The right to an indemnity can be lost in numerous ways and professionals must take great care in their dealings with insurers. Failing to disclose a material fact or committing a misrepresentation on a proposal form can entitle an insurer to avoid the policy *ab initio*, such that the entire policy is treated as never having existed. This is extremely serious and leaves the professional uninsured and potentially uninsurable in future. Often proposal forms have a 'basis of contract' wording which also turns the information provided into warranties, breach of which would discharge an insurer from all liability as from the date of breach.

Policies will also contain numerous claims conditions, many of which will be conditions precedent, such as the duty to notify claims or circumstances promptly or the fact that no admissions are to be made without the insurer's authority. A breach would mean that the insurer would be discharged from liability for the claim in question.

As a general rule in dealing with insurers it is important to err on the side of caution and if in doubt to disclose or notify. In some policies insurers will waive some of their rights, by including, for example, an innocent non-disclosure clause that will only entitle them to avoid for a non-disclosure made fraudulently. Some of these are required by a minimum set of approved terms imposed by the relevant professional body but in respect of construction professionals only RICS members have this advantage.

Even very minor breaches can have disproportionate effects. Thus a condition requiring notification of a claim within 14 days is likely to mean precisely that. If notification takes place on day 15, the insurer may be entitled to reject cover for the claim (see, e.g., *Aspen UK Limited* v. *Pectel Limited (2008)*, where there was a requirement for immediate notice of any circumstance but there was a long delay). It is crucial that policies are read and the obligations understood, not just by the individual responsible for such matters but also much more widely throughout the organisation.

15.6. Limitations of professional indemnity insurance

Some of the most significant risks to professionals are those arising from claims that are outside insurance cover or within excesses where these are substantial, or which might arise because as explained above certain obligations of the insurance contract between the professional and the insurer have not been met.

The following features relating to typical indemnity insurance policies should be noted.

- Fines and penalties: insurers will not indemnify an insured against illegal acts or non-compliance with certain statutory requirements (e.g., the Health and Safety at Work, etc., Act). Thus they will not pay any fines or penalties imposed on the professional but some policies will pay for some or all of the defence costs of a prosecution because of the likelihood that a civil action will be brought against the insured if the prosecution is successful. It is against public policy to insure against criminal penalties.
- Liquidated damages: insurers will not indemnify an insured against liquidated damages (or penalties) as such. It is possible under some policies for an indemnity to be provided if the liquidated damages payable do not exceed the damages that would have been payable if assessed by the courts on the usual common law basis.
- Deliberate acts and omissions: insurers will not indemnify an insured against the consequences of deliberately risky actions. They must be fortuitous events.
- Legal liability: insurers are not concerned with events for which an insured is not legally liable. Legal liability is a matter determined by law and proving liability will be a matter for whoever is making the claim against the insured. So if a payment is made for commercial reasons where there is no liability, insurers will not cover this.
- The course of business: insurers will generally only consider claims arising out of the insured's pre-agreed recognised business. This is because insurers will have one set of rates, terms and conditions for, say, consulting engineers, and another for a firm undertaking contracting activities. Therefore, a professional should obtain the insurer's approval, usually through the insurance broker, for any proposed work outside what could be considered normal activities. Examples might be the direct hiring of plant, or works usually undertaken by civil engineering or building contractors such as the physical opening up of a structure by the professional for inspection purposes. The alternative is to act as an agent for a client in the engagement of such services, or for the client to employ contractors directly. If the professional agrees to employ a sub-consultant of a different discipline from his own, again the insurer's approval should be obtained. Insurers will normally require notice to be given of any joint venture arrangement and will decide in each case whether or not they are prepared to cover the arrangement. (Joint ventures, agency risks and vicarious liability are discussed elsewhere in the text. See also Section 15.7 on subrogation rights.)
- Warranties for fitness for purpose: clients (a term which here can include design and build contractors) can seek to impose on professionals a warranty for 'fitness

for purpose' – see Section 4.2. For the reasons explained in that section, professional indemnity insurers usually exclude such warranties specifically from cover or agree to indemnify the insured only to the extent that damages would have been payable in the absence of such a warranty. In 'soft' insurance market conditions (i.e., when business is slack and insurance premiums are low) it may be available for certain professionals, but there is never any guarantee that it will be included in future annual renewals, particularly if insurance market conditions harden. This does not mean that a professional cannot sometimes accept an uninsured risk, but that would be a commercial decision.

- Pollution and contamination: construction activities may result directly in creating pollution or, more indirectly, in disturbing natural pollution, or pollution from earlier activities which is already in the ground. There may be no risks, or another party other than the professional may accept the risks and/or a client may indemnify the professional. Where that is not the case, the professional should seek insurance cover. Unfortunately, there can be difficulty in professionals obtaining full cover for what is quoted typically as 'injury, damage or loss of use of property arising from seepage, pollution or contamination (including the costs of removing, neutralising and cleaning up afterwards), arising out of professional activities'. Professional indemnity policies, if they do not exclude pollution and contamination claims, often include a specific aggregate annual sub-limit on indemnity for all claims relating to such matters whether these arise in connection with one project or several. The professional therefore needs to be careful not to accept liability under an agreement for a particular project for an amount which, when combined with earlier liability exposures for similar projects that year, would exceed the annual aggregate liability insured under the policy. This requires careful management. What is called an 'evaporation clause' can sometimes be incorporated into professional services agreements. This seeks to limit the professional's contractual liability to whatever remains of the aggregate cover at the time of any claim. ACE Agreement 1 includes such an 'evaporation clause' as an option which limits liability for pollution and contamination claims, *inter alia*, to 'the amount, if any, recoverable by the Consultant under any professional indemnity insurance policy taken out by the Consulting Engineer' – see Section 16.9.
- Asbestos: in recent years large asbestos claims have been made against employers and previous employers. As far as professional indemnity insurance of construction professionals is concerned, insurers reacted by either excluding such claims or providing very limited cover. However, the risk of such claims is now perceived to have receded and many professional indemnity insurance policies now provide cover on an aggregate basis for asbestos claims but the total aggregate limit may be less than that available for pollution and contamination claims and/or exclude personal injury or cover particular costs only.
- Date recognition claims: as has been said, limited aggregate cover only may be available or no cover at all.
- Terrorism claims: again as a result of acts of terrorism in recent years, most professional indemnity insurance will exclude all claims arising out of terrorism and acts of terrorism, which can be very widely defined.

- Fungal or toxic mould: there have been a spate of claims in the USA and Canada arising from this problem, and again insurers reacted by excluding such claims from professional indemnity insurance policies in the UK. However, as at present such claims have not materialised, some professional indemnity insurance policies no longer have such an exclusion.
- Collateral warranties: some professional indemnity insurance policies have specific restrictions concerning the collateral warranties that can be entered into by the insured. For example, they may not provide cover in respect of any collateral warranty which permits more than two assignments or provides a warranty that goes beyond the obligations undertaken in the underlying appointment.
- Territorial limits: there can be territorial limits to the scope of the cover. For example, some insurers will not provide insurance or will provide it on a different basis for claims made in the USA or Canada.

15.7. Subrogation rights

Subrogation permits an insurer which has paid a claim (up to the limit of its indemnity to an insured) to pursue recovery of its payment from any third party it considers partially or wholly responsible for the claim arising in the first place. Professional indemnity policies are usually composite policies but a potential problem can arise for professionals in joint name policies where the parties are called 'co-insured', or where they are joint venture partners, because the respective interests of, say, a main contractor and a professional are different. The problem arises because in theory co-insureds are parties to the insurance contract, not third parties. The Courts have increasingly been reluctant to permit recovery, either because a clause in the policy expressly waives the right or because they will imply a term to that effect. This is usually circumvented by a 'cross-liability' clause inserted into a joint name insurance policy which permits the co-insureds to be treated as separately insured; that is, one can be treated as if it were a third party to the other. In such circumstances subrogation rights will apply. Attempts to delete or omit cross-liability in joint name policies arranged by a contractor or other joint venture partner in order to prevent the professional's insurers seeking recompense from them for monies paid out to the professional, should be recognised as unusual and advice should be sought before accepting such a deletion or omission. It is also very important to recognise that the underlying contracts between the joint insureds can prevent a recovery on the basis that an insurer is only subrogated to whatever rights an insured has.

15.8. Liability of partners

The liability of partners is unlimited. They are jointly liable for the debts and other obligations of the partnership (Section 9 of the Partnership Act 1890) and jointly and severally liable for the partners' wrongful acts and omissions such as negligence (Section 12 of the Partnership Act 1890). See Section 12.7 concerning joint and several liability generally.

15.9. Liability of directors

There is no joint and several liability between directors.

An individual director cannot be liable in contract to those who contract with the director's company, as the company is the other party to the contract. However, if a director gives a personal guarantee in respect of the company's obligation under any contract, the director could, of course, become personally liable.

A director can be sued individually in negligence but the claimant would have to show that the director himself owed a duty to take reasonable skill and care in the actions he performed or advice he gave. It would therefore be necessary to establish a close relationship or direct contact between the director and the events resulting in the negligent act.

15.10. Liability of limited liability partnerships

The Limited Liability Partnership Act 2000 allowing the creation of limited liability partnerships (LLPs) came into force in 2001. LLPs lie between a company, with its relatively inflexible structure and elaborate rules, but with limited liability, and a partnership, which has much more flexibility but where liability is unlimited. An LLP is registered at Companies House and is a legal entity separate from its partners. The partners are called 'members', and each is an agent of the LLP. They are able to organise their business on any basis they agree.

Measures to safeguard creditors are included in the Act. For example, there is a 'clawback' provision against members of withdrawals taken out of the LLP in the two years prior to liquidation. Members can also be liable for wrongful trading in the same way as a director of a limited company. Knowledge or deemed knowledge of the insolvency or imminent insolvency of the LLP is required to establish liability, so if a firm has been well run but meets an unexpected and sudden end, the risk of members having to contribute to the assets of the LLP through the operation of these provisions should be limited.

The minimum number of members is two, which is the number also needed to create a partnership.

There is no requirement in the legislation for compulsory professional indemnity insurance, nor is there compulsory professional regulation. An LLP is therefore available to any two individuals wishing to carry on any type of profession or trade together.

As with companies, LLPs have to make public certain financial information, including accounts, which must be audited if the size of the LLP is above certain thresholds.

The members are treated as self-employed (unless evidently in an employment relationship) and continue to benefit from the fact that PAYE does not operate on drawings, so their drawings from the LLP are not subject to payroll deductions for tax and national insurance. LLPs are also able to grant security over their assets by fixed and floating charges.

Members of LLPs have, as members, limited personal liability. Thus, creditors and claimants only have access to the assets of the business, which include any professional indemnity insurance taken out by the LLP, but not to the personal assets of any members. A member of a professional firm may be liable if they have been personally negligent, for example, in the advice he has given. This is comparable to the position of directors.

15.11. Liability of individual employees

Individual employees can become liable to third parties, whether this is the client or other people involved in a construction project in tort under the principles described in Chapter 14. They can also in certain circumstances owe a direct duty of care in negligence if they carried out the work or were in direct control or managed the work involved. Generally, they will not be sued direct because the claim will be made against the employer, for example the professional's company, partnership or limited liability partnership, and if there is sufficient professional indemnity insurance the claimant will not need to join in the individuals concerned. Also, for the most part, the employee will have an express or implied indemnity against claims from his employer unless he was acting outside the scope of his employment. Where this can result in direct claims against an employee is where there is no or insufficient professional indemnity insurance. This was illustrated by the case of *Merrett* v. *Babb (2001)*. Mrs Merrett asked a firm to carry out a survey on the property she proposed to buy and the survey was done by an employee of the firm, Mr Babb. The survey was negligent but when Mrs Merrett brought her claim the firm had become insolvent, there was no professional indemnity insurance and the administrators had not bought any run-off insurance. Mrs Merrett successfully sued Mr Babb in tort for negligence and he had to meet the claim from his own assets.

As a result of this case ACE Agreement 1 and RIBA 2010 provide that no claims can be brought direct against the professional's employees – see Section 16.9.

15.12 Public liability insurance

Public liability insurance policies are indemnity policies. They provide protection to the insured against any liability to the members of the general public, as opposed to employees of the insured, and cover claims resulting from physical actions causing accidental injury and damage. They are usually written on a 'claim occurring' or 'occurrence' basis: the relevant insurance cover will be that in existence at the time the event occurred.

It is again usual for such policies to contain limitations on the indemnity provided, so that the policy could stipulate a pecuniary maximum payment in relation to any claim or any period.

A public liability insurance policy will, typically, restrict the cover by reference to the business carried on by the insured, so that if an accident occurs the insured will only be covered if the accident arises out of the insured's stated activities. The policy will also typically specify the particular liability covered so that an insured incident must be of the specified type.

The importance of specifying liabilities in the policy is that the premiums payable under the policy depend on the probability of accidents occurring and the occupation and size of the insured's business.

Claims brought by an insured's employee when acting in the course of business will usually be covered by employer's liability insurance. If, however, an employee is not carrying out any identifiable professional activity when he is injured or injures a member of the public, the professional's public liability insurance may cover the liability.

It is common for professional indemnity insurance and public liability insurance to be taken out at the same time through the same broker, which, taken in conjunction with statutory compulsory employer's liability insurance, will assist to ensure there are fewer gaps or overlaps in the liability cover provided.

An accident occurring to a third party while an employee is driving a company vehicle in working hours would be covered by a motor policy, not by professional indemnity or public liability insurance. However, an employee of a professional organisation monitoring work on site, and who in the course of that monitoring causes injury to a contractor's employee or a member of the public, may be covered by professional indemnity insurance, as the professional's employee was engaged in professional activities at that time.

As both professional indemnity insurance and public liability insurance cover personal injury and damage to property, clients can become confused as to which policy might respond and can seek indemnities in relation to matters covered by a public liability policy when they intend to refer to professional indemnity insurance policies, and vice versa. Also, some limitation of liability clauses can exclude claims for loss and damage to property, as under PSC3 – see Section 9.14 – or seek separate limits for such claims. These claims would be covered by the professional indemnity insurance policy if the claim arises as a result of professional activities carried out by the insured.

15.13. Disclosure of insurance details

Most insurance policies will prohibit the disclosure of any detail of that insurance to clients or other third parties. However, it has become common for clients and contractors to request confirmation of insurance. Such details should only be released with the relevant insurer's permission and insurance brokers may well have a standard certificate giving details of the limit of indemnity (sometimes on a 'not less than' basis) and of the main limitations to that insurance such as aggregate limits for pollution and contamination claims only. Even though the professional may have substantial professional indemnity insurance the amount disclosed should be that appropriate for the particular commission and the fee payable. If a professional has £10 million professional indemnity insurance, because that is the amount he believes his practice requires because some of the firm's projects are very large, there is no need for him to disclose (and usually as a result agree to maintain) professional indemnity insurance at that level for a commission where, for example, his fees are £100 000 or less and the project worth £2 million. In such circumstances, £1 million or £2 million professional indemnity insurance should be sufficient.

15.14. Provision of Services Regulations 2009

The Provision of Services Regulations came into force on 28 December 2009 and requires the majority of private sector business providing services in the UK to make certain information available to clients and to deal with client complaints promptly. They also prohibit discrimination against individual clients on the basis of place of residence.

The services provided by engineers, architects and surveyors fall within the scope of the Regulations.

The Regulations include some requirements concerning professional indemnity insurance, but it is not entirely clear whether these apply to all construction professionals. They certainly apply to architects.

The relevant Regulations only apply to business services where the business is 'required' to hold professional indemnity insurance. This is obviously the case with architects and surveyors but is not so obviously the case with consulting engineers. If they were required by their professional code of conduct to hold professional indemnity insurance, they would be within the Regulations. It might also be the case where they have agreed through their professional appointments to have and maintain professional indemnity insurance.

It is probably best to adopt a cautious approach and assume that the insurance requirements do apply to all construction professionals.

The requirements are to give information about the cover, the contact details of the insurers and the territorial coverage. This information can be made available in one of four places, one of which is to 'make it easily available to the client at the place where you provide the service'.

A professional would therefore comply with this requirement if he had a notice in the reception area of his offices giving this information: that the firm has professional indemnity insurance, that it covers work carried out for clients from the relevant offices in the UK, and details of the territorial scope of the cover and the insurer's name and postal address.

It would be prudent to check that the relevant insurer is happy with this approach. The other ways in which the information can be given are

- 'to provide the client with such information on your own initiative'
- 'make it easily available to the client electronically'
- 'include it in any information or documents you supply to the client which set out a detailed description of the services you supply'.

These would require the information to be given for each proposed commission.

Certificates from insurance brokers may include more information than is required by the Regulations and so it is not recommended that these are used. However, if the information is made available at the professional's offices, any information that insurance brokers given to a professional's client, whether by a certificate about cover or otherwise, would be in addition to the requirements of the Regulations with which the professional has already complied and can therefore be in whatever terms are acceptable to insurers. In particular, there would be no need to refer to worldwide or any other territorial cover.

15.15. Insurance requirements of standard conditions of engagement

All main standard conditions now contain obligations to maintain insurance. ACE Agreement 1 includes both professional indemnity insurance and public liability insurance. It provides:

> F8.1 The Consultant shall maintain public liability and professional indemnity insurances in the amounts and for the length of time sufficient to cover the Consultant's liabilities under this Agreement provided always in either case that such insurances are available at commercially reasonable rates and subject to all exceptions, exclusions and limitations to the scope of cover that are commonly included in such insurances at the time the insurances are taken out or renewed as the case may be. The Consultant shall immediately inform the Client if such professional indemnity insurance ceases to be available on the terms required by this clause at commercially reasonable rates in order that the Client and the Consultant can discuss means of best protecting their respective positions in the absence of such professional indemnity insurance.
>
> F8.6 As and when reasonably requested to do so by the Client, the Consultant shall produce for inspection brokers' certificates to show that the insurance cover required by clause F8.1 is being maintained.

This is a simpler version of earlier clauses. As it now refers to 'the amounts and for the length of time sufficient to cover the Consultant's liabilities under this Agreement' there is no need for a specific amount nor for a time period to be put in the Particulars. Nor is there any longer a reference to pollution and contamination claims, nor to the absence of insurance for terrorism claims. The words 'subject to all exceptions, exclusions and limitations to the scope of cover' would cover all these and allow for different or additional limitations or exclusions in future professional indemnity insurance policies.

When considering what is sufficient time: the professional will need to maintain professional indemnity insurance until the expiry of the agreed time limit for claims under clause F7.8. Otherwise claims could be brought in contract for up to six years from the last possible breach of contract if the appointment is signed under hand and 12 years if executed as a deed. The professional might also consider whether 15 years would be prudent as this is the longstop for claims in negligence.

It is usual to include an obligation to inform the client if professional indemnity insurance or public liability insurance policies are no longer available, which could happen if the firm is wound up or goes into liquidation or if the premiums being demanded were simply prohibitive. This could give the client an opportunity to seek project insurance or perhaps to buy run-off insurance.

It is also usual for a professional to agree to give evidence of his professional indemnity insurance and public liability insurance from time to time by way of insurance broker's certificates as set out here.

RIBA 2010 deals only with professional indemnity insurance but in similar terms, providing:

> 7.4 The Architect shall maintain until at least the expiry of the period specified in clause 7.1 professional indemnity insurance with a limit of indemnity of not less than the amount or amounts specified in the Project Data provided such insurance continues to be offered on commercially reasonable terms to the Architect at the time when the insurance is taken out or renewed.
>
> Such insurance shall be subject only to such other limitations, exceptions and exclusions as are commonly included in such policies.
>
> 7.5 The Architect, when reasonably requested by the Client, shall produce for inspection a broker's letter or certificate confirming that such insurance is being maintained.
>
> 7.6 The Architect shall inform the Client if such insurance ceases to be available at commercially reasonable terms or subsequent to the date of this Agreement an aggregate limit applies to any matters other than those specified in the Project Data in order that the Architect and Client can discuss the best means of protecting their respective positions.

It will be seen that this wording is comparable to that in ACE Agreement 1 save that it only relates to professional indemnity insurance and the professional has to put the amount and the period in the project data. It is assumed in the project data that the professional's professional indemnity insurance will be on a per claim basis with annual aggregate limits for pollution and contamination and/or asbestos or fungal mould, which may or may not be the case. As to the continuing obligation to inform the client if the insurance ceases to be available or if an aggregate limit applies other than those specified in the project data, there does not seem to be an obligation to inform the client if any other restrictions become attached to the policy.

The date specified in clause 7.1 is the date of the last services performed or if earlier practical completion of construction of the project.

RICS 2011 in relation to professional indemnity insurance only provides:

> 12.1 Without prejudice to its obligations, the Consultant effects and maintains professional indemnity insurance as specified in the Appendix for the Relevant Period subject to such insurance being available in the insurance market on reasonable terms and rates. Such insurance is to be with a well-established insurance office or underwriter of repute.

12.2 If the specified insurance is not available on reasonable terms and rates or is not maintained in accordance with this Appointment, the Consultant notifies the Client immediately and the parties discuss the best way to protect their respective positions, having regard to the provisions of this Appointment and the status of the Project at the time.

12.3 The Consultant produces written evidence that the specified insurance is being maintained whenever reasonably requested to do so by the Client.

The appendix refers to 'Such amount as is maintained by the Consultant on the date of this Appointment' which, although brief, does not give any indication as to what that might be so as to be able to ascertain whether any later changes have been made.

The 'Relevant Period' is to be six years from practical completion if the appointment is signed under hand and twelve years from practical completion if executed as a deed. This may not always be an appropriate starting point from which to determine the period.

As can be seen, all obligations in the standard conditions of engagement (and also in the CIC forms of collateral warranties) are limited so that the insurance is to be maintained 'as long as such insurance is available at commercially reasonable rates' or similar wording. This protects the professional against swings in the professional indemnity market when premiums could become prohibitively expensive, or even not available at all.

Professional indemnity policies held by many professional firms are usually professional 'group' policies underwritten by several lead insurers covering a pool of similar insureds. The policy is not available for inspection by those outside the pool. It is normal to satisfy clients and others by means of a broker's certificate – the policy should rarely, if ever, be shown to third parties.

15.16. Insurance obligations in non-standard conditions of engagement

The clauses in non-standard conditions of engagement may be more detailed than those in the standard forms and need to be considered on their merits. First, they should accurately describe the professional's current professional indemnity insurance. The obligation should not last an unreasonable length of time, nor should the production of a policy be the only acceptable evidence of satisfactory insurance.

The limit(s) of indemnity included in the clause should be as low as is possible, commensurate with the risks for that particular project should the professional be found to have been negligent. For example, the risks of poor design on a short length of estate road are different from those on several miles of motorway. The limit of indemnity should not be the amount that the professional happens to carry at the time of the appointment. That is irrelevant and, indeed, it may not be the amount available when the claim is made.

An obligation to maintain insurance cover to a certain level for a project is not the same as limiting liability to that amount. If liability is to be limited, this needs to be agreed and recorded separately in the appointment – see Chapter 16.

Other potentially onerous elements of clauses requiring insurance to be maintained are as follows.

(*a*) Those stipulating the amount of excesses. These can vary and are a matter for negotiation between the insured and the insurer. If the excess required is lower than that which the professional would normally agree with insurers, this could result in a higher premium.
(*b*) Those requiring different cover from that normally contained in the insurance policy. The professional will be obliged to find and pay for such insurance or be in breach of contract.
(*c*) Those stipulating the use of a particular insurer or requiring the client's consent to an insurer. This might restrict a professional's ability to find the most suitable and economic professional indemnity insurance. The same is true of a prohibition on obtaining insurance from outside the UK, for example.
(*d*) Those stipulating an excessive length of time for which the insurance is to be maintained – see Section 15.15.
(*e*) Those restricting the professional's rights in relation to any claim which has been made under the policy; for example, an obligation not to settle or compromise any claim or enter into any agreement with the insurer without the client's consent. This is usually expressly prohibited by professional indemnity insurance.
(*f*) Those requiring the professional to hold any money in respect of any claim from the insurer on trust for the client. Other parties, such as co-defendants and the professional's legal advisers or expert witnesses, may have an entitlement to that money.
(*g*) Those saying that the insurance should not have any unusually onerous terms and conditions. Who is to determine whether any terms are 'unusually onerous'?
(*h*) Those stipulating that the insurance has to be taken out at rates other than those that are commercially reasonable. The test should be what is available to that professional and what is commercially reasonable for the professional to pay as it is he who has to buy the professional indemnity insurance. The test should not be, for example, whether professional indemnity insurance would be available to other similar professional firms or the professional generally at different rates. Sometimes the clause may provide that when determining whether or not professional indemnity insurance is available at commercially reasonable rates, no account is to be taken of any effect on that premium of the professional's claims record. That is not acceptable. First, the professional indemnity insurance available to a professional is inevitably going to reflect his claims record. Secondly, the insurers may find it impossible to separate out that part of the premium that does not reflect the claims record. All sorts of factors go into determining the level of premium.

All of these points become more important when the professional indemnity market hardens and premiums generally increase. This could happen at a time when the profitability of the professional is falling.

Some non-standard conditions of engagement allow the client to obtain insurance if the professional cannot provide evidence of continuing cover, and deduct any premium from amounts due to the professional or recover it immediately as a debt. This could result in the professional paying much more than might otherwise be necessary. (The client will not be able to take out professional indemnity insurance as such, but it may be able to take out a project policy or decennial policy to achieve a similar result.)

15.17. Run-off insurance

When a firm ceases to practise, winds itself up or becomes insolvent, it will not be able to continue to pay for professional indemnity insurance on an annual basis, but can still face claims for as long as the law allows. This could be 15 years or more under the Limitation Act 1980. In the case of a company or limited liability partnership, this may not matter as neither will have any assets to meet any claim, but it is possible that claims could be made against former directors or members, or even former employees. As for partners, they or their estates will continue to be liable individually even though they may have retired or died.

In such circumstances, it is prudent to consider whether a professional indemnity insurance policy could be bought to cover any claims that might come in after the firm ceases to practise. These are known as run-off policies, and typically for a single premium will provide a certain level of cover for a certain number of years. They can also be renewed annually.

Partners who retire or leave a firm which is continuing to practise will need to consider obtaining the agreement of the continuing partners to maintain sufficient professional indemnity insurance for a sufficient number of years to protect the retiring or former partners should claims be made against the partners arising out of matters when he was a partner.

15.18. Value of reducing number of claims against the professional

Insurance is not cheap. Generally, in each area of insurance (e.g., professional indemnity, public liability, motor, etc.) what is paid in as premiums represents what is paid out as claims, but with insurer's administration costs, profit margins, brokers' fees and so on added. Overall any 'pool' of insureds pays in substantially more in premiums than it gets back through the settlement or payment of claims. The fewer claims there are, the lower the premiums needed to be paid. Professionals with good claims records may individually be able to negotiate lower premiums. As already described, insured claims have non-insured elements (that is, excesses and upper limits) which have to be met by the professional. In addition, even successfully defended insured claims will involve considerable time and effort, usually by senior members of staff that will not be recoverable in whole or in part. Once again, the fewer insured claims there are the better for the profession as a whole and for professional organisations individually. It is generally accepted to be cost-effective to attempt to reduce the possibility of insured claims, and to rely on insurance solely as an ultimate safety net.

15.19. Choice of insurers

The choice of insurers is usually made by a professional in conjunction with a broker. Several professional and 'trade' organisations have their own schemes of insurance which can more closely match the requirements of their particular members and provide the commercial advantage of cheaper premiums as a result of their large memberships and resultant premium income. However, insurers are not 'bottomless pits' of money. In the recent recession, some insurance companies have folded. In 1996, one specialist professional indemnity insurer ceased trading when its claims became consistently larger than its premium revenue. In effect, it was a 'mutual' and was self-financing, and to pay its claims it legitimately approached its (mainly) structural engineering insureds for more money. This mutual status and the associated obligation to pay more money was unexpected for many of those concerned and caused difficulties for both insured and insurers. The moral must be: choose insurers with care and read and understand any small print!

Professional Services Agreements
ISBN 978-0-7277-4158-5

http://dx.doi.org/10.1680/psa.41585.261

Chapter 16
Limitation and exclusion of liability

It is important to understand that the liability of professionals in respect of claims in contract and in tort (which are the most likely basis for claims against professionals) is unlimited: that is, the professional is liable to the full extent of his assets including (but not restricted to) his professional indemnity insurance. In the case of a company, this will be all the company's assets as well as its professional indemnity insurance. The liability of individual partners, directors, etc. is discussed in Sections 15.8 to 15.11. Having or agreeing to maintain professional indemnity insurance of a certain amount does not limit the professional's liability to that amount.

Most standard conditions of engagement therefore include clauses limiting or excluding liability. It is becoming more usual for the parties to agree limits of liability in non-standard conditions of engagement. Such clauses can be vital where the project is very risky or where the risk is out of all proportion to the fee, as can happen, for example, in environmental assessments. It is first necessary to set out some background before considering the detail of some clauses.

16.1. Limiting or excluding liability generally and the Unfair Contract Terms Act 1977

Parties to a contract can never effectively agree to limit or exclude liability for damages for personal injury or death because this is expressly prohibited by Section 2 of the Unfair Contract Terms Act 1977 – see below. They are, however, able to agree to limit or exclude any other liability they may incur to each other (subject to some restrictions imposed by the Unfair Contract Terms Act 1977 discussed later in this section). It is possible to exclude or restrict liability in relation to third party claims by notice, again subject to the Act, but not through contractual terms because there is no contract with that third party. Thus, until recently, if a claim was made against a professional by a tenant of a building the professional has designed under a contract with the building owner, the professional was not able to rely on any limitation of liability agreed with the building owner because that limitation was not contained in a contract with the tenant or agreed by the tenant. The only way a professional could be protected against such third party claims was if the client agreed to give the professional an indemnity against them. The Contracts (Rights of Third Parties) Act 1999, however, allows the parties to a contract to limit liability to third parties who have been given the right within that contract to enforce a term of that contract – see Section 2.3.

Where the professional enters into an appointment direct with third parties, for example, in a collateral warranty, a clause limiting or excluding the liability of the professional could be included and could be effective.

Clauses excluding or limiting liability are governed by Sections 2 and 3 of the Unfair Contract Terms Act 1977.

Section 2 says that a person cannot by contract or notice exclude or restrict liability for death or personal injury and that a person's awareness of or agreement to a contract term or notice purporting to exclude or restrict liability for negligence is not of itself to be taken as indicating the voluntary acceptance of any risk. Its most important provision as far as professionals are concerned is that in cases other than those concerning liability for personal injury or death 'a person cannot so exclude or restrict his liability for negligence except insofar as the term or notice satisfies the requirement of reasonableness'. What could constitute a reasonable exclusion or limitation of liability in professionals' appointments is considered later in this chapter.

Section 3 states that a party in breach of contract cannot rely on a clause excluding or limiting liability in respect of the breach unless the term is 'reasonable'. Section 3, however, only applies where one of the contracting parties is a consumer or where a party is dealing on the other party's 'written standard terms of business'. A 'consumer' is defined in Section 12 of the Act as someone who does not make the contract in the course of a business, nor holds himself out as doing so.

There is no definition of the phrase 'written standard terms of business' in the Act, and the courts have considered the situation where there have been negotiations over some standard terms.

The courts (*Salvage Association* v. *Cap Financial Services (1995)*) decided in that case that Section 3 would not apply where the standard terms were used as a starting point for negotiations, and the contracting parties had equal bargaining power.

The courts (*St Albans City and District Council* v. *International Computers Limited (1996)*) also decided that the terms of business were essentially a question of fact, although the Court of Appeal indicated that any negotiations that take place before the contract is made should be disregarded. It was held that the word 'deal' meant 'makes a deal', irrespective of any prior negotiations and it is only necessary for the contract to be made on those terms.

Where the parties use standard conditions or a model form issued by a professional institution or other body, this will not necessarily constitute 'the written standard terms of business' of the party relying on them. The courts held (*British Fermentation Products Limited* v. *Compare Reavell Limited (1999)*) that, although an agreement incorporated the Institution of Mechanical Engineers Model Form of Contract Form C, it was not enough to show that the standard form was used in one instance. The form had to be used 'invariably or at least usually' by the party concerned. Since this could not be established the Model Form was not, in that case, 'written standard terms of business' for the purposes of the Act.

The Act is predominantly aimed at the supply of goods and hire purchase agreements, which is why there are few cases that help in establishing principles for construction contracts.

The Act also applies to a restriction of the duty which gives rise to a liability or which would render a contractual performance substantially different or no performance at all. For example, it would be unreasonable if a professional prepared a report for a client and then said it took no responsibility for its accuracy. However, stating that the report is for a particular purpose (e.g., it is for valuation purposes and is not a structural survey) is not restricting the duty of care undertaken by the professional in relation to that report.

As to the terms which could affect liability and which the Act says must satisfy the test of 'reasonableness', these include conditions.

(*a*) As to time limits within which to make claims – see Section 16.6.
(*b*) Excluding or restricting other rights, for example, to reject goods.
(*c*) Excluding or restricting evidence or procedure, for example, 'if you accept these goods this is conclusive evidence as to their fitness for purpose'.

Terms in professional appointments which may be assessed for 'reasonableness' in accordance with the Act include the following.

(*a*) Terms which exclude or restrict liability in negligence – see Section 4.1 – this will cover both a contractual term 'to use reasonable skill and care' (this is not restricted to where a party is dealing with a consumer or on standard terms) and where there is a common law duty to use reasonable skill and care.
(*b*) Indemnities by a person dealing as a consumer to indemnify another party in respect of a liability which may be incurred by that other party for negligence or breach of contract. However, indemnities given by persons who do not deal as a consumer are unaffected by the Act.
(*c*) Terms which limit liability to a certain amount.

The Act says in relation to such terms that a term has to be 'a fair and reasonable one to be included having regard to the circumstances which were or ought to have been known or in contemplation of the parties when the contract was made'.

It is therefore very important that clauses limiting liability are considered at the time the appointment is being negotiated.

Certain criteria have been put forward for deciding in particular cases whether or not a clause restricting or excluding liability is 'reasonable'.

(*a*) It has to be clear because, if it is not, it is construed very strictly against the party relying on the clause to restrict or exclude its liability.

(*b*) A clause is more likely to be considered 'reasonable' if it restricts liability rather than excludes it. Liability, therefore, should be totally excluded only in rare cases – see Section 16.5. However, a clause limiting liability to a small sum may be treated by the courts as if it was an exclusion clause.

(*c*) A clause restricting or excluding liability should be discussed with a client before an appointment is finalised. The risks and the level of limitation proposed need to be discussed and it is prudent to document the discussions. This will help to establish that the client understood the sort of liability the professional could incur when agreeing to the limitation and therefore accepted that the clause was reasonable.

(*d*) Where a limitation clause is in a body of conditions drafted by a professional body and widely used, this could show it was 'reasonable', particularly where the other party (i.e. not the one relying on the limitation) has showed a willingness to contract on these conditions.

(*e*) The availability and amount of insurance is very important. In some cases it has been thought that if there was a correlation between the limitation clause and the insurance available, this would demonstrate that the clause was reasonable. For example, the courts held (*George Mitchell* v. *Finney Lock Seeds (1983)*) that: 'Breathtaking sums of money may turn on professional advice for which it is impossible to obtain adequate insurance cover and which would ruin him personally if liable'. As explained in Section 16.2, the availability of insurance will be considered by the court in assessing the reasonableness of a clause limiting liability to a specific monetary sum. Thus, it helps if a clause limiting liability is combined with an obligation to maintain professional indemnity insurance for that sum.

(*f*) If liability for claims in negligence are to be limited as well as claims in contract, including the contractual duty of care, then it is advisable for 'negligence' to be stated specifically and not refer, for example, only to claims in tort. The case of *The Canada Steamship Lines* v. *R (1952)* set out three principles:

  (*i*) If an exclusion clause expressly refers to negligence, effect will be given to its terms.

  (*ii*) If the clause does not expressly refer to negligence, the court will consider whether the words used are wide enough, in their ordinary meaning, to cover negligence.

  (*iii*) the clause will generally not exclude liability for negligence if the words cover a possible head of damage other than negligence. This is based on *dicta* from Lord Greene in *Alderslade* v. *Hendon Laundry Limited (1945)* that 'if a contracting party wishes to limit his liability in respect of negligence he must do so in clear terms, in the absence of which the clause is construed as relating to liability not based on negligence'.

Clauses which seek to limit or exclude liability for fraudulent misrepresentation have been held to be ineffective and it is common to exclude from limitation clauses in bespoke appointments claims arising from fraud or gross negligence.

As to whether a limitation clause can cover claims by third parties as well as the client, see comments on the CIC limitation provisions in Section 16.9.

16.2. Limiting liability by reference to a monetary amount

It is perfectly permissible to limit liability by reference to a monetary amount, but in determining whether or not the limit is 'reasonable' the courts are required by the Unfair Contract Terms Act 1977 (Section 11(4)) to have regard to

(*a*) the resources that the professional could be expected to have available for the purpose of meeting the liability should it arise, for example, the assets of the company or partners
(*b*) how far it is open to the professional to be covered by insurance.

As it is usually open to a professional to obtain professional indemnity insurance, a failure to do so would probably make any clause limiting that professional's liability 'unreasonable'.

It is important that the amount to which liability is to be limited is not calculated by reference to a formula, such as ten times the fee, or to the same amount for every appointment. Each project needs to be considered separately on its merits and the amount should be an agreed monetary amount appropriate for that project. It also needs to be remembered that an aggregate limit for all claims under that contract will only apply if this is expressly stated in the limitation clause.

The validity of specific limitation clauses was considered in detail by the Court of Appeal (*St Albans City and District Council* v. *International Computers Limited (1996)*). In that case, the St Albans council ordered a computer system from International Computers Limited to calculate the level of community charge that should be set by the authority and to carry out the administration and collection of the charge. Due to an error in the system, the computer over-estimated the number of chargepayers in the region with the result that the council collected through the community charge significantly less than required, but still had to pay out the amounts that should have been recovered. The council claimed damages, although a limitation clause in the contract purported to restrict the level of damages payable by International Computers to £100 000.

At first instance the court applied the provisions of Section 3 of the Unfair Contract Terms Act 1977 and, as the contract was made on the defendant's written standard terms of business, the limitation clause would only be valid if it was 'reasonable'. The court applied the Section 11(4) guidelines to assess this, noting that the defendant was a substantial company with ample resources to meet any liability, and that the defendant held product liability cover of £50 million worldwide. The court further noted that the defendant had called no evidence to show that it was fair and reasonable to limit its liability under this contract to £100 000. It therefore appears that any monetary figure used in a limitation clause cannot simply be chosen at random; it must be a figure that can be justified objectively.

The court, when considering whether it was better that the loss should fall on a local authority or on an international computer company, stated:

> The latter is well able to insure . . . and pass on the premium cost to the customers. . . . I do not think it unreasonable that he who stands to make the profit should carry the risk.

The court therefore decided in favour of St Albans City and District Council and the Court of Appeal refused to interfere with this finding.

In deciding the maximum liability it is prepared to accept, a professional should assess the nature of the risks for each particular project and also the damages that could be payable if the professional is in breach of contract or is negligent. These will include not only the costs of putting things right but also all the other losses that could be caused to the client, for example inability to use or rent its building. An assessment would also need to be made of the damages that could flow from any particular matters, such as pollution and contamination claims.

The amount of professional indemnity insurance cover is a relevant factor but this does not mean that if the professional has £5 million insurance cover, the limit of liability for every appointment should be £5 million. The monetary amount included in the limitation clause should be appropriate for the type of commission being undertaken.

An obligation in an appointment to maintain professional indemnity insurance does not limit a professional's liability to the amount of cover provided by that insurance. A separate, express clause is needed, agreeing that liability should be limited. It needs to be remembered also that it is the policy in force when a claim is made that will provide the indemnity (if any) for that claim, not the policy in force at the time of making the original agreement with a client – see Section 15.3.

The courts (*Moores* v. *Yakeley Associates (1999)*) have considered the effect of a monetary limitation in the RIBA's standard conditions SFA/92. The court held that a limit of £250 000 was reasonable because it was not an arbitrary figure but based on the architect's assessment of the likely cost of the works; the fees were in the order of £20 000 and the ceiling was ten times that amount; the client was in a stronger bargaining position than the architect – he could have instructed any architect; the client and his solicitor had both been aware of the clause and had had an opportunity to object and a comparison of their respective resources showed that the architect had none and the client was very wealthy. It did not matter that the architect had professional indemnity insurance in excess of £250 000. This decision was upheld on appeal.

16.3. Limiting liability by reference to 'net contribution'

Where joint liability can arise a net contribution clause can be helpful – see Sections 12.11 and 12.12. There is no reason why a clause limiting liability cannot use the amount of the professional's net contribution as a means of establishing a maximum that should be paid in relation to a claim. It needs to be remembered, however, that such clauses have not been tested by the English courts, either as to their efficacy or their 'reasonableness'.

The Scottish courts considered whether an engineer would be excluded by those parts of the Unfair Contract Terms Act 1977 which apply in Scotland from relying on the net contribution clause in the ACE Agreement 1998 B(1) in the case of *Langstane Housing Association Limited* v. *Riverside Construction (Aberdeen) Limited (2009)*.

The pursuer argued that the net contribution clause fell within Section 16(1)b of the Scottish Act. Section 16(1)b of that Act states that 'Where the term of the contract purports to exclude or restrict liability for breach of duty arising in the course of any business . . . , that term . . . shall . . . have no effect if it was not fair and reasonable to incorporate the term in the contract'. The court adopted the argument put forward by the engineer that a net contribution clause does not seek to exclude or restrict liability for breach of duty within the meaning of Section 16(1)(b): the net contribution clause seeks to ensure that the engineer is only held liable for the consequence of his own breach of duty and was not held liable by joint and several liability for the breaches of duty by others and thus Section 16(1) did not apply so as to impose a 'fair and reasonable' test in this case.

As the decision comes from the Scottish Outer House Court of Session, it is not directly applicable in the English courts but it will have weight.

The ACE Conditions which were applicable in that case were the 1998 version. The wording for the net contribution clause in Clause B8.2 of the 1998 version is not the same as Clause 7.5 of ACE Agreement 1. Therefore, any weight given to the decision cannot be assumed to apply equally to other versions of the clause.

Although there was no detailed consideration of the wording of the net contribution clause itself, the judge did appear to assume it worked in the manner it is generally thought such a clause would work. Also, while not deciding that net contribution clauses generally will be enforceable and would be outside the Unfair Contract Terms Act, it is useful in showing that in a particular case it was enforceable.

Sometimes non-standard conditions of engagement include a provision that where there is a design/build contractor the professional's entitlement to rely on a net contribution clause is reduced or otherwise limited. This is misconceived – see the comments on the proviso to RICS 2011 net contribution clause below.

It also needs to be remembered that a net contribution clause can only apply where the claim is such that two or more parties are jointly 'liable with the professional for the same damage'. If the professional is found to be solely responsible for the damage arising out of that claim, such a clause will not be effective in limiting liability at all and a general limitation of liability clause would be needed.

16.4. Limiting liability by reference to particular losses or direct, indirect or consequential losses

Limiting liability by reference to other matters can be effective if the limitation is reasonable. For example, liability could be limited to the costs of repair or of cleaning up a site, or to the amount recoverable under professional indemnity insurance (an 'evaporation clause') – see Section 16.9. Certain types of damages can be excluded or limited, such as relocation costs, loss of profits or consequential losses.

If the liability is to be limited by reference to the sort of damages payable, this needs to be done carefully with appropriate advice. It is worth referring to two of the cases which have dealt with 'direct' and 'consequential' losses. The courts (*British Sugar plc* v. *NEI Power Projects Limited and Another (1997)*) decided that 'consequential' losses were those other than 'normal' losses; that is, the 'normal' loss is that which every claimant in a like situation will suffer, and 'consequential' loss is that loss which is special to the circumstances of the particular claimant. Thus, 'consequential' losses would cover indirect damages and, therefore, might not cover loss of profits or loss of rent, for example. If a professional agreed with its client that 'consequential' losses would be excluded, indirect damages only would be excluded.

This approach was confirmed by the courts (*Deepak Fertilizers & Petrochemical Corporation* v. *Davy Kee* (1) *ICI Chemicals* (2) (1998)), when loss of profits and wasted overheads were found to be 'direct' damages and not indirect or consequential damages.

Another Scottish case has considered the effect of a provision limiting 'liability for costs' in a collateral warranty given by professionals employed by the contractor to the ultimate client – *Glasgow Airport* v. *Kirkman & Bradford (2007)*. Clause 1 of the warranty stated:

> In the event of any breach of this warranty:
> The Sub-Consultant's liability for costs under this Agreement shall be limited to that proportion of such costs which it would be just and equitable to require the Sub-Consultant to pay having regard to the extent of the Sub-Consultant's responsibility for the same . . .

There were problems with the floor slab and when the contractor became insolvent Glasgow Airport sought to recover its outlay under the collateral warranty provided by the consultants, Kirkman and Bradford, for failure to exercise reasonable skill, care and diligence in the performance of their duties.

Kirkman and Bradford denied liability, maintaining that clause 1(a) of the warranty acted as a net contribution clause and the effect of 'liability for costs', referred to in clause 1(a), was that the pursuers could only recover 'costs' of repair or reinstatement of the floor slab. Kirkman and Bradford maintained that their liability did not extend to consequential loses. Glasgow Airport contended that Kirkman and Bradford's liability under the warranty was unrestricted and that they were therefore entitled to recover all losses arising from its breach (subject to the rules of remoteness).

It was held that it made commercial sense that 'liability for costs' described Kirkman and Bradford's whole liability under the warranty. The common terms used in the net contribution clause did not suggest an intention to restrict Kirkman and Bradford's liability from that which would otherwise flow from the terms of the agreement and there was no express provision that Kirkman and Bradford's liability was limited to costs.

It will be seen from this that reliance on general terminology is dangerous if the aim of a clause is to exclude liability for consequential losses.

It might also be 'reasonable' to exclude certain types of losses.

In *Regus (UK) Limited* v. *Epcot Solutions Limited (2008)*, the Court of Appeal gave a decision holding a standard business term excluding liability to be reasonable under the Unfair Contract Terms Act. The contract in question was for the supply of serviced office accommodation, including air conditioning. The contract term in question excluded liability for all business losses and loss of profits and consequential losses generally.

The judge at first instance had held this term to be unreasonable, because it effectively excluded any liability for the services, such as the air conditioning (which was alleged to be defective in this case). It was conceded, however, on the appeal, that the judge had been wrong on this point, since the term did not exclude liability for the diminution in value of the services, which would normally be the main measure of loss for defective services.

Having established that, the Court of Appeal held that the exclusion of liability for the particular losses mentioned in the term was reasonable.

This finding turned, of course, on the particular circumstances of this contract and the relationship between the parties. One point made was that it is probably easier for the customer to insure against its own business losses caused by defective equipment than for the supplier of the equipment to attempt to insure against those losses through liability insurance.

A full discussion of the damages that may flow from any breach of a professional's appointment is outside the scope of this text.

16.5. Exclusion of liability

In exceptional cases it can be 'reasonable' to exclude liability in respect of some particular claims or matters. For example, if a professional has not been asked to consider pollution or contamination in relation to a particular project because the client is taking separate specialist advice about this, it could be reasonable to state in the appointment that the client is doing this and any liability of the professional for any claim arising out of, or in connection with, pollution and contamination is excluded. However, if the professional has undertaken any services in relation to any particular matter it may well not be thought to be reasonable to exclude liability even if the professional has limited or no professional indemnity insurance. In such circumstances, the professional should seek to agree separate lower limits for such claims.

16.6. Limiting the time within which claims can be brought and time limits for claims

Time limits must also satisfy the requirement of 'reasonableness' and this may prove difficult because there are different time periods for bringing claims in contract and in negligence.

Time limits which relate only to claims in contract will more readily satisfy the test of reasonableness – six or 12 years might be reasonable, depending on whether the contract is under hand or executed as a deed – see Section 3.14 – because such claims can only be brought within six or 12 years from the date the cause of action accrues. If the work to be carried out under a contract is of short duration then a shorter time limit might be reasonable.

It is possible for claims in negligence to be brought after a claim in contract would be time-barred. A limitation to bringing claims in negligence to, say, six years from completion of the services might not be considered 'reasonable' as it could be a sizeable reduction in the position under the law at present, which is

(*a*) six years from the date of the accrual of the cause of action; or
(*b*) if longer, three years from the date when the claimant knew or should reasonably have known of the damage and its being caused by the defendant's negligence; with
(*c*) a long stop of 15 years from the date of the last act of negligence to which the damage is attributable (except in relation to fraud or deliberate concealment).

Thus, if a client only became aware of the defects in year 12 after completion of a project and it had been agreed that claims could only be brought at the end of year six, that claim would be excluded before the client had any chance of knowing about it. (However, 15 years is itself an arbitrary time limit and it may not be unreasonable to reduce the length of time within which claims can be brought if the long-term availability of certain types of insurance, including professional indemnity insurance, is in doubt.)

Again, if claims in negligence in addition to claims for breach of contract, including, in the case of professionals, a breach of the contractual duty of care, are to be subject to a time limit, it is advisable to refer to such claims specifically – see Section 16.1. It may also be advisable to include any claims of set off in any clause limiting the time for claims.

The recent case of *Robinson* v. *P E Jones (Contractors) Limited (2011)* illustrates the arbitrary effect of such time limits as well as how liability can be incurred.

In that case, in December 1991 the contractors agreed to construct and sell a house to the claimant and his wife. The transaction was completed in April 1992.

In September 2004, the claimant was advised that the construction of flues for the gas fires was defective and that they needed to be reconstructed. The claimant complained to the contractors and, in December 2006, commenced proceedings against them.

The problem for the claimant was that a claim against the contractors for breach of contract was obviously too late: the period of liability for the contractors under the NHBC's standard form agreement had long since expired; and, under the general law of limitation, the time for bringing a claim for breach of contract is, at most, 12 years from the date of the breach.

However, the claimant might have been able to avoid this problem if he could have brought a claim against the contractors based on the tort of negligence. For such claims, as has been said, there is a three-year limitation period that runs from the date the claimant knew of the existence of a potential claim, subject to the 15-year-long stop. The claimant's claim was brought within both the three-year and 15-year period.

The contract provided that the contractors could only be liable to the claimant under the provisions of the NHBC agreement. The contractors argued that this excluded the right of the claimant to bring a claim in tort. The claimant argued that this was not the case and that this provision was anyway unenforceable due to the Unfair Contract Terms Act. The court found in favour of the contractors on both these points, which meant that the claimant's case failed. A more fundamental issue was whether the claimant could bring a claim in tort at all.

The court said the contractors had not assumed responsibility to the claimant in the Hedley Byrne sense – see Section 14.4. It was a 'normal contract', whereby the contractors would construct a house to an agreed specification and the warranties of quality and the claimant's remedies were set out.

The reasoning is not altogether clear. The court placed emphasis on cases in the House of Lords in which it was held that contractors do not owe a duty in tort in respect of economic loss. However, those cases were concerned with loss suffered by third parties and not by the other party to a contract. One would have thought that contracting parties would assume responsibility to discharge contractual obligations to each other and thus potentially become liable for economic loss.

Perhaps the key point from this case arises from the different obligations of a contractor and a professional. The duty in tort is a duty of care – see Section 4.1. The contractual duty owed by contractors to prevent economic loss (which in effect means the duty to construct the building without defects) is not a duty of care as such: it is a strict duty and so it is actually more onerous than a duty of care. But if it is not a duty of care in contract, it follows that it cannot also be a duty owed in tort. By contrast, the contractual duty owed by professionals is a duty of care and the courts seem to indicate that professionals do normally owe their client a duty in tort concurrently with their contractual duty.

The issue was and is only ever likely to be of practical relevance because of the anomalous quirk in the law of limitation that provides difference limitation periods for a claim for economic loss brought under a contract and for the same claim if it can be brought in tort.

This anomaly would have been swept away by reforms recommended by the Law Commission more than ten years ago. Proposals to implement those reforms were finally brought forward in 2010, but were then dropped.

For professionals, the case underlines the importance not merely of having a time limit in their appointments, but of making it clear that the time limit applies to claims in tort (including those in negligence) as well as contract.

Where claims involve fraud or deliberate concealment discovered many years later, the long stop provision would not apply. So if a professional sought to apply a 15-year-(or similar) long stop on claims involving fraud or deliberate concealment, this would almost certainly be considered unreasonable.

Non-standard conditions of engagement often include a provision limiting the time within which claims can be brought to 12 years from practical completion or some other event, notwithstanding that if the appointment is to be signed as a deed a 12-year time limit for claims in contract will apply in any event. Such a provision could well extend the professional's liability. Time begins to run from the breach of contract and if this is a negligent mistake by the professional which happened early in a project a time limit that ran from completion of the services or the project could add years to the time limits prescribed by law. If accepted, it is best that such a clause also says 'or such earlier date as may be prescribed by law' so that time limits are not excluded, although in a recent case the judge decided that the normal meaning of the words 'not later than [a certain time]' would not prevent the running of a period that expired earlier and it would therefore need to be made very clear in a clause drafted by a client that the earlier expiry of periods was not to apply.

16.7. The Unfair Terms in Consumer Contracts Regulations 1994

In addition to the provisions of the Unfair Contract Terms Act 1977, the terms of contracts entered into by consumers are governed by the Unfair Terms in Consumer Contracts Regulations 1994, which came into force on 1 July 1995. For the purposes of the Regulations, a consumer is defined as a 'natural person who is acting for purposes which are outside his business' when making the relevant contract. The Regulations will furthermore only apply to standard conditions which have been drafted in advance, so that individually negotiated contracts are excluded. The Regulations only apply to contracts for the supply of goods and services.

The purpose of the Regulations is to strike out any term which is held to be unfair, and the rest of the contract is interpreted without the offending term unless the contract then makes no sense. The Regulations define an unfair term as any term which causes a 'significant imbalance in the rights of the parties to the contract to the disadvantage of the consumer'. There is a schedule containing terms which might be considered to be unfair, but it is emphasised that the list is for illustrative purposes only, and the test for unfairness must therefore be applied to every contract.

The Regulations are wider in scope than the Unfair Contract Terms Act 1977 since they can potentially apply to every term of a contract which could be argued to be unfair, whereas the Unfair Contract Terms Act 1977 only applies to certain terms, for example, unreasonably limiting liability or limiting the time in which a claim can be made. The Regulations are furthermore concerned with the unfairness of terms, whereas the Act only invalidates terms if they are unreasonable. A term could be reasonable and yet be unfair, so the Act will not necessarily invalidate unfair terms.

It is, however, unlikely that the Regulations will be of concern to many professionals, given that they would only apply to the professional's standard form contracts for services to consumers. Most contracts concerned with building and construction projects will be made with people in the course of their business, so the Regulations would not apply. However, two recent cases are of some relevance.

In *Munckenbeck and Marshall* v. *Harold (2005)* the court considered the indemnity in the SFA/99 form of agreement produced by the RIBA. This related to the architect's costs of legal proceedings and the provision requiring interest at 8% over base rate on late payment in a contract with a consumer. Both terms were considered 'unfair' and thus unenforceable against the consumer – regardless of the fact that they were in a standard professional agreement. They had not been drawn to the consumer's attention and caused a significant imbalance between the parties, even though they were there to protect the architect against unfair treatment and the consumer had negotiated a reduction in the architect's fees. RIBA 2010 does not contain such an indemnity. The interest rate on late payment is still 8% over the dealing rate of the Bank of England as this is the rate allowed by the Late Payment Act (Late Payment of Commercial Debts (Interest) Act 1998) between businesses. It would have to be amended, however, in contracts with a consumer as a result of this case.

In *Domsalla (t/a Domsalla Building Services)* v. *Dyason (2007)* the court considered some terms in the JCT minor works contract. This allowed for adjudication even though the Construction Act does not apply to residential occupiers. The court held this was not unfair in this case, as it did not alter the balance of the parties' rights and obligations but the Unfair Terms Regulations are potentially applicable to contractual/agreed adjudication.

The court also held that the withholding notice clause in the contract was unfair as the consumer had no hand in selecting the clause, no advice as to its existence, meaning or effect, and no means of ascertaining that it was in the contract – as the consumer was not shown the contract when signing it or given an opportunity to read it at that time.

Finally, an adjudicator does not have jurisdiction to make an erroneous decision as to the application of the Unfair Terms Regulations to a consumer contract.

16.8. Arguments in favour of including limits of liability

Although, as has been said, there has in recent years been a marked increase in the percentage of non-standard conditions of engagement that include limitations of liability to amounts per claim or to total aggregate claims, it can sometimes still be difficult to persuade clients to include a reasonable limitation. They can be under the misapprehension that unlimited liability will in fact give them greater recovery in the event of a claim.

However, unlimited liability is in practice a myth. What the client can recover will depend on the assets of the professional at the time the claim is made and the major component of this in most cases will be the professional indemnity insurance he has available to meet the claim at the time and other assets such as any property and/or office furniture, etc.

The existence of financial caps can protect the client while at the same time providing professionals with an assurance that their potential liability will not be beyond what is reasonable taking all relevant circumstances into account and regardless of whether or not the professional has the insurance to meet the claim.

If financial caps were not used, a single successful claim could result in that firm's liquidation. What is the position then for the other clients who have engaged the professional (or indeed the same claimant but on another project), and what about the holders of collateral warranties or third party rights? There may well be no insurance and certainly no other assets to meet these other claims, leaving claimants to bear the entire burden of their loss and damage, possibly leading to their own liquidation. If all these claims were subject to a reasonable cap and the professional has professional indemnity insurance, there is a better chance of these claims being met, at least in part.

Financial caps therefore provide a greater chance that claims may be satisfied.

Unlimited exposure is also a factor in the levels of insurance premiums. If the industry adopted a sensible capping regime, insurers may lower premiums. As premiums are reflected in professionals' fees, lower premiums could mean lower fees.

It is sensible for clients and professionals to consider an appropriate limit at the beginning of a project and discuss then what the risks are, who is best placed to manage them, and the likely damages in the event of a claim, in order to arrive at a reasonable cap on liability.

16.9. Limitation of liability in standard conditions of engagement including evaporation clauses and net contribution clauses

The limitation provisions in ACE Agreement 1 follow the main limitations to be found in professional indemnity insurance but, as has been said, it is possible that when any claim is made the restrictions and exclusions in the professional's professional indemnity insurance will have changed and this needs to be borne in mind when agreeing limits of liability.

First, ACE Agreement 1 limits liability for all claims to an aggregate amount to be set out in the Particulars. This is a change from earlier editions which limit liability to an amount per claim.

Clause F7.1 reads:

> F7.1 Notwithstanding anything to the contrary contained in this Agreement and without prejudice to any provision in this Agreement whereby liability is excluded or limited to a lesser amount, the total liability of the Consultant under or in connection with this Agreement whether in contract, in tort, in negligence, for breach of statutory duty or otherwise shall not exceed the sum set out or the amount referred to, as the case may be, in clause B12 of Part B: *The Particulars of Agreement*.

The Particulars at B12 say that '. . . the total liability of the Consultant shall not exceed in aggregate the lesser of (i)' (a sum to be set out) 'or (ii) a multiple of ten times the total of the fees payable to the Consultant by the Client. If no amount is inserted in paragraph (i) paragraph (ii) only shall apply'. It was thought advisable to include a fall-back or default position in case professionals forgot to include an amount. Although it is possible that having a standard fall-back of ten times the fee might be challenged as 'unreasonable', this will depend on the circumstances of the case in which it is sought to apply this limit. Such a limit can be 'reasonable' as in the *Moores* v. *Yakely* case referred to in Section 16.2.

ACE Agreement 1 then goes on to deal with liability for pollution and contamination claims on an alternative basis – the parties can either agree a limit of liability for such claims or, in a commission where the professional is not responsible for advising on pollution and contamination matters, liability can be excluded. This second option in the Particulars excluding liability is as follows:

> B13 Notwithstanding anything to the contrary in this Agreement and without prejudice to any provision in this Agreement whereby liability is excluded or limited to a lesser amount the Consultant is not responsible under this Agreement or otherwise for advising as to the actual or possible presence of pollution and contamination or as to the risks of such matters having occurred, being present or occurring in the future and the liability of the Consultant whether in contract, in tort, in negligence, for breach of statutory duty or otherwise for any claim or claims arising out of or in connection with pollution and contamination is excluded.

As to limiting liability for claims for pollution and contamination, B13 (first option) states:

> B13 Notwithstanding anything to the contrary in this Agreement and without prejudice to any provision in this Agreement whereby liability is excluded or limited to a lesser amount the total liability of the Consultant whether

> in contract, in tort, in negligence, for breach of statutory duty or otherwise in respect of all claims under or in connection with this Agreement arising out of or in connection with pollution and contamination shall not exceed in aggregate the sum of £[ ] provided always that such pollution and contamination liability as limited by the aggregate or balance thereof shall not exceed in respect of any one claim or series of claims arising out of the same occurrence or series of occurrences the amount, if any, recoverable by the Consultant by way of indemnity against the claim or claims in question under any professional indemnity insurance taken out by the Consultant and in force at the time that the circumstances that might give rise to the claim or claims or if later the claim or claims is or are reported to the insurer in question. This proviso shall not apply if no such amount is recoverable due to the Consultant having been in breach of the Consultant's obligations under clause F8.1 of Part F: *The Terms of Contract* or of the terms of any insurance maintained in accordance therewith or having failed to report such circumstances or the claim to the insurer in question timeously.

It will be noted that the amount to be stated is an aggregate amount for such claims. This is because many professional indemnity insurance policies, even though they may provide a limit of indemnity per claim for other types of claims, have an aggregate limit for pollution and contamination claims. The aggregate limit to be inserted in B13 can be the same amount as the per claim limit, or a lesser amount, depending as always on the risks for that particular project. It is also prudent to agree a lesser amount than the full amount of the professional indemnity cover to allow something for other pollution and contamination claims as it needs to be appreciated that this aggregate limit will apply to all pollution and contamination claims that arise in any one year (that is, the year covered by the professional indemnity insurance policy) in respect of any and all projects the professional has been involved with, not just the project in respect of which the parties are agreeing this cap on liability. Thus if, when a claim comes in in respect of pollution and contamination, the professional has already had a claim or claims which have exhausted or eaten into that limit of indemnity, there will be nothing or only the balance of the cover available for this new claim. So the aggregate amount that is agreed for pollution and contamination claims, while it may be the amount of the aggregate cover available at the time to the professional for such claims, needs to be qualified by the proviso set out in B13. This is known as an 'evaporation' clause – that is, it seeks to limit the amount payable by the professional to the amount if any available for such claims under his professional indemnity insurance at the time. This could be something or nothing, depending on how many pollution and contamination claims have been met in that year.

B13 goes on to say that the professional will not be able to take advantage of this proviso, and therefore will be liable to pay regardless of whether he has an indemnity under his professional indemnity insurance policy, if he fails to maintain professional indemnity insurance in accordance with clause F8.1 of the Terms or if he breaches any term of that insurance or fails to report the claim in the proper time. These latter two events could result in professional indemnity insurers being entitled to make no payment under the policy. It is thought that these exceptions to the 'evaporation' clause will help in establishing its 'reasonableness'.

These types of limitations have been common for pollution and contamination claims for some years now, and although they have not been tested in the courts the general principles do seem to have been accepted where such limitations are agreed.

ACE Agreement 1 deals with claims arising out of asbestos in the same ways as those for pollution and contamination.

B14 (first option) says:

> B14 Notwithstanding anything to the contrary in this Agreement and without prejudice to any provision in this Agreement whereby liability is excluded or limited to a lesser amount the total liability of the Consultant whether in contract, in tort, in negligence, for breach of statutory duty or otherwise in respect of all claims under or in connection with this Agreement arising out of or in connection with asbestos or any product or waste that contains asbestos shall not exceed in aggregate the sum of £[ ] provided always that such asbestos liability as limited by the aggregate or balance thereof shall not exceed in respect of any one claim or series of claims arising out of the same occurrence or series of occurrences the amount, if any, recoverable by the Consultant by way of indemnity against the claim or claims in question under any professional indemnity insurance taken out by the Consultant and in force at the time that the circumstances that might give rise to the claim or claims or if later the claim or claims is or are reported to the insurer in question. This proviso shall not apply if no such amount is recoverable due to the Consultant having been in breach of the Consultant's obligations under clause F8.1 of Part F: *The Terms of Contract* or of the terms of any insurance maintained in accordance therewith or having failed to report such circumstances or the claim to the insurer in question timeously.

B14 (second option) says:

> B14 Notwithstanding anything to the contrary in this Agreement and without prejudice to any provision in this Agreement whereby liability is excluded or limited to a lesser amount the Consultant is not responsible under this Agreement or otherwise for advising on matters that wholly, partly, directly or indirectly arise out of or result from asbestos (including without limitation the costs of testing for, monitoring, abatement, mitigation, removal, remediation or disposal of any asbestos or product or waste that contains asbestos) and the liability of the Consultant whether in contract, in tort, in negligence, for breach of statutory duty or otherwise for any

> claim or claims arising out of or in connection with asbestos or any product or waste that contains asbestos is excluded.

It was thought necessary to split asbestos claims from pollution and contamination claims because professional indemnity insurers introduced some years ago specific exclusions and very severe restrictions on the cover available to professionals for such claims in the wake of the large claims during the period 2000 to 2005 for personal injuries arising out of the presence of asbestos.

Some policies provide no cover at all for such personal injury claims, and it will be recalled that the professional cannot exclude or restrict liability for such claims. As to other damages, limited cover can be available, for example, on a costs-inclusive, aggregate basis, or for negligence claims only, or to certain types of losses such as remediation or rectification costs or diminution in value of buildings or structures up to a certain amount.

The 2004 ACE Agreements responded to this by limiting liability to the amount recoverable in respect of asbestos claims under the professional's professional indemnity insurance, that is on the 'evaporation' basis, and contained an indemnity back from the client to cover all claims, including personal injury claims, which exceeded the amount of professional indemnity insurance the professional had available for such claims.

The fear of a substantial number of asbestos claims being brought against professionals, however, has not been realised, and thus the ACE has adopted in its 2009 Agreements the approach described above and no longer includes an indemnity from the client. It is likely, however, that the professional's professional indemnity insurance cover for asbestos claims will be more restricted than that for pollution and contamination claims and therefore that any aggregate limit agreed for such claims will be substantially lower than that for pollution and contamination claims.

A separate limit or exclusion is also provided for terrorism claims. This again was necessitated by professional indemnity insurers withdrawing all cover for claims arising out of terrorism, which is generally defined very widely, in the early 2000s.

B15 (option 1) provides:

> B15 Notwithstanding anything to the contrary in this Agreement and without prejudice to any provision in this Agreement whereby liability is excluded or limited to a lesser amount the total liability of the Consultant whether in contract, in tort, in negligence, for breach of statutory duty or otherwise in respect of all claims under or in connection with this Agreement arising out of or in connection with the designing or advising on or otherwise taking measures to prevent or mitigate the effect of any act of terrorism or any action that may be taken in controlling preventing suppressing or in any way relating to any act of terrorism shall not exceed in aggregate the sum of £[ ].

In the absence of any professional indemnity insurance, it is thought it would be 'reasonable' to restrict this to a relatively small amount as it will be the professional's own assets that would have to meet such claims. This could also be used to assist in discussions with the client as to who is to take the risk of such claims where no insurance is available.

Liability can also be excluded if the professional does not advise on such matters in the second option of B15:

> B15 Notwithstanding anything to the contrary in this Agreement and without prejudice to any provision in this Agreement whereby liability is excluded or limited to a lesser amount the Consultant is not responsible under this Agreement or otherwise for designing or advising on or otherwise taking measures to prevent or mitigate the effect of any act of terrorism or any action that may be taken in controlling preventing suppressing or in any way relating to an act of terrorism and the liability of the Consultant whether in contract, in tort, in negligence, for breach of statutory duty or otherwise for any claim or claims arising out of or in connection with such matters is excluded.

ACE Agreement 1 also includes a net contribution clause in the following terms:

> F7.5 Further and notwithstanding anything to the contrary contained in this Agreement and without prejudice to any provision in this Agreement whereby liability is excluded or limited to a lesser amount, the liability of the Consultant, if any, for any loss or damage ("the loss or damage") in respect of any claim or claims shall not exceed such sum as it would be just and equitable for the Consultant to pay having regard to the extent of the Consultant's responsibility for the loss or damage and on the assumptions that:
>
> (*i*) all other consultants and advisers, contractors and sub-contractors involved in the Project shall have provided contractual undertakings to the Client on terms no less onerous than those set out in clause F2.1 in respect of the carrying out of their obligations in connection with the Project;
>
> (*ii*) there are no exclusions of or limitations of liability nor joint insurance or co-insurance provisions between the Client and any other party referred to in this clause F7.5 and that any such other party who is responsible to any extent for the loss or damage is contractually liable to the Client for the loss or damage; and
>
> (*iii*) all the parties referred to in clause F7.5 have paid to the Client such proportion of the loss or damage which it would be just and equitable for them to pay having regard to the extent of their responsibility for the loss or damage.

It will be noted that the other parties with whom the professional could be jointly liable (i.e., the others within the 'net') are described widely as 'all other consultants and advisers, contractors and sub-contractors involved in the Project'. Clause F7.5(ii) also deals with the possibility that these others may have agreed limitations or exclusions of liability with the client which prevents them from being liable or liable to the extent they would have been without such provisions. It also deals with the effect of any of the other parties being joint or co-insured with the client which could again prevent liability from arising as in the Co-op case described in Section 12.12.

ACE Agreement 1 makes it plain that liability for personal injury or death is not excluded or limited:

> F7.6 If the Client is an individual, no exclusion or limitation of liability in this Agreement shall affect any liability of the Consultant for death or personal injury suffered by the Client.

As to possible claims against employees of the professional as illustrated by the *Merrett* v. *Babb* case described in Section 15.11, ACE Agreement 1 seeks to protect employees in such circumstances as follows:

> F7.7 Save in respect of death or personal injury the Client shall only look to the Consultant (and not to any of the Consultant's Personnel) for redress if the Client considers that there has been a breach of this Agreement. The Client agrees not to pursue any claims in contract, in tort or statute (including negligence) against any of the Consultant's Personnel as a result of carrying out the Consultant's obligations under or in connection with this Agreement at any time and whether or not any of the Consultant's Personnel is named expressly in this Agreement.

'Consultant's Personnel' are defined as 'any employee or member of the Consultant including any officer or director of a company or a member of a limited liability partnership or self-employed or agency personnel working for the Consultant'. It will be noted that partners are not included as they are not employees and will be liable to the client on the basis described in Section 15.8.

Finally, in ACE Agreement 1 there is a limit for the time within which claims can be brought:

> F7.8 No action or proceedings under or in respect of this Agreement whether in contract or in tort or in negligence or for breach of statutory duty or otherwise shall be commenced against the Consultant after the expiry of the period of liability stated in clause B17 of Part B: *The Particulars of Agreement* or such earlier date as may be prescribed by law.

Note this refers to all types of claim, not just those for breach of contract – see comments on *Robinson* v. *Jones* case in Section 16.6.

The period of liability is to be stipulated in B17:

B17 The period of the Consultant's liability is from the effective date hereof to [ ] years after the completion of the Services or the termination of the Services if earlier.

If, however, the time limit under the general law of limitation for the time for claims has expired, the provision in F7.8 – 'or such earlier date as may be prescribed by law' – would mean that earlier time bar would apply – see Section 16.6.

ACE Agreement 1 includes an obligation to maintain professional indemnity insurance for the amount and length of time sufficient to cover the professional's liabilities under the Agreement. It is hoped this will help establish the 'reasonableness' of the limitations of liability.

RIBA 2010 limitation provisions begin with a limit on the time to bring claims in terms not dissimilar to ACE Agreement 1, save that there is no express reference to claims in negligence – see comments at Section 16.1.

7.1 No action or proceedings arising out of or in connection with this Agreement whether in contract, in tort, for breach of statutory duty or otherwise shall be commenced after the expiry of the period specified in the Project Data from the date of the last Services performed under this Agreement or, if earlier, the date of practical completion of construction of the Project or such earlier date as prescribed by law.

As to the limit of liability, RIBA 2010 provides:

7.2 In any such actions or proceedings:

7.2.1 The Architect's liability for loss or damage shall not exceed the amount of the Architect's professional indemnity insurance specified in the Project Data, providing the Architect has notified the insurers of the relevant claim or claims as required by the terms of such insurance.

7.2.2 No employee of the Architect, including any officer or director of a company or a member of a limited liability partnership or any agent of the Architect, shall be personally liable to the Client for any negligence, default or any other liability whatsoever arising from the performance of the Services.

This fixes the professional's liability in relation to such actions or proceedings as are described in clause 7.1. Unfortunately, as has been said, the actions or proceedings referred to in clause 7.1 do not expressly include actions in negligence. It may not be sufficient simply to refer to tort – negligence should be referred to specifically in a limitation clause and there is therefore a real possibility that the professional's liability in negligence will not be limited by clause 7.2.1 nor subject to the time limits in clause 7.1.

The basis of the professional's limit of liability is to the amount of the professional's professional indemnity insurance specified in the project data. Such an overall limit ought to be made subject to the net contribution clause. The limitation only applies if the professional has notified insurers of the relevant claim or claims as required by the terms of such insurance. If that notice is not given, the professional's liability will be unlimited. This may well help the limitation pass the 'reasonableness' test under the Unfair Contract Terms Act 1977.

By allowing for the professional to fill in in the project data the amount of aggregate cover for pollution and contamination and asbestos or fungal mould, RIBA 2010 allows the professional's liability to be limited in line with his insurance if that is its current basis. Amendments will be needed if it is not. It also needs to be remembered that the professional's professional indemnity insurance basis could change during the time within which claims can be brought. RIBA 2010 also seems to presuppose that the professional will put in the amount of professional indemnity insurance he actually has (and thus limit liability to that amount) when he should put in an appropriate amount for the project which could be significantly less.

Clause 7.2.2 is the provision protecting employees of the professional from direct claims by the client, although this clause does refer specifically to negligence as well as the defaults or any other liabilities.

The net contribution clause is in clause 7.3:

> 7.3 Without prejudice to the provisions of clause 7.2, the liability of the Architect shall not exceed such sum as it is just and equitable for the Architect to pay having regard to the extent of the Architect's responsibility for the loss and/or damage in question and on the assumptions that:
>
> 7.3.1 all other consultants and contractors providing work or services for the Project have provided to the Client contractual undertakings on terms no less onerous than those of the Architect under this Agreement;
>
> 7.3.2 there are no exclusions of or limitations of liability nor joint insurance or co-insurance provisions between the Client and any other person referred to in this clause; and
>
> 7.3.3 all the persons referred to in this clause have paid to the Client such sum as it would be just and equitable for them to pay having regard to the extent of their responsibility for that loss and/or damage.

The scope of this clause, however, may be less wide than that under earlier RIBA Agreements or indeed ACE Agreement 1 because it only refers to 'consultants and contractors'. The 2007 RIBA Conditions included advisers and sub-consultants and the SFA/99 was wider still – 'all other consultants, contractors, sub-contractors and other persons providing the services'. There is a danger, therefore, that some of the other people with whom the professional could be jointly liable are not brought within this clause, leaving the professional at risk of paying more than his proportionate share if those people become insolvent or have insufficient assets to meet a claim.

RICS 2011 again provides for a maximum aggregate limit of liability to an amount to be specified in the Appendix as follows:

> 5.1 Except for liability for death and/or personal injury, the maximum aggregate liability of the Consultant to the Client under or in connection with this Appointment and all other documents relating to the Services (whether in contract, tort (including negligence), for breach of statutory duty or otherwise) is limited to the amount specified in the Appendix.

The Appendix provides for the amount to be the amount of the professional's professional indemnity insurance referred to in clause 12.1.

A net contribution clause can be added by agreement – it does not apply automatically:

> 5.2 Where the Appendix states that this Clause is applicable, and without prejudice to any other exclusion or limitation of liability available to the Consultant, the Consultant's liability for loss or damage suffered by the Client in the event of any breach of this Appointment is limited to the proportion of such loss or damage that it would be just and equitable to require the Consultant to pay having regard to the extent of the Consultant's responsibility for the same and on the assumptions that:
> (*a*) the persons identified in the Appendix have provided contractual undertakings to the Client on terms no less onerous than those set out in this Appointment in respect of the carrying out of their obligations in connection with the Project;
> (*b*) there are no exclusions or limitations of liability nor joint insurance or co-insurance provisions between the Client and any person referred to in Clause 5.2(a); and
> (*c*) all the persons referred to in Clause 5.2(a) have paid to the Client such sums as it is just and equitable for them to pay having regard to the extent of their responsibility for such loss or damage.
> Provided always that the Consultant does not assert that it has no or a reduced liability to the Client under this Appointment solely by reason of the fact that the Contractor accepts responsibility for the design of the Project, to the extent such design is undertaken by the Consultant.

If this is agreed upon the parties have to indicate this in the appendix and also set out the persons to come within the net. The appendix provides for the building contractor, any sub-contractor and other members of the professional team to be included. The 'Professional Team' is defined as 'the Consultant, the professionals listed in the Appendix and any other professionals engaged by the Client in relation to the Project and which the Client notifies to the Consultant'. It seems curious, in the context of such a clause, that whether or not another professional is within the scope of the net contribution clause should depend upon notification by the client. There might be a professional employed by the client on some crucial element whom he does not notify to the professional and yet who would otherwise have been jointly liable with the professional. This part of the appendix needs careful consideration and may need amendment.

Otherwise the net contribution clause is substantially similar to that in ACE Agreement 1 except for the added proviso. It is thought that this proviso is based on a misapprehension as to the way net contribution works where there is a design/build contractor, and ought to be deleted – see Section 16.10.

RICS 2011 also contains a clause limiting the time within which claims can be brought:

> 5.3 No action or proceedings for any breach of this Appointment may be commenced against the Consultant after the expiry of the Relevant Period.

It will be noted that this is limited to claims for breach of contract only so no other claims will be subject to this limitation.

The 'Relevant Period' is six years from practical completion if the appointment is signed under hand and 12 years from practical completion if signed as a deed. This might well have extended the time limits for claims – see Section 16.6.

The drafting does beg the question as to what the time period will be if in fact there is no practical completion under the building contract.

CIC provides for an aggregate limit of liability for all claims including those in negligence to be set out in the contract details but unusually this covers claims from both the client and 'any Interested Party', that is anyone to whom third party rights are to be given, so that the professional is protected from claims from any or all of them which would take the professional over that agreed amount. It is debatable whether the reference to 'Interested Party' is necessary as the Contracts (Rights of Third Parties) Act provides that the third party takes subject to the same limitations as the client. This wording, however, makes this clear. The amount agreed here could be quite high, perhaps even a figure that exceeds the professional's professional indemnity insurance for a single claim, so consideration would need to be given in such circumstances as to whether an inner limit to an amount per claim was advisable. The CIC limitation in any event provides for the parties to provide a separate (and presumably lesser) aggregate limit, within the overall limit, for asbestos claims arising out of the presence of asbestos on site.

CIC does not incorporate a net contribution clause but the guidance sets out the circumstances in which the parties might wish to include one and provides some wording which is substantially the same as in ACE Agreement 1, except that there is a reference to the contractor's responsibility for design.

The limit of liability provisions in PSC3 are discussed in Section 9.14.

16.10. Net contribution where there is a contractor with design responsibility

The argument appears to be that where a professional acts for the contractor and the contractor has full design responsibility to the client under his contract, a net contribution clause is not appropriate or if it is the contractor should not be included, or that wording like that in the RICS 2011 in Section 16.9 needs to be added.

In the author's view, the correct position is that although the contractor may have 100% design responsibility through his contract with the client, this does not exclude other people having responsibility as well. The purpose of the net contribution clause is to take each person's responsibility into account and then make an apportionment, just as in normal contribution proceedings. The contractor should therefore be included.

Indeed it could be argued that any contractor, in order to be a contributor at all, is 100% liable for the damage in question, as are all contributors. The reduction in liability as a result of the apportionment occurs only after the initial 100% liability has been established. (Having said that, net contribution clauses are wide enough to cover people who may only be responsible for part of the loss and damage.)

A net contribution clause does not have the effect that any contractual responsibility for design on the part of any party referred to in the clause, including the contractor, extinguishes entirely the liability of the professional for any loss or damage attributable to such a design; nor does it preclude the responsibility of the professional and other parties for such loss and damage being taken into account for the purposes of such a clause.

Professional Services Agreements
ISBN 978-0-7277-4158-5

http://dx.doi.org/10.1680/psa.41585.287

Chapter 17
Indemnities

17.1. Indemnities and guarantees generally

Indemnity clauses are sometimes included in professional services agreements whereby a professional provides an indemnity to a client against certain liabilities set out in that indemnity. Indemnities have particular characteristics of which the professional should be aware.

A contract of indemnity can be defined in several ways; in its widest sense it is any contract under which a party underwrites as a primary liability the loss or liabilities of another party. A contract of indemnity need not be in writing and can arise through statute or the relationship between the parties.

A contract of indemnity creates its own independent original obligations and the indemnifying party will be primarily liable for the obligations under the contract of indemnity throughout its term.

An 'indemnity' can also describe certain contracts of insurance or guarantee (usually described as 'bonds' in construction contracts), as the purpose of those contracts is to provide some compensation to an innocent party, for example, when a third party fails to perform its obligations under its contract with the innocent party.

A contract of indemnity is not the same as a contract of guarantee (or bond). The latter is often described as being a collateral contract, existing in parallel with a main contract, under which a guarantor (or surety) assumes a secondary liability for the obligations of one of the parties under the main contract. If the party under the main contract performs its contractual obligations completely and perfectly, the guarantor (or surety) is released from its liability and the contract of guarantee ends. A contract of guarantee must be in writing. Guarantees are discussed in Chapter 22.

In this chapter, the indemnities being considered are those which professionals may be required to give to clients under their appointments. These tend to appear most frequently where the professional is being appointed by a contractor, possibly because the contractor has had to sign a building contract which contains indemnities and wants his professionals to be liable to him on the same basis as the contractor will be liable to his employer.

Indemnities are not unlawful and are enforceable as the courts take the view that the parties are free to enter into any bargain they chose (subject only to some specific exclusions such as contracts which are against public policy).

17.2. Problems with indemnities

Unless indemnities are very carefully worded, there is a risk that the amount that the professional will have to pay under the indemnity in respect of some default on the professional's part will be higher than the damages awarded if the matters covered by the indemnity had been the subject of a claim and decided by the courts. This can apply to amounts payable as a result of the indemnity to the client and also to third parties, because of the professional's default.

An indemnity could, for example, allow a client to recover damages which would

(*a*) not normally be recoverable because they are too remote
(*b*) otherwise be reduced by the client's duty to mitigate its loss or by reason of the client's contributory negligence
(*c*) not be recoverable because they may not have been properly or reasonably incurred.

It is also possible that an indemnity would allow legal costs and expenses to be recovered which would otherwise be disallowed by the court.

Depending on the wording of the indemnity a professional may not necessarily have to have been negligent before being liable to pay under an indemnity. If a client has suffered the type of loss for which the professional is to provide an indemnity, the professional has to pay. The client would then be under no obligation to prove that the professional failed to exercise reasonable skill and care; it merely has to show that it has suffered an indemnified loss. For example, there might be an indemnity in a contract by a professional to its client against any damage to property owned by the client, even though damage to that property was inevitable when performing the services. Because of the indemnity the client could claim for the damage to its property, even though the professional had not been negligent.

Indemnities are not only asked for by clients to cover their own losses, but can extend to claims brought by third parties against the client in respect of some default of the professional. For example, a contractor could sue the client because of some error in design which is alleged to have caused additional work and delay. It is possible in this situation that a design professional will not know anything about the claim between the contractor and the client nor be involved in any court or arbitration hearing about it, and will only be informed once the amount has been agreed or determined by the court or arbitrator when the client seeks to collect that amount from the professional under its indemnity. The problem is likely to become more prevalent as a result of faster alternative dispute resolution procedures, such as adjudication.

Settlements of claims can also give rise to particular difficulties. For example, a client may decide for commercial reasons to settle a claim with a third party for £500 000 when the damages properly payable by the professional if decided by the court would be £350 000. There may be no opportunity, if the indemnity is drafted in appropriate terms, to argue that the amount the professional should pay should be limited to £350 000.

The length of time within which a client can collect the amount due under an indemnity can also be a problem. Although in practice the client will probably not wait to claim under the indemnity once a claimable event occurs, the client has in fact between six and 12 years to do so, depending on whether the indemnity is contained in a contract under hand or signed as a deed.

The limitation period for claims made in respect of loss or damages or expenses, etc. covered by an indemnity given in a contract only begins when the loss or damage which is to be indemnified has been established: for example, when the court has decided the amount of the loss or the monies the client is claiming have been expended. By contrast, the limitation period for claims for breach of contract begins when the contract is breached, which could be significantly earlier. Where the indemnity is against 'liability', however, time will run from the time the cause of action accrues, i.e. from when the client incurs his liability to the third party as a result of something the professional has done. It is therefore possible that a claim can be made under an indemnity contract after the limitation period in respect of claims under the principal contract has expired.

If indemnities are given to specified third parties as well as the client, those third parties may be entitled to claim under the indemnity directly as a result of the Contracts (Rights of Third Parties) Act 1999 – see Section 2.3. This could happen if the client asked for an indemnity not only for himself but for his funder or other parties, such as purchasers and tenants who had an interest in the project. Unless the Act is excluded, indemnity clauses will need to be checked to see whether such rights have been given to third parties.

Indemnities are not needed and should therefore be resisted. If a professional is in breach of the terms of a professional services agreement, a client will be entitled to the damages prescribed by law and he may also be able to sue the professional in negligence. If third parties sue the client because of something the professional has done, the professional can be joined in the proceedings or damages can be claimed from the professional by the client in contribution proceedings. In the absence of an indemnity, the law does not say that professionals should be liable to their clients for damages, etc. on an indemnity basis, nor that there should be no safeguards or fetters on the amount clients can spend in rectifying a default of the professional and then seek to recover from professionals, nor that professionals should pay whatever clients decide to pay to third parties in respect of any claims brought by third parties arising out of something the professional has done.

An indemnity therefore only makes it easier for a client to collect money from a professional and increases the damages he can recover. There are no advantages to the professional.

If a professional agrees to give an indemnity, an additional fee for the additional risk should be considered. It will be important to consider the extent and nature of the indemnity which is being proposed and what types of losses are to be covered; for example, are they to include personal injury or death, property damage or even all losses arising out of the contract? A widely drafted indemnity clause can create onerous liabilities and it may be called upon in situations which the professional had not envisaged. Unacceptably wide indemnities should not be accepted.

None of the standard conditions of engagement referred to in this text contain indemnities from the professional, save for PSC3 which has an indemnity to the client against claims for copyright infringement and possibly more – see Section 9.13.

17.3. Position under professional indemnity insurance policies

Most professional indemnity insurance policies cover indemnities, but 'negligence-only' policies may only do so if the indemnity is in respect of negligence and not, for example, other breaches of contract and/or other acts, omissions or defaults where negligence is not in issue.

Where payment under an indemnity includes an *ex gratia* element as described in the preceding section, that may not be recoverable under professional indemnity insurance. This will depend on the terms of the relevant policy. Even if it is recoverable, this may increase the professional's premiums for future insurance because the size of the payment made under the policy has been increased by the *ex gratia* element.

Further, if professional indemnity insurers find they are paying out claims more regularly on an indemnity basis, rather than the common law basis, they may in future exclude or limit liability for such claims.

Checking whether a particular indemnity falls within the professional indemnity insurance policy will provide limited reassurance as this will only apply while that policy is in force. If later policies exclude indemnities, the professional could find he has no cover when the claim is made.

17.4. Some specific indemnities

It is impossible to include every indemnity wording that might appear in non-standard conditions of engagement and to comment on them. The words 'indemnify' or 'indemnity' are usually used, but sometimes it is expressed as 'hold harmless'. The following example illustrates some of the foregoing general principles:

> The Consultant shall indemnify the Client against any damages, costs, losses, expenses, actions, claims or proceedings arising out of or caused by any breach of the terms hereof or any negligent act or omission by the Consultant in carrying out or failing to carry out its obligations hereunder, whether express or implied.

This illustrates some of the problems outlined in Section 17.2 as follows:

(*a*) As has been said, if a professional is in breach of his obligations or has been negligent, he does not become liable for any and all damages and losses, etc. suffered by the client. Restrictions apply to these. However, payment of all damages, costs, losses, etc. is provided for in this clause.

(*b*) Equally, a professional is not necessarily liable for all the costs and expenses suffered by a client. The courts have a system of assessment to establish what costs and expenses are properly recoverable. In this clause, 'any costs or expenses' would not necessarily be restricted to those the court would order to be paid and thus a professional could pay more than those for which he might otherwise be legally liable.

(*c*) A breach of the terms of the appointment may not result in any particular loss or damage to the client, but the client could incur costs and expenses in finding this out. These costs and expenses could be recoverable under the quoted indemnity.

(*d*) The indemnity refers not only to a 'negligent act or omission' but also to 'any breach of the terms hereof'. So any breach of a contractual obligation could give rise to payment of damages, etc. on an indemnity basis and the professional need not have been negligent to incur such a liability. This is the case where the term is a strict obligation – see Section 4.3.

(*e*) The indemnity refers to the 'carrying out or failing to carry out its obligations hereunder whether express or implied'. It may be acceptable to include 'express' obligations because those will be specifically set out in the appointment and the professional will therefore know what they are – subject to the point that failure to carry out an express obligation may not necessarily result in any damages. However, to extend the indemnity to 'implied obligations' could be extending the liability of the professional beyond what is reasonable. Who can say what those implied terms or obligations will be – this is an issue the court would have to decide. A professional should therefore resist such a specific inclusion.

Another example of an indemnity is:

> The Consultant shall be responsible for any discrepancies, inadequacies, errors or omissions in the documents and information provided by it under this Agreement and the consequences thereof whether or not the Client shall have approved the same. The Consultant shall at its sole expense rectify promptly any such deficiency by amending, replacing or supplementing the deficient document or information as appropriate and shall bear the cost of any consequences thereof and/or shall indemnify the Client against all costs, losses, expenses, damages, claims, actions or proceedings brought or claimed by or on behalf of third parties in respect of any such consequences thereof.

The problem with the first part of this clause is as follows. The obligation is to 'rectify promptly any such deficiency . . .'. If the deficiency is such that, for example, it affects the professional's designs, the professional will have such an obligation in any event. What if the deficiency is a mathematical or detailing error which, in fact, has no effect on any of the professional's designs or information? In the latter circumstances the professional has no obligation to amend and no liability for 'the cost of any consequences thereof'. The same would be true if the deficiency did not amount to a breach of contract or negligence. The wording seeks to pre-empt what the consequences of any error might be, instead of leaving it to be sorted out under the common law, which requires that there must be a breach of contract or negligence and that a client has suffered loss or damage which is properly recoverable. The 'costs of any consequences' could be outside the scope of the professional's indemnity insurance cover.

The indemnity in the latter part of this clause is limited to amounts claimed by third parties, but it could make the professional liable to third parties when, without such an indemnity, the professional would not incur any liability, or liability to such an extent, to these third parties. The professional may not be liable under the common law for 'any such consequences', but he could become liable for them under the terms of this indemnity, for the reasons stated above.

There is also the fact that once the professional is liable under this part of the clause, he is going to have to pay all costs, losses, expenses, damages, etc., not just those which are recoverable under the common law.

Where the indemnity is against 'liability' only, the amounts recoverable should only be those recoverable under the common law.

Some indemnity clauses are not so onerous, for example, they may expressly limit the amounts recoverable to those that are reasonably foreseeable or properly recoverable in law, reduce the amount payable because of any contributory negligence, require the client to have mitigated his loss, exclude the *ex gratia* element of any settlement, and make the liability arise only if the professional has been negligent. However, although the objective may well be to make the amounts payable comparable to those recoverable under common law, there must be some doubt as to how the courts would interpret such an indemnity. If the parties were trying to provide that the professional would only be liable to the extent prescribed by law in the event of any breach of contract and/or negligence, why was there any need to have such an indemnity at all? Also, the rules for the payment of damages, etc. under the common law are complex and extensive and any limitation to the extent of an indemnity is almost certainly not going to be able to cover them all.

Some indemnities, although this is rare, will provide for notice to be given to the professional of any third party claim received which the client considers is covered by the indemnity and/or will allow the professional to take over the conduct of any negotiations and any litigation or arbitration.

It is stressed again that indemnities by a professional should be resisted as the law already provides adequate remedies to a client, both in respect of its own claims and any claims coming from third parties.

17.5. Provisions equivalent to an indemnity – what will be paid, or the extent of liability

Some clauses in non-standard conditions of engagement can impose an obligation on the professional to pay losses or damages greater than those payable under the common law and/or which are the equivalent of an indemnity.

For example, a professional can be required to pay or state that he will be liable for all the losses or other damages suffered by the client and/or third parties. If 'losses' are widely defined, such as 'all losses, costs, claims, demands, actions, damages, awards, liabilities and expenses howsoever arising (including any legal expense)' that is an agreement in effect to pay losses, damages and costs on an indemnity basis. 'All losses' by itself may be wide enough for the client to recover losses on an indemnity basis.

17.6. Where indemnities may be acceptable

There are, possibly, three situations in which indemnities may be acceptable.

(*a*) An indemnity to a client in respect of copyright infringement by a professional. If it is correctly worded most professionals will accept such an indemnity because copyright is a matter over which the professional generally has control – that is, the professional will usually be aware whether or not it is infringing someone's copyright when producing a design, report, information, etc.

(*b*) Because one joint venturer can become liable for acts or omissions of a co-joint venturer it is usual for cross-indemnities to be given to enable an innocent joint venturer to collect any amount that it has paid to a third party from his co-joint venturer(s).

(*c*) Where a limitation or exclusion of liability has been agreed between a client and a professional, this will apply to third party claims brought against the client which the client then seeks to pass on to the professional, but it will not protect the professional against third party claims made directly against the professional. If a limitation of liability is agreed, a professional should not have any liability to third parties who claim directly, where those claims would exceed the amount of the liability agreed with the client. The professional needs an indemnity from the client against such claims, if the professional is to be protected. As with all indemnities, this protection will only be of any worth if the client has sufficient insurance and/or assets at the time the claim is made to meet the amount due under the indemnity. None of the standard conditions referred to in this text contains such an indemnity, even though all the standard conditions have provisions to limit or exclude liability – see Section 16.9.

The ACE Conditions used to have such an indemnity from the client in respect of asbestos claims by third parties, as the professional indemnity insurance cover available to professionals for such claims is severely restricted. However, the indemnity has been dropped from ACE Agreement 1, partly because insurers have found that such claims are being brought primarily against employers or previous employers and are therefore under employers' liability policies, and partly because clients were objecting to the inclusion of the clause.

17.7. Effect of limitations of liability on indemnity clauses

Clauses in a contract which limit liability will also set a limit on the amount payable under an indemnity included in that contract unless claims under indemnities are specifically excluded from the scope of the limitation. For example, the indemnity given in PSC3 in respect (it is assumed) of copyright infringement is expressly excluded from the general aggregate limit of liability in clause 82.1. If the contract which includes the indemnity provision also limits the time in which claims can be brought, this limit will also apply to claims in respect of amounts due under the indemnity.

Professional Services Agreements
ISBN 978-0-7277-4158-5

http://dx.doi.org/10.1680/psa.41585.295

Chapter 18
Set-off, abatement, liens, retentions and liquidated damages

18.1. Set-off and abatement

Set-offs have been discussed in Section 7.15.

The Construction Act 1996 and the amended Construction Act recognise the right to set-off and formalise it by requiring the payer to identify what is being set off and the ground for doing so in the notice of intention to withhold payment or notice of intention to pay less – see Section 17.15.

Abatement is technically distinct from set-off. It is usually raised as a defence where an innocent party to a breach of contract is entitled to deduct from the cost of the contract the reduction in value of the relevant goods or services caused by the breach. This could be determined by comparing the current market value of the item as constructed with the market value it ought to have; or the diminution in value could be determined by reference to the cost of remedial work. The sum claimed can never exceed the sum the professional would otherwise have been entitled to be paid for his services.

Abatement is a common law right and has been available as a defence at common law since the beginning of the twentieth century. It is restricted to claims arising out of contracts for the sale of goods or services and materials. Unlike set-off, it is therefore not available under all contracts. Also, it is only available where there are defects in the quality of the works or services. It cannot be used as a defence, for example, where there is for a claim for delay or damage to other property.

18.2. Liens

In the absence of express contractual provisions to the contrary, a professional has a legal right to retain documents, such as the designs and drawings the professional has prepared or significantly improved, until he has been paid for them (at which point they must be handed over). This is known as a 'particular lien'.

There are various rules covering this right of retention. The lien only covers the costs of preparing the documents concerned, so that any other amounts of money that might be part of the total owed to the professional do not have to be paid in order to secure the release of the documents. Payment for the professional's work must be due. The professional cannot use a lien to obtain an early payment.

The lien only comes into existence when the professional has completed the documents or drawings, although a lien can be claimed for work actually done if the contract is terminated before the work is completed. A lien cannot be exercised in relation to intellectual property as it is intangible.

The lien only exists while the professional still possesses the work, so that the lien is lost once the work is passed on to a third party. In this respect, a lien is a personal right of the professional.

A lien will often arise in connection with work which will also attract copyright. Copyright, licences and the effect of payment on any licence are discussed in Chapter 10.

In practice, there may be a risk of being sued if the exercise of a lien causes delay to the project. For example, the client may seek to recover from the professional the costs of a contractor's claims for delay under the building contract if this was due to the delayed delivery of drawings as a result of exercising the lien. A lien, therefore, may only be valuable as a means of obtaining payment when withholding 'as-built' or final drawings or documents that a client might need, say, to complete its health and safety file under the CDM regulations.

18.3. Retentions

Some non-standard conditions of engagement allow for retention of a specified amount of the fee until certain events or circumstances have occurred. The reason for a retention is to provide security for the client for due performance of the contract and as a fund to be drawn upon either to complete the services or rectify defects if the professional fails to do so. They are common in contractors' contracts but not professional appointments. The amount of the retention in construction contracts is usually between 3% and 5% of the contract sum – one half released at practical completion and the balance at the end of the defects period.

Professionals are sometimes engaged on the basis that no fees will be payable until planning permission for the project is obtained. Problems can result if the client abandons the scheme, withdraws the scheme from consideration by the planning committee or if planning permission is available but not 'collected' by the client – see also comments on conditional payment in Section 7.5.

The risks for the professional will depend on how precisely the amount to be retained can be calculated and the definition of the circumstances in which it can be released. The professional will need to consider whether the provisions concerning the retention in the agreement set out clearly when the retained monies will be released. If the test is subjective, for example, 'when the client is satisfied that all the services have been completely and correctly carried out', it might be difficult to determine when the amount is due. If it is fixed to a certain date, such as 'practical completion' or 'the Certificate of Making Good', this may be more certain provided, of course, these events actually occur. The former occurs in a project that proceeds (although problems could occur if the first contractor becomes insolvent) but the latter frequently does not occur at all.

The Olympic Delivery Authority took a stand against the deduction of retentions in its contracts for Olympics sites.

18.4. Liquidated damages

The parties to a contract can agree that if the contract is breached by either party, a fixed amount of damages will automatically be recoverable by the innocent party. The purpose of such an agreement is to allow the innocent party to be compensated without the need to show that it has suffered loss and to avoid the costly and often difficult procedures involved in doing so. Sometimes the parties will fix a sum as liquidated damages if a specific breach occurs. Should any other breach occur, the innocent party still needs to sue for unliquidated damages. If the parties have used a liquidated damages clause, and the innocent party's losses exceed the stipulated sum, it cannot disregard the clause and sue for more compensation; similarly the party in default cannot reduce the sum payable even if the innocent party has incurred losses of less than the agreed amount.

In general, the courts will not interfere in any agreement freely entered into by the parties, and they will not prevent a party from making a bad bargain. However, in the area of liquidated damages clauses, the courts have been reluctant to uphold clauses which are expressed to be liquidated damages clauses, but which in practice operate as a penalty to encourage the parties to perform the contract. Penalties are unenforceable.

A liquidated damages provision must be a genuine pre-estimate of the losses that would be suffered by the innocent party if the obligation under consideration is breached. The courts will treat the clause as a penalty if the damages payable exceed these estimated losses, or if the same amount is payable in circumstances that have different financial consequences, or if the sum payable is extravagant or out of all proportion to the range of possible losses which might be incurred. Also, if the parties put 'nil' in a liquidated damages clause, the courts have decided that this means the parties have agreed no damages will be paid in respect of a failure to complete on time.

Although penalty clauses are unenforceable, the courts (Court of Appeal in *Jobson* v. *Johnson (1989)*) have held that an innocent party can still recover its actual losses.

Liquidated damages may be included in non-standard conditions of engagement where a professional does not complete services within a pre-agreed timescale or where certain drawings or documents are not provided by a certain time. They are usually expressed in terms of so much money per day of delay.

Liquidated damages are inappropriate for professional services agreements and are not needed. The reason is the same as that given for indemnities in Chapter 14. If a professional is in breach of contract the client will be entitled to the damages prescribed by law. The client may also have a remedy in negligence. A client does not need a liquidated damages clause in order to recover damages. Generally, too, unlike contractors' liabilities, a professional's liability will be underwritten by professional indemnity insurance, not the assets of the company.

The problems for a professional in assessing a liquidated damages clause tied to the programme for completion of the services are as follows.

(*a*) Whether compliance with the programme is something that is entirely within the professional's control and, if not, whether any extensions of time can be given for changes by the client, delay by others, etc., and whether these extensions can be challenged if not correctly given.
(*b*) Whether the proposed rate of liquidated damages is a genuine pre-estimate of the client's loss. If liquidated damages are a genuine pre-estimate and there is a total aggregate limit in place, exposure can be limited. If it is not, the professional may have a good chance of arguing that it is a penalty and only the actual damage or loss would be recoverable.

In the case of late delivery of documents, it is difficult to see how the same rate of liquidated damages could apply to all documentation supplied late. Late delivery of documents may have no effect or very little effect, or have severe consequences for the whole project.

In one non-standard professional services agreement, liquidated damages for late delivery of documents were combined with liquidated damages for erroneous or incomplete documents. However, as with delays, the damages that flow from some errors could be very little or none, whereas others could result in very grave consequences. A fixed level of damages would not distinguish between the two. There was also a problem in the interaction in the two provisions – how are the liquidated damages calculated if at the same date the professional is late with some documents and those that are delivered on time contain errors? Are the liquidated damages paid until all late documents are delivered, and then only a proportion paid until errors have been put right, or do both sets of liquidated damages run until everything is put right?

The professional's professional indemnity insurance needs to be considered. Liquidated damages are often expressly excluded altogether and the indemnity limited to the amount that would have been payable if damages had been assessed on the common law basis. Also, the cover provided by the policy could be limited to acts of negligence only and the damages which result from any breach of the duty of care. As a result, if the liquidated damages agreed are more than the damages that would have been payable if the claim had been made on the basis of negligence or breach of the duty of care, there would not be a full indemnity to the professional from the insurer.

PSC3 contains an option for the payment of delay damages fixed by reference to completion of a section of the services or for the whole of the services. ACE Agreement 1, RIBA 2010 and RICS 2011 have no such provisions.

Some professionals may decide that liquidated damages are a way of limiting liability for delay if a reasonable sum can be agreed together with a cap and if there is a possibility of very substantial damages in the worst case of default.

Professional Services Agreements
ISBN 978-0-7277-4158-5

http://dx.doi.org/10.1680/psa.41585.299

Chapter 19
Assignments and novations

19.1. Assignments generally

This section is concerned with the transfer of 'rights' or 'benefits' under contracts and of 'obligations' and 'burdens' under contracts to a third party. This is achieved by assignment. In the context of a professional services appointment, 'rights' or 'benefits' are generally the right of the client to sue the professional if he has been negligent or in breach of contract, and the right to payment in the case of the professional. As to 'obligations' and 'burdens', these are generally the obligation of the client to pay the professional and the obligation of the professional to perform the services.

19.2. Legal assignments and the 'no loss' argument

The Law of Property Act 1925 provides that a legal right (or 'benefit') may be assigned provided the assignment is made in writing by the assignor and the other party to the contract is notified of the assignment in writing. The assignment must also transfer the entire benefit under the contract, so that the assignor cannot retain some interest in the contract. The assignment is therefore described as 'absolute', in the sense that the assignee becomes the new owner of all the rights. The assignment cannot be stated to be by way of charge. If the assignment is valid, it passes the legal rights under the contract to the assignee, as well as the legal and other remedies relating to those rights. These are transferred, however, 'subject to equities', that is, subject to any rights of set-off or other defences which would have been available to the other party to the contract against the assignor. Once contractual rights have been assigned, therefore, the assignee has the right to bring legal proceedings to enforce those rights in its own name, without the need to join the assignor as a party to the action.

Thus a professional should be in no worse a position by virtue of the assignment, if faced with a claim for breach of contract, than he would have been without the assignment. The claim will simply be brought by the third party instead of the client. As well as the same defences being available, the statutory limitation period will run from the same date; that is, the date of the alleged breach of contract.

The consent of the other party to the contract to the assignment of the benefits under that contract is not required unless the contract otherwise provides.

It must be noted, however, that the obligations (or 'burdens') under a contract cannot be assigned without the consent of the other party, again unless the contract otherwise provides. If consent is given, this would constitute a novation of the original contract – see Section 19.7.

It is also possible for that assignee to assign on the same rights unless expressly prohibited, as a case decided that even when the assignee or a number of assignments were expressly permitted that did not fetter the general unqualified right to assign 'rights' without consent. It is for this reason that many assignment clauses that permit some assignment of the rights add expressly 'but not further or otherwise'.

It should also be noted that although the 1925 Act sets out a mechanism for assigning contractual rights, it has not destroyed the common law and equitable principles, so that an assignment which is void under the 1925 Act, because the formalities have not been complied with, may still take effect as an equitable assignment.

An oral assignment of legal rights could therefore still take effect in equity – see Section 19.3.

An assignment cannot be used to change the nature of the contractual obligations which can only be changed by agreement. The effect of an assignment is to give the assignee the same rights that the client would have had – no more and no less. The damages recoverable would probably be no more than those that would have been recoverable by the client. An assignee cannot recover losses which could not be recovered by the assignor if there had been no assignment, and the assignee cannot as a result of the assignment claim for losses it itself has suffered. The unresolved issue is whether the assignee could recover losses of a kind that the assignor would not have suffered because, for example, the assignor and assignee are in different businesses and therefore could suffer different kinds of losses. The better view at present is that he could not. The assignment therefore only operates as a mere substitution of the contracting parties.

In July 2006, the Court of Appeal rejected the argument that an assignee who takes over an assignor's interest is precluded from recovering substantial damages on the basis that the assignor has suffered no loss. This was known as the 'no loss' argument which arose where the assignor had transferred its interest for full value (i.e. without knowledge of the loss) and it was argued had thus only a claim for nominal damages; however the Court held that the assignor had still suffered loss. Where a client owns property and the breach of contract occurs during the project when the client is still the owner, the client has incurred a loss even though it may later sell or lease the property before realising this. In such a case, the loss may fall on another party but could still be recovered by the client if he brought the case.

Many non-standard conditions of engagement and novation agreements, however, still include wording designed to get around the 'no loss argument', but this is now otiose.

Many contracts expressly prohibit the assignment of any rights of action in respect of accrued breaches without consent. The House of Lords (*St Martins Corporation Limited* v. *Sir Robert McAlpine & Sons Limited (1994)*) held that an assignment which was made in breach of a prohibition on assignment clause in the contract made after the employer had transferred its interest in the property but before the relevant breach of contract by the contractors did not give the assignee the rights to sue the original contractor. The House of Lords held that the rights to sue do not vest in the assignee independently of the invalid assignment. They did, however, hold that in such a situation the original contracting parties could still sue each other, even where the owners of a defective building had sold the property and no longer had an interest in it. However, at the time of the breach the original owners no longer had an interest in the building so this was a genuine case of no loss being suffered by the original owner. The House of Lords held, nevertheless, that the original owner was entitled to recover for the losses suffered by others (i.e. the new owner) on the basis that it could be foreseen that any damages caused by a breach of contract would cause loss to a future owner of the property and that the parties made the contract on the footing that they would be able to enforce the contract on behalf of any party suffering the loss.

As with other contractual provisions, however, such prohibitions on assignment may be waived by the parties expressly or by their conduct so that an assignment which was prohibited by the appointment is in fact valid.

Some professional indemnity insurers are concerned that assignment provisions in professional appointments (or silence on assignment) could allow the client to assign the benefit of one or more of the client's professionals' contracts but not all and can object to such provisions on that ground. This is because if the original client does not choose to sue the unassigned professionals, for example because the client is insolvent or no longer has an interest in the project, only one or some of the professionals (i.e. the assigned professionals) may be sued by the assignee where the unassigned professionals are not. So in these circumstances insurers may advise their insured to agree to assignment clauses, provided that such an assignment is only effective or permitted if all the professionals' appointments are similarly assigned.

From the client's point of view, it would seem unwise for there to be an unqualified prohibition of assignments by the client since it could amount to a fetter on selling the subject of the project.

(See also Section 2.3, dealing with the implications of a client acting on behalf of, or for the benefit of, third parties, and Sections 14.15 and 14.16 concerning passing on reports and the Technotrade case.)

19.3. Equitable assignments

In equity, an assignment of whole or part of the benefit of a contract is valid even if the legal formalities have not been complied with, provided the assignee paid for the transfer under the contract of the rights to the assignee. The effect of a valid equitable assignment is the same as for a legal assignment.

In contrast to the provisions of the Law of Property Act 1925, no formalities are required to make an equitable assignment of rights. No specific form of assignment is necessary, and the assignment need not be in writing. The assignor must, however, indicate that the rights have been transferred to another person, whether by request or permission. The equitable assignee cannot, however, sue the other party in its own name – it must join the assignor as a party to action.

As with legal assignment, no equitable assignment of the obligations or burdens of any contract is permitted without the consent of the other party. If consent is given, this could constitute a novation of the original contract – see Section 19.7.

19.4. Extracts from standard conditions of engagement

The position under the standard conditions is as follows.

ACE Agreement 1 provides:

> F11.1 Neither the Consultant nor the Client shall, without the written consent of the other, assign or transfer any benefit under this Agreement save that the Client may assign the benefit of this Agreement at any time up to twelve weeks after the Services have been completed without the consent of the Consultant and provided that all fees properly due to the Consultant up to the date of the assignment have been paid.

The exception in this clause to the prohibition on the assignment of rights would allow the client, for example, to assign the benefit of the appointment to a group company or another company or third party up to twelve weeks after the services have been completed provided all the fees to date have been paid.

RIBA 2010 prohibits assignment without consent.

> 4.1 Neither the Architect nor the Client shall at any time assign the benefit of this Agreement or any rights arising under it without the prior written consent of the other, which consent shall not be unreasonably withheld or delayed.

Neither ACE Agreement 1 nor RIBA 2010 deal with 'obligations'. Thus under these forms the obligations cannot be assigned without the other party's consent. It would be surprising, however, if they allowed the assignment of obligations because that would mean a new contract. Also, the client will not want to find that a professional he had not chosen and appointed is carrying out the services to be performed. Equally. the professional may be concerned if the obligation to pay has been passed to another client who may be a bad payer or one with whom he has had unhappy experiences in the past or who may have few assets.

RICS 2011 deals with both in more detail:

> 14.1 The Consultant does not assign or transfer any of its rights or obligations under this Appointment to any person without the prior written consent of the Client (which consent is not to be unreasonably withheld or delayed).

14.2 The Client is entitled to assign the benefit of this Appointment to any third party specified in the Appendix.

14.3 The Client is entitled to transfer all of its rights and obligations under this Appointment to any Funder and/or to any Client Group Company provided that, where such rights and obligations are transferred to a Client Group Company, such Client Group Company establishes to the Consultant's reasonable satisfaction that it is as able as the Client to perform and discharge the Client's obligations under this Appointment and, if it is reasonable in all the circumstances for the Consultant to request it to do so, the Client guarantees the obligations of the Client Group Company under this Appointment.

14.4 The Consultant does not contend that any assignee or transferee is precluded from recovering any loss it has incurred due to any breach of this Appointment because the assignee or transferee is not a named promisee under this Appointment.

It is possible that clause 14.3 would allow the client, if he so wished, to assign the rights to any third party specified. Alternatively, it may merely permit the client to have its obligations performed by another company but not release the original client.

The rights under RICS 2011 of the client to assign or transfer its obligations to a client group company are qualified to some extent so as to meet some of the objections the professional may have to such an assignment or transfer. The client group company has to satisfy the professional that it is able to perform the client's obligations under the appointment and the professional can 'if it is reasonable in all the circumstances' require the client to guarantee the obligations of the client group company. This would or should address any concerns about payment, for example, from a shell company, but it would not deal with any reservations the professional may have about the client *per se*.

The last point, in clause 14.4, is aimed at the 'no loss' argument – see Section 19.2.

PSC3 does not mention assignment and thus because the contract is silent assignment of the benefits of the contract would be permitted.

19.5. Assignments in collateral warranties

The assignment clauses in collateral warranties require special consideration. Briefly (this is discussed more fully in Chapter 21) a collateral warranty entered into by a professional with a third party gives that third party a right to sue the professional when it might not otherwise have been able to do so. Generally, the terms of the collateral warranty will set out the damages recoverable by the third party (or set some limit on these) and stipulate the time within which claims can be brought. Sometimes, however, collateral warranties are unlimited as to damages and silent as to the time within which claims can be brought. If assignment of the collateral warranty by the third party is prohibited, the risk to the professional of being sued upon the collateral warranty would come to an end once the third party had ceased to have any interest in the building or project which is the subject of the collateral warranty. If, however, the benefit of the collateral warranty can be assigned (as will be the case if nothing is said), the professional is at risk of being sued by the assignees until the time limit for bringing claims under the collateral warranty has expired. This could mean that many subsequent assignees may have the right to bring claims against the professional, although any restrictions in the original collateral warranty as to the amount that can be recovered will also apply to those claims. Professionals will therefore usually seek either expressly to exclude any right to assign the benefit of a collateral warranty or restrict the number of such assignments. Further, some professional indemnity insurance policies have specific provisions as to the number of assignments that can be permitted under any collateral warranty, and professionals must ensure they check these before agreeing to assignment provisions in collateral warranties and include any such restrictions in the collateral warranty.

It also needs to be remembered that even if a claim by an assignee is unsuccessful, the professional will incur time and expense in dealing with that claim which may not be recoverable.

19.6. Effect of the Contracts (Rights of Third Parties) Act 1999 on assignment

Prohibitions on the right to assign do not necessarily prevent claims by parties who might otherwise have been assignees. Third parties may be able to bring claims in tort or as a result of the Contracts (Rights of Third Parties) Act 1999. Under the Act a third party may be entitled to enforce a term of a contract in certain circumstances – see Section 2.3. The Act provides that the third party must be expressly identified by name or as a member of a class or as answering a particular description. Thus, a generic description of third parties can be used in the contract between the client and the professional, such as 'occupiers', 'funders', 'tenants' or 'purchasers'. Third parties coming within these descriptions have the right to enforce a term of the contract against the professional, such as the right expressed to be for the benefit of all tenants to have an air-conditioning system designed with reasonable skill and care. There would be no limit on the number of tenants who can sue the professional if the air-conditioning system has been negligently designed, provided they come within the description and subject to the statutory time limits for bringing their claim. In the example given, any 'tenant' of a building described as such in the appointment, be it the first or any subsequent tenant, would be able to sue the professional, even though the same tenant would not have been able to do so unless it was a permitted assignee under the appointment. If the tenant had an assignment of a collateral warranty, again it would not have been able to sue the professional if it was outside the permitted number of assignments – see Section 19.5.

Conversely, where rights of assignment have been permitted to certain third parties, those rights can be excluded from the general exclusion of rights under the Contracts (Rights of Third Parties) Act which is usually found in the professional's appointments.

19.7. Novations generally

As has been said, the only way in which contractual obligations or burdens can be assigned, either in law or in equity, is with the consent of the other contracting party unless otherwise provided in the contract. This is achieved by the parties to the original contract and the third party to whom the obligations are being assigned entering into a new appointment – a novation agreement – whereby it is agreed that the obligations will now be undertaken by the third party.

Strictly, a novation agreement is one which substitutes one party for another and substitutes the original contract for one on the same terms. It should not change the terms of the original contract.

As a novation agreement replaces the original contract, the usual requirements for the validity of contracts will apply on the making of the agreement. See also Chapter 20 on novation to a design and build contractor, which raises particular considerations.

19.8. Novation by a client to another client

Unless provided otherwise in the appointment a professional can withhold consent to an assignment to a third party of its client's obligations under the professional's appointment. This would include the obligation to pay the professional's fees. As has been said, there could be real objections to this. For example, the professional may not wish to work for that particular client, the client might be close to insolvency or there may be a conflict of interest. Some appointments may provide for the assignment of the burdens and benefits of the contract without consent, or may oblige the professional to agree in advance to a novation in a certain form to a specified third party. This is often the case where there is to be a design and build contractor and the client wishes to novate its professionals to that contractor – see Chapter 20.

The novation agreement should be drafted on the basis that the retiring client is discharged from any further liability and thus steps out of the contract completely, and the new client assumes responsibility for all the contractual obligations, including past obligations.

However, most people use 'novation' somewhat loosely to cover, for example, an arrangement where the whole benefit and only part of the burden is assigned with consent. For example, when a new client takes over an appointment, the new client usually agrees to pay future fees but often does not wish to take on the burden of paying fees accrued but not paid, or responsibility for certain past decisions of the original client.

This is not satisfactory. The new client is going to want to sue the professional for any breaches that may have occurred before the new client took over, and it should work both ways. The professional will not want to sue the original client for some things (and particularly not for non-payment of fees if the reason the appointment is being novated is that the client is in financial difficulties) and the new client for others.

There should also be no variation to the time within which claims can be brought. The time limits to apply should be those which would have applied if the original client had remained the professional's client throughout. Some novation agreements try to give new clients extra time for bringing claims in respect of pre-existing breaches. This should not be accepted by a professional. Also, if the original appointment was under hand, a novation agreement should not be signed as a deed.

Any novation agreement should also be checked for any unusual or onerous terms in the same way that any other contract would be checked.

19.9. Novation by a professional to another professional

Most appointments state expressly that a professional must not assign its obligations, for example, the performance of the services, without the client's consent. The client then has the opportunity to check and approve the new proposed professional's abilities to carry out those obligations. Neither the ACE Agreements nor RIBA 2010 mention the assignment of obligations and so the usual common law rules apply: neither party can assign the obligations under the appointment without the other party's consent (unless otherwise provided, which it is not).

The RICS 2011 provisions on assignment are discussed in Section 19.4.

A prohibition on assigning obligations would also apply to any proposed transfer of burdens or obligations from one legally separate part of the professional's organisation to another.

19.10. Interaction with sub-contracting

Although a party to a contract may not, without the consent of the other party, transfer its obligations to a third party, it is generally entitled to arrange for those obligations to be performed on its behalf, say by a sub-consultant, while remaining liable for any defect in such performance – see Section 12.3. However, where a professional is employed for the professional's special skill and knowledge, there may be an implied prohibition against sub-contracting. It is best, therefore, to obtain consent to any sub-contracting, in advance, in the original appointment document. It is now usually the case that non-standard conditions of engagement prohibit sub-contracting without consent.

Professional Services Agreements
ISBN 978-0-7277-4158-5

http://dx.doi.org/10.1680/psa.41585.307

Chapter 20
Novation of the professional by a client to a design and build contractor

20.1. Introduction

The construction industry has in recent years seen a substantial increase in projects carried out on a design and build basis, and with some or all of the professionals originally employed by the client being novated to the building contractor usually at the time the building contract is let.

The main reason why clients like the design/build route is that this gives one-point responsibility – in broad terms, as the design and build contractor is responsible for both design and construction, if anything goes wrong the client does not have to ascertain whether it has been caused by a design error or a construction error: the design and build contractor is responsible for both and it is the design and build contractor who will have to sue any of the professionals if he feels they have been at fault. However, clients were concerned about the quality of the professionals who might be employed by the design and build contractor and so they employ their own professionals to prepare the preliminary design and the employer's requirements (or sometimes much more than this) and then provide that the design and build contractor is to employ some or all of these same professionals to fulfil the contractor's obligations to complete the design. This brings with it the obligation to examine the design at the point at which responsibility was taken over, to understand the principles underlining the work done thus far and to form a view as to its sufficiency.

There are in the construction industry at the time of writing two types of novation – the *ab initio* and the 'switch' forms.

In order to understand the differences between them, it is helpful to understand what is meant by novation and the situations for which it was originally intended.

20.2. The meaning of 'novation' and the situations for which it was intended

Novation is usually discussed in connection with assignment, that is the assignment of rights and obligations under a contract – see Chapter 19. However, the point usually made is that technically novation is not the same thing as assignment.

Novation is a concept that was borrowed from Roman law in the mid- to late-nineteenth century because of the problem of assigning such rights and obligations. In those days there was no right in law (although there was in equity) to assign rights and there was no right at all to assign obligations, without the consent of the party to whom the obligations were owed.

This caused problems at a time when companies were amalgamating and partnerships were changing with more frequency. When a company amalgamated with or was taken over by another company, did a person who had a contract with the original company now have a contract with the new company? If the composition of a partnership changed – if a partner retired or a new partner joined – did a person who had a contract with the old partnership now have a contract with the new partnership?

The answers in these cases depended on whether the person contracting with the company or partnership had clearly accepted that the obligation of the original company or original partnership could now be performed by the new company or new partnership. If he had, then the contract with the old company or old partnership was extinguished.

Another issue concerned the transfer of debts. In the classic example, A owed £100 to B and B owed £100 to C. All three parties may be happy to resolve this situation simply by agreeing that A will pay the £100 to C. But if A does not pay, can C then recover from A? Has there been an effective novation? Only if it was clearly intended that the debt due from A to B was to be extinguished – otherwise the novation would fail for want of consideration.

These were the sorts of situations for which novation was intended – where it made obvious commercial sense for the contract to be transferred, but, because of the technicalities of the law, this could only be achieved by the agreement of the parties, with one contract being replaced by another.

Today, as discussed in Chapter 19, one can assign the benefit of a contract without the consent of the other party, but one still cannot assign the burden or obligations of the contract, without consent, unless the contract provides otherwise.

So no party to a contract can assign the entirety of a contract. The consent of both parties is still needed, which means that one contract is extinguished and replaced with another, and there is no actual assignment.

Novation, although it is not in fact an assignment, is intended to achieve the effect of an assignment.

Thus the person who is a party to the first contract but not to the second, that is the person in the position of the assignor, drops right out of the picture. The person who is a party to the second contract but not to the first, that is the person in the position of assignee, inherits the position of the assignor and should be in no different or better position than the assignor was at the time of the assignment. As for the other person, who remains a party to both contracts, his position should remain unaltered.

It is open to parties to agree variations to this, but if they do they are departing from the notion of a pure novation.

A novation, like an assignment, works best in straightforward situations, where the obligations that are novated, like payment of a debt, are not likely to be affected by the identity of the new party.

Payment of a debt would be a one-off obligation. The obligation may, though, be an ongoing one. X may be engaged to Y to build a car. Y then decides that he wants X to build the car for Z instead. So, halfway through the building of the car, X enters into a new agreement, whereby X will now build it for Z.

In a case like this, where a contract has been partly performed at the time of novation, what is being novated? This can only be the performance of the remainder of the contract, and any liabilities arising out of the performance of the contract to date.

One cannot novate obligations that have already been performed. Take the simple example of a debt of £100, where the creditor agrees to accept the obligation of another party to pay the debt in place of the original debtor. If, before novation of the debt takes place, £50 of it is paid, then in reality the debt novated will be a debt of the remaining £50, and not the original £100.

In the construction context the professional is not going to start his design all over again when employed by the contractor.

But while novation cannot cover obligations already performed, it will cover any liabilities arising out of the past performance of those obligations. However, those will, of course, be liabilities that have accrued in favour of the original party.

20.3. Differences between novation to another client and to a design and build contractor

It needs to be appreciated that there is a very real difference where the novation is to be to a second client and where the novation is to be to a design and build contractor.

In design and build novation, the original client does not drop out of the picture – he continues to carry out the same project. The contractor is to construct that project for the original client and only takes over the professional's appointment, not the project itself. The client and the design and build contractor have very different interests.

- The client wants the project carried out to high standards within his budget and timescale; then he will sell or use the building.
- The contractor wants to construct the project as efficiently and economically as possible, make a profit and then leave.

In all this, the tender price of the contractor is crucial. The client is looking for a competitive price. The contractor wants to know as accurately as possible what it is going to cost him to build it.

20.4. When novation takes place to a design and build contractor

Generally, novation takes place when the building contract is entered into and therefore *after* the contractor has tendered for the work, but it could take place earlier than that. Also, since it is to be a design and build contract, the contractor at that stage will be taking over responsibility for some or all of the design. Equally, novation takes place after many of the professional's obligations have already been performed.

What is being novated at that time therefore are only the obligations under the original appointment of the professional that have yet to be performed by the professional. This could be, at one end of the scale, when nearly all the design has been completed, or at the other end when the design is still at a preliminary stage – for example, only that necessary to put together the employer's requirements.

As has been said, obligations that have already been performed cannot be novated unless the professional is to do them again.

20.5. Nature of liability arising out of services performed prior to novation

As to the liabilities arising out of the design done prior to novation, that is in respect of obligations already performed by the professional, these obligations can only ever have been performed for the original client even if the design and build contractor may now take the benefit of them – that is, will now be able to sue the professional in place of the original client.

As has been said, a professional does not owe duties in a vacuum – see Section 4.1. He owes a duty to a particular person and his relationship to that person determines the scope of his duty and thus the kind of loss for which he is liable (and therefore for which he is liable to that person's assignee: the design and build contractor in a design and build novation).

So the relevant question in law relating to the liability of a professional for loss arising from his performance prior to novation is 'Was it the kind of loss that the professional owed a duty *to the original client* to use reasonable skill and care to prevent?' If so, he will be liable to the design and build contractor.

On novation, the professional does not escape all liability for services performed prior to the novation. The design and build contractor will be able to sue the professional in respect of errors committed prior to novation if they resulted in the kind of loss for which the professional was under a duty of care to his original client to avoid, for example, the cost of repairing defective work resulting from a defect in the professional's design prior to the novation. This offers some protection to a design and build contractor but does not help him if he has got his tender price in relation to the professional's design wrong, because that is not the sort of loss a professional is under a duty to his original client to avoid.

20.6. The Blyth & Blyth case

In the Blyth & Blyth case, the engineers had done some design for the client. The contractor used the professional's information in the employer's requirements prior to tender to estimate how much steel reinforcement would be needed, and costed on that basis. Thereafter, the building contract was let and the professional novated to the contractor. The contractor then discovered that much more steel reinforcement was needed and sued the professional on the basis of alleged deficiencies in the employer's requirements which resulted in their tender price being too low. As the design and build contractor had accepted responsibility for the design he had to bear this loss and could not recover it from the client.

The novation agreement by which Blyth & Blyth were novated to the contractor contained the following clause:

> The liability of the Consultant under the Appointment whether accruing before or after the date of this Novation shall be to the Contractor and the Consultant agrees to perform the Appointment and to be bound by the terms of the Appointment in all respects as if the Contractor had always been named as a party to the Appointment in place of the Employer.

There may be room for debate over what this clause means, or what is the effect of an agreement containing such a clause. It may simply refer to the performance of the professional's services following novation. If so, that ought to be made clear.

However, if it means that the performance of the professional prior to novation and his potential liability for work done prior to novation will be judged on the fictitious basis that during that time the professional had in fact been engaged by the contractor, it would involve a retrospective variation in the scope of the professional's duty. That would be unacceptable to the professional and its professional indemnity insurers, and it would not make commercial sense from the professional's point of view. It would also be departing from the principles of a novation.

The novation agreement continued as follows:

> . . . the Consultant agrees that any services performed under the Appointment by the Consultant or payments made pursuant to the Appointment by the Employer to the Consultant before the date of this Novation will be treated as services performed for or payments made by the Contractor and the Consultant agrees to be liable to the Contractor in respect of all such services and in respect of any breach of the Appointment occurring before the date of this Novation as if the Contractor had always been named as a party to the Appointment in place of the Employer.

It appears from this that the intention was to alter the scope of the professional's liability retrospectively, but the judge was not convinced. He pointed out that it would produce nonsensical results and a conflict of interest, and said:

> It would in any event be inherently unlikely that the parties should intend the effect of the Novation agreement to be that of re-casting a duty owed and performed to the employer as being a duty owed to the contractor retrospectively.

Counsel acting for the contractor conceded that that must be so. However, this effectively doomed the contractor's claim.

The responsibility for getting the tender price too low was not the kind of loss which the professional was under any duty to the client to use reasonable skill and care to avoid, so no claim could be made against the professional, without re-casting his duty retrospectively.

Professionals may advise clients on aspects concerning tenders but they do not become responsible to the client if the design and build contractor makes insufficient allowance for any particular item in his tender as this is not a loss the client will suffer.

The whole debate about *ab initio* and 'switch' novation agreements is really an argument about to whom and in respect of what the professional should be liable in respect of services performed prior to the actual novation.

20.7. *Ab initio* novation

Design and build novation *ab initio* is based on the traditional form of novation, when the first client transfers his interest in the project to a second client and the second client steps into the shoes of the first client as far as the professional's appointment is concerned and is treated as if he was always a party to that contract, in place of the first client. The clauses they contain are often similar to those in the Blyth & Blyth case quoted in the previous section. Bespoke forms of *ab initio* novation agreements can have additional clauses which seek to make clear for what sort of losses the professional will become liable to the design and build contractor.

But the way in which these *ab initio* forms seek to make the professional liable to the contractor for services carried out for the client prior to the novation is to create the fiction that the professional's client has been the contractor from the beginning – the retrospective re-casting of the duty of care the judge referred to in the Blyth & Blyth case.

This does not make any sense in reality. How can a professional agree to the legal fiction that he has always worked for the contractor? Up until novation the professional is working to the requirements of his client and he should have his interests only in mind.

He was not in fact employed by the contractor at that time and was therefore under no duty to advise him as to the sufficiency of his tender price, nor as for what contingencies and allowances he should provide in his price. Indeed, had he done so at the time, he would have been in conflict with the duties he owed to his client and would be acting in a manner which was adverse to that client's interests.

If, however, the professional agrees to this retrospective duty of care the contractor may have a remedy if he can show that if he had been properly advised by the professional, on the basis that the professional was acting for him at that time, the contractor would not have got his tender price wrong.

It is not necessary for the words *ab initio* to be used to create this sort of liability.

Examples where it could be created are set out below. These too are all fictions.

(*a*) An undertaking or warranty that the professional owed a duty of care to the contractor from the outset.
A professional should not owe a duty to the contractor prior to novation, even if it could be argued that the scope of the duty owed to the contractor is no different to the scope of the duty owed to the client. A professional should not owe a duty to the client and contractor at the same time.

(*b*) An acknowledgement or undertaking that the contractor has 'relied' on the information and advice provided by the professional to the client – despite the fact that the information and advice was provided with the client's interests in mind and may indeed never have been passed or given to the contractor.
In accepting this, the professional might be accepting a tortious duty to the contractor in respect of work done whilst his client was the original client. The scope of any such duty may be unclear, but such provisions should be avoided by professionals.

(*c*) An acknowledgement or undertaking that any losses suffered by the contractor arising out of the services provided by the professional for the client 'were within the contemplation' of the professional at the date that the professional first entered into a contract with the client – even though at that stage the professional may have no idea who the contractor will be.
Again, in accepting this, the professional could be making itself vulnerable to claims for losses over and above those for which he would be liable to his client. This is exactly what the contractor may want. Whether words such as these would actually achieve that may be open to question. A professional, however, should not take the risk.

20.8. Client's position under *ab initio* novation

Clients are not concerned about the liabilities their professionals incur to third parties – in this case the design and build contractor before he becomes the professional's employer. They do, however, want their design and build contractor to take on full one-point responsibility for all design and construction, and if an *ab initio* form of novation agreement helps to achieve this the client will try to procure one from their professionals.

The client's position is not affected either way – having passed over all responsibility to the design and build contractor he has no need to reserve any rights to sue his professional in respect of any errors that may have been committed prior to novation. He may of course want a warranty back from the professional in respect of the post-novation services, although, having opted for one-point responsibility, having a collateral warranty under which the client can sue the professional direct, for example if the design and build contractor becomes insolvent, smacks of having your cake and eating it.

20.9. Switch novation

This is not a novation in the traditional sense. The 'switch' recognises the reality of the design and build situation. The professional works for his client at the outset and then from a certain stage works for the contractor. The same result could be achieved by bringing to an end the original appointment by the client comprising the services to be performed up to novation, and having a new appointment between the design and build contractor and the professional comprising the services the professional is to perform thereafter for the design and build contractor. Under a 'switch' form of novation the professional switches from acting for one client – the original client – to acting for another client – the design and build contractor at the point of novation.

There is a review of the standard form for *ab initio* novation – the CLLS (City of London Law Society) Agreement – and the switch novation – the CIC novation agreement – at Sections 20.15 and 20.16, in this chapter.

20.10. Post-novation services

While there must be no retrospective change in the scope of the professional's duty prior to novation, there is an inevitable change following novation. The professional's duty is now owed to the contractor, and the contractor's interests and concerns are different from those of the client. The professional must take this into account in the performance of his duties following novation.

This may not only affect the manner in which the professional performs his duties but may call into question the content of those duties. The services that a contractor requires may be different from those that the client would have required had there been no novation. Some of the services set out in the original appointment may not be appropriate for an engagement by a contractor. An obvious example would be to assist the client in any dispute with the contractor. Clearly, the professional could not now perform that service. Could he assist the contractor in disputes with the client? It is thought not, given that he was formerly engaged by the client.

There may also be services in the original appointment that should be differently worded in circumstances where they are being performed for the contractor rather than the client and there may be additional services not mentioned in the original appointment that the contractor would like the professional to perform.

This situation has to be addressed at the time of novation. There is no reason in principle why the parties should not agree to a variation in the services remaining to be performed. Whether this is necessary will depend on what is in the original appointment and how the problem was addressed, if at all, at that stage.

20.11. Avoiding conflicts of interest

If there is going to be a novation to a design and build contractor, it is crucial to avoid conflicts of interest. This is why it is vital to distinguish between the situation prior to novation and the situation after novation. One must avoid any attempt to make the professional serve two masters during either stage.

When originally appointed, the professional may agree at that stage to the appointment eventually being novated to the contractor. However, he should not owe a duty to the contractor until he is actually engaged by the contractor and released from his obligations to the client. Otherwise there is a real risk for all parties concerned that he will not be able to perform his services properly.

It is important to remember that where the contractor is formulating his tender price before the professional is novated to him and at a time therefore when the professional is engaged by his original client, the professional owes his duty of care to his client and should be protecting his interests. The professional should owe no duty at that time to the contractor, nor should he be seeking to protect the contractor's interests by, for example, putting warnings on his tender drawings or otherwise as to the stage of the design nor as to what assumptions should be made by the contractor or what risks the contractor should allow for. This would be seeking to protect conflicting interests for which the professional could properly be criticised.

Some non-standard conditions of engagement may also contain duties that the professional is to continue to perform for the client after the professional has been novated to the contractor, for example, reporting to the client on the contractor's performance during construction. This again puts the professional in the position where a serious conflict of interest could arise and should be resisted.

20.12. Confidentiality

It also needs to be remembered that the professional may well, while acting for the client, be given confidential information which he will be under a duty to the client not to divulge to the contractor once employed by him. As to how he is to reconcile this with the duty he will owe to the contractor after novation to inform him of matters relevant to his interests which could include that confidential information – see Section 11.4.

20.13. When liability to the contractor can arise outside any novation

Should a professional volunteer advice to the design and build contractor during the tender period and/or should he answer queries from the design and build contractor on which the contractor relies, then the professional could become liable to the contractor if he does so negligently under the law of tort, regardless of any terms in his appointment or any novation agreement – see Section 14.3. This could happen quite outside and independently of any novation agreement, for example, if the professional discusses his design and its stage of development with any of the contractors during the tender process, or if he gives specific assurances to the contractor concerning his design or elements of it. Whether this would in fact give rise to liability for the losses actually suffered would depend on the scope of the duty assumed.

The *Galliford Try Infrastructure Ltd* v. *Mott MacDonald Ltd (2008)* case was put on this basis in the absence of any agreed novation – that Motts had made representations to Galliford Try as to the prop forces that could be taken by the concrete slab part of the floor and/or that the steelwork would not be required to perform this function. The judge found there was no representation of this sort by Motts, no reliance by Galliford Try and that Motts had not been negligent.

20.14. 'Equitable' or implied novation

'Equitable novation' is not a recognised legal term and in fact it is probably better described as 'implied novation'. The expression is used as a shorthand to describe the situation where the three parties involved, the client, the professional and the design and build contractor, knowing that novation of the professional to the contractor has been agreed in principle, act as if the novation has been effected without any formal deed of novation or new appointment ever being entered into.

The way this usually happens is that the professional stops acting for his original client and performs his remaining services for the contractor and the client stops paying the professional and the contractor pays the professional the remaining fee. It will be remembered that 'novation' is in effect the ending or extinguishing of the original contract and the formation of a new contract whereby the contractor and the professional agree to perform the outstanding obligations under the original contract for the other. There is no reason in law why this cannot be implied by or inferred from the conduct of the parties.

In the absence of express terms (which could be in writing or oral), the terms inferred for the new contract between the contractor and the professional are likely to be restricted to basic matters. The court may have regard to what is regularly agreed on novations and assume that what the parties intended, applying an objective test, was that the terms of the original contract applied to the new contract, subject to any necessary amendments and any expressly agreed variations, for example, to the services to be performed for the contractor. Whether there is in fact a new contract will be determined on the usual principles.

Since the parties have in effect agreed by their conduct that the professional should now act for the design and build contractor under a new contract, it is likely that the courts will hold that the original contract has come to an end by mutual consent and both parties will be estopped from asserting that there are still outstanding obligations to be performed by the other. In a formal novation agreement, this is dealt with by the client and the professional formally releasing and discharging the other from the further performance of obligations under the original contract, with perhaps some agreed exceptions.

This raises the question as to whom the professional is to be liable if he has been negligent in the performance of the services for the client prior to ceasing to act for him.

Normally, this would be dealt with in the novation agreement itself. If it is not, the right to sue the professional would probably remain with the client. The contractor might try to argue that under the new contract the professional had undertaken liability on the basis of having been employed by the contractor *ab initio* and it could be that such an argument might succeed. However, it is not thought that a court would readily infer, without more evidence, that the professional had agreed with the contractor to be liable to the contractor for breaches of the appointment with the client prior to novation or that the client had agreed to release and discharge the professional from any such liability to the client.

If the contractor is to have a formal right to sue the professional for breaches prior to the novation, the client would have to assign those rights to him and might in fact do so at a later date. If it is to be a legal assignment, it has to be done in writing and notice given to the other party: the professional. There could be an equitable assignment of those rights, in which case the contractor could sue the professional, but he would have to do so in the client's name. The court would have to be satisfied that such an assignment had been effected. It is unlikely, however, that a court would readily infer an implied assignment of his rights from the client to the contractor and thus that this was an implied term in the new contract.

As to the professional, what would his position be if the original client still owed him fees? Again, it would probably be difficult to imply a term that the contractor had become responsible for payment of outstanding fees due from the client in the absence of evidence to the contrary and so the professional's right to sue the client for those outstanding fees would remain.

20.15. The CLLS novation agreement

The CLLS published in 2004 a standard form of novation agreement for the circumstances where the professional is to be novated from the employer to a design and build contractor.

This form adopts what might be called the *ab initio* approach, in that it states that:

> 1.2 The Consultant undertakes to perform the Appointment and to be bound by its terms in every way as if the Contractor were, and had been from the inception, a party to the Appointment in lieu of the Employer.

This clause, by itself, may be interpreted as referring only to performance of the appointment following novation, but there is a risk that it could be interpreted as varying retrospectively the scope of the professional's duties prior to novation.

The Agreement continues:

> 1.4 ... the Consultant warrants to the Contractor that it shall be liable for any loss or damage suffered or incurred by the Contractor arising out of any negligent act, default or breach by the Consultant in the performance of its obligations under the Appointment prior to the date of this Agreement. Subject to any limitation of liability in the Appointment, the Consultant

> shall be liable for such loss or damage notwithstanding that such loss or damage would not have been suffered or incurred by the Employer (or suffered or incurred to the same extent by the Employer).

Whether the professional has committed a 'negligent act, default or breach in the performance of its obligations under the Appointment' will depend, in any given situation, on the nature and scope of the duty owed by the professional at the time of performance and the kind of loss that the professional was then under a duty to take reasonable skill and care to avoid. That duty will have been owed to the client and the professional will, therefore, only be liable for the kind of loss that the professional was under a duty to the client to take reasonable skill and care to avoid.

However, this clause could be interpreted as making the professional liable for losses outside the scope of his duty of care, especially when read in conjunction with the earlier *ab initio* wording. This would again involve a retrospective variation in the scope of the professional's duty, which would be unfair to the professional, as it would to any professional person. This agreement should therefore be avoided by professionals.

The contractor too undertakes to perform the appointment and be bound by its terms as if it had been a party to the appointment in lieu of the client from 'inception'.

20.16. The CIC novation agreement

The CIC does not endorse novation in the design and build situation, but its standard form novation agreement, published in 2004 and reproduced at Appendix 2, aims to take into account the realities of the situation and address some of the practical problems that arise.

The CIC novation agreement is a 'switch' type of novation. The way the CIC novation agreement achieves this is by the professional remaining liable to the client in respect of the work done for him prior to the novation. Post-novation, the professional agrees to work for the contractor and to be liable to him. The contractor agrees to step into the shoes of the original client post-novation and thenceforth act as the client. Thus the 'switch' happens at the point of novation.

It is worth looking at the provisions of the CIC novation agreement in some detail.

20.16.1 Clause 1 – mutual release

Under clause 1, the client and the professional release each other from further performance of their respective obligations under the appointment.

This release does not extend to accrued liabilities; that is, liabilities arising out of services already performed. These are therefore preserved.

There are also two specific performance obligations on the part of the professional that are excepted from the release, if of course they are terms of the original appointment. The first is any obligation that there may be to give collateral warranties at the request of the client. The other is any obligation of confidentiality in the appointment. However, this is only so far as this does not conflict with the duties now owed by the professional to the contractor. This recognises that the contractor is now the professional's client.

20.16.2 Clause 2 – future performance by the professional

Under clause 2 the professional agrees to perform all outstanding services and other obligations for the contractor.

However, there is provision for these services and other obligations to be varied if the parties so agree.

Subject to that, the terms of the original appointment apply to the performance of the remaining services. This includes any limitation of liability in the original appointment.

20.16.3 Clause 3 – future performance by the contractor

Under clause 3 the contractor agrees to perform all the client's outstanding obligations under the appointment, from which the client has of course been released.

Again there is provision for the parties to vary these obligations by agreement. For example, in the light of any variations under clause 2 to the services, the parties might want to agree a variation to the fees to be paid for these services.

Subject thereto, the outstanding obligations are again to be performed on the terms of the original appointment.

20.16.4 Clause 4 – professional's warranty to the contractor

Under clause 4 the professional gives a warranty to the contractor concerning the services carried out by the professional prior to novation. This is because the contractor will accept some responsibility for design under a design and build contract and this will generally include an element of the design done prior to novation (when the professional was acting for the client). The contractor will want a remedy against the professional who did the design work to cover him in the event there is an error in that design and he suffers loss as a result.

Note, first, that this is confined to services for which the contractor has now taken responsibility. This may not be all the services. For example, under the JCT Major Projects Form, the contractor is not responsible for the adequacy of the design contained within the employer's requirements. So where the main contract is on that form, the warranty will not extend to the professional's designs in the employer's requirements.

The next point to note is that the professional merely warrants that all such services have been performed 'for the employer' in accordance with the terms of the appointment. The intention here is that the professional can only be liable to the contractor, as far as pre-novation services are concerned, for acts or omissions that constitute a breach of the duty that was owed to the client.

The point is underlined by clause 4(b), which expressly states that the professional can raise the equivalent rights in defence of liability as if the claim were being brought by the client.

The last part of clause 4(b) reads:

> . . . save that the Consultant shall not be absolved from liability to the Contractor for such loss merely by virtue of the fact that the loss has not been suffered by the Employer.

This is an exception to the 'equivalent rights' rule and was included to meet a concern that some people felt about the effect of the Blyth & Blyth decision. Their view was that, in a situation where the client had passed all the risk on to the contractor, the contractor would never be able to claim for any loss arising out of a breach of duty by the professional prior to novation, because such loss would not have been suffered by the client.

Thus, taking the earlier example of the cost of repairs caused by negligence in the professional's design, could it be argued that, because the client is no longer, following novation, having to bear these costs, the client has suffered no loss, and as this would give the professional a complete answer to a claim by the client, a claim by the contractor must therefore fail? This is sometimes referred to as the 'black hole' or 'no loss' argument.

It is not thought that the Blyth & Blyth decision went as far as that. Even if it did, it is not thought that sort of reasoning would succeed in an English court.

What Blyth & Blyth was concerned with was not whether the client had actually suffered the loss in question but whether the loss in question was of a kind that the professional was under a duty to the client to use reasonable skill and care to prevent. The question of whether the client has actually suffered any such loss may be an indicator as to whether this was loss of that sort. In Blyth & Blyth the loss in question was of a kind that could never, by its very nature, be suffered by the client, and could hardly therefore be categorised as the sort of loss that fell within the scope of the professional's duty to the client.

Nonetheless, in order to avoid 'no loss' arguments, if the loss is of a nature for which the professional is or would have been liable to the client, the last few words of this clause provide that the professional could not defeat a claim brought by the contractor merely by virtue of the fact that the loss has not been suffered by the client, although such wording is probably no longer required – see discussion of the 'no loss' argument in Section 19.2.

It still has to be shown that the loss in question is loss that the professional was under a duty to the client to use reasonable skill and care to prevent. Anything else would involve a retrospective variation of the scope of the professional's duty.

The CIC agreement does not contain a warranty to be given to the client in respect of services performed following novation. If such a warranty is given, then the converse should apply. The professional can only be liable to the client for the kind of loss that it is within the scope of his duty to the contractor to use reasonable skill and care to prevent. Such a warranty may be of only limited value to the client. Although it is not part of the agreement, the CIC has drafted such a warranty for use with the novation agreement. This can be found on the CIC website in the liability and contracts section at www.cic.org.uk as a supplemental agreement.

20.17. Conclusion

The CIC novation agreement should be the only one recommended to professionals for the design and build situation. It has been endorsed by all the relevant consultant bodies. This will also mean that the professionals will be engaged on a proper basis and not put in a position of conflict of interest and/or where their duties of confidentiality could be broken. Several projects have used it.

A final cautionary note: we are already seeing adjudication and claims arising out of novations of professionals to design and build contractor and these will only get worse (i) as time goes on and more projects are carried out on this basis; and (ii) in a recession people (particularly contractors) sue to try and recover the monies they have lost.

There is attached at Appendix 3 to this book a short note on the switch and *ab initio* novations which may be helpful for passing to clients and contractors and their lawyers to explain why the CIC novation agreement is the one professionals are advised to use.

Professional Services Agreements
ISBN 978-0-7277-4158-5

http://dx.doi.org/10.1680/psa.41585.323

Chapter 21
Collateral warranties

21.1. What are collateral warranties?

To understand collateral warranties, it is necessary to go back briefly to the law of negligence and the question of to whom it is that the duty of care is owed. The common law duty of a professional is to take reasonable skill and care in the performance of professional services. A professional owes a client a concurrent duty in contract and in tort – see Section 4.1. The impact of the case of *Murphy* v. *Brentwood District Council (1991)* was to restrict liability for economic loss in tort – defined to include the cost of rectifying defects and damage caused by defects to the property containing defects. However, there will be situations where there will be a duty not to cause economic loss and therefore where the professional can become liable in tort for economic losses. These cases often occur where professional advice is given – see Sections 14.3 and 14.4. It is now unlikely that designers would in most cases be liable in tort for economic loss, as defined above, to future occupiers. If a duty of care could be established then the recovery would be limited to personal injury and damage to third party property.

In the early- to mid-1980s the courts were becoming reluctant to impose a duty of care in negligence by a professional to anyone other than to the client with whom the professional had a contract. This gave the client problems with prospective building occupiers (such as users, purchasers and tenants) who could be taking over a potentially defective building without the possibility of being able to sue the contractor or designer who might have been at fault. This was because the construction team had no direct contract with the occupiers who in turn would probably not be able to show they were owed a duty of care by the construction team in negligence. They could not claim against the client because the client had not designed or constructed the building. The client would probably not be able to bring a claim on the occupier's behalf (even if the client could be persuaded to do so) because, having sold the building or let it on full repairing leases, the client had suffered no loss and thus had no basis for a claim. Funders were also concerned because, if the client had financial difficulties and the funder exercised a right to take over and complete the building (or sold it in a half constructed state), the funder might have no legal cause of action against the construction team.

The concept of a collateral warranty was evolved to try to create a direct contract, a 'collateral warranty', between the funder, purchasers/tenants, future occupiers, etc. on the one hand and the contractor and the rest of the construction team on the other. If there was poor construction, defective design or even inspection, those third parties could then sue the contractor and/or the rest of the construction team, including professionals, directly in contract.

Although it is possible that purchasers/tenants, etc. could be given rights under the professional's appointment by the Contracts (Rights of Third Parties) Act 1999 – see Section 2.3 – and this is sometimes used by clients, many appointments still provide that the professional is to enter into collateral warranties, and the Contracts (Rights of Third Parties) Act rights are specifically excluded.

A collateral warranty is described as such because it is a contract which is collateral to the construction contract or the professional appointment. It is also a statement or representation about what a contractor or professional did, since it is often signed after the works are completed.

A collateral warranty gives third parties the right to sue the giver of the collateral warranty when they might not otherwise have that right in law. The question of what matters the third party can sue in respect of and the extent of the damages recoverable will depend on the wording of the collateral warranty itself. A professional should consider carefully before agreeing to give a collateral warranty, especially if the professional is not being paid extra to undertake this additional liability. If there are no overriding commercial reasons, why should the professional extend his risks for nothing? A professional will be under no obligation to enter into a collateral warranty if the client has not included in the professional services agreement an undertaking or agreement that the professional would do so or has failed to agree an acceptable form of wording with the professional. If the client then approaches the professional after the services have been completed or the end of the project, and asks for collateral warranties in favour of future owners/occupiers, the professional is entitled to refuse or if, for commercial reasons, the professional wishes to give one, it can at least stipulate the terms of the collateral warranty and obtain a fee which adequately reflects the additional risks and additional administrative costs incurred.

In some situations, however, there could be a potential advantage in agreeing to give a collateral warranty in that it provides an opportunity for the professional to define and restrict his liability to the party in whose favour the collateral warranty is being given. In the absence of a collateral warranty, that party might have rights as a third party against the professional in tort or under the Contracts (Rights of Third Parties) Act – see Section 2.3. A collateral warranty provides an opportunity to exclude those third party rights and substitute rights that are more restrictive.

21.2. Reducing risks under collateral warranties

If collateral warranties cannot be avoided they should be limited in at least three ways to reduce the risks to the professional.

(*a*) They should be limited in number. Many appointments oblige the professional to give collateral warranties to any funders, future owners, occupiers or users and all would potentially have a right to sue the professional for an unlimited amount.

Even if the professional's liability is limited to an amount within the professional's professional indemnity insurance when the claims are made – see point (*c*) below – the administrative and legal costs that could be incurred if the professional were sued by, say, 20 collateral warranty holders could be substantial. It is usual to limit the number either by agreeing that the first purchasers and/or first tenants only are entitled to a collateral warranty or by stipulating that only tenants who take a certain number of square feet or a floor of a development are entitled to a collateral warranty, perhaps also with a numerical limit.

(*b*) There should be limitations on the time within which the collateral warranty has to be entered into or on when claims can be brought. Many appointments simply say that the collateral warranties must be given 'at any time'. This could mean that if, years after a project is completed, a client finds a tenant/purchaser, the client can ask the professional to enter into the collateral warranty. This is a particular risk in a recession when developers can take some considerable time to let a building. If the collateral warranty is signed as a deed, the tenant/purchaser will then have 12 years from the date of the collateral warranty to sue the professional, and this could be 18 or more years after the original duties have been completed. If such a clause is agreed to, the professional should ensure that the collateral warranty itself has a finite time limit, fixed to a date such as practical completion, within which claims can be brought under the collateral warranty, preferably limited to, say, six years. (It should be noted, however, that such a limitation may be subject to the reasonableness test under the Unfair Contract Terms Act 1977 – see Chapter 16.) Alternatively, the appointment should state that there is a limited time after the services have been completed within which the professional can be asked to sign a collateral warranty. If there is no time limit for claims in the collateral warranty, the professional needs to remember the implications of signing the collateral warranty under hand or signed as a deed – see Section 3.14. If the appointment was signed under hand, the collateral warranty should not be signed as a deed.

(*c*) The liability of the professional under the collateral warranty should be limited. (It would otherwise be unlimited.) The professional should consider the obligations contained in a collateral warranty and assess the risk of any breach of those obligations and what damages could be paid as a result to determine the appropriate limit of liability, in exactly the same way as the professional would decide on the appropriate limitation of liability in the appointment itself. Generally, all collateral warranties (except possibly to a funder) include some limitation of liability. Because a funder's collateral warranty often permits the funder to step in and become the client on certain terms, many professionals take the view that the funder should be treated as the client and thus will agree to become liable to a funder for matters he would not accept for a purchaser or tenant. Net contribution clauses are often accepted in purchaser and tenant collateral warranties – see Sections 12.10 and 12.11 and also Sections 16.9 and 16.10. Ideally, a limitation of liability and a net contribution clause should be included in all the collateral warranties or an aggregate limit of liability agreed which covers all the collateral warranties given and the original appointment. Thus, if the client and all the tenants of a building brought claims totalling

£10 million, an agreed aggregate limitation of liability of £5 million which covered the appointment and the collateral warranties would restrict the amount recoverable to £5 million (subject to the 'reasonableness' test – see Chapter 16). Also, particularly for purchaser and tenant collateral warranties, some limitation on the damages recoverable should be included, such as the maximum liability being no more than the cost of repairs. This is particularly important because of the way professional indemnity insurance responds to claims under collateral warranties – see Section 21.3.

21.3. Professional indemnity insurance for claims under collateral warranties

Most professional indemnity insurance policies will cover claims under collateral warranties. However, where the policy is on a per claim or occurrence basis, it needs to be appreciated that only one limit of liability may be available to meet the claims of all the collateral warranty holders. This is because if the claims arise out of one event or one occurrence, for example, the professional's negligent design to the foundations, leading to claims by all the collateral warranty holders of whom there are, say, 20, with an average claim of £500 000 each, the professional indemnity insurance policy limit applies to all these claims, not each claim. Thus, if the professional's limit of indemnity under his professional indemnity insurance policy is £5 million, the claims described above totalling £10 million will not be met and the professional will be facing uninsured claims of £5 million.

This is another reason why purchaser/tenant collateral warranties are often limited to the cost of repairs. There will only be one cost of repairs and a view will have to be taken as to whether in the most serious claim those costs will still be within the professional's professional indemnity insurance. This is also why professionals cannot be advised to accept the forms of collateral warranty for purchasers/tenants where, in addition to the cost of repairs, it is anticipated that the purchaser/tenant will be paid an agreed monetary amount to compensate him for his other losses, such as relocation costs, additional rent, disruption to his business, etc.

It follows from what has been said above that the cost of repairs plus a monetary amount for a number of purchasers/tenants could quickly take the professional outside his professional indemnity insurance.

Some professional indemnity insurance insurers have particular conditions for collateral warranties if cover is to be provided. This is often a restriction to a specific number of assignments only, or that the collateral warranty must not impose any additional liability or obligation beyond that in the underlying appointment. It is therefore very important that professionals check their professional indemnity insurance policies before agreeing to give collateral warranties. It also needs to be remembered that it is the policy when the claim comes in that responds and the restrictions or conditions of the policy may have altered by then – another reason for the professional to agree as limited a collateral warranty as possible.

21.4. Collateral warranties for residential property

As a general rule no collateral warranties should be given to the owners/occupiers of residential properties. These should be covered by guarantees/insurance schemes like those of the NHBC.

21.5. Collateral warranties for contractors

Where a professional is employed by the client for the project, no collateral warranties should be given to the contractor (or any sub-contractors). The professional owes a duty of care to the client and does not undertake any direct obligations for a contractor, nor indeed does he owe him a duty of care. If he were to do so, he would be placed in a position of a conflict of interest. It follows from this that no client should ask his professional to give a collateral warranty to a contractor.

21.6. Extracts from standard conditions of engagement

ACE Agreement 1 provides:

> F10.1 Where the Consultant has agreed to provide a collateral warranty to third parties as specified in clause B18 of Part B: *The Particulars of Agreement*, the Consultant shall enter into such collateral warranties in the Construction Industry Council's relevant standard form or such other form agreed with the Consultant, provided that such form shall not give any greater benefit to those to whom they are issued in quantum, duration or otherwise than is given to the Client under the terms of this Agreement. It shall be a condition of the provision of such collateral warranties that all fees due to the Consultant at the date of execution of the collateral warranty have been paid.

This makes it clear that whether or not a collateral warranty is to be given is to be agreed before the appointment is entered into. This allows the professional to assess the risks attached thereto and if appropriate to make an adjustment to the fee. The CIC's form of collateral warranty is referred to as this is recognised as a reasonable industry standard, but other forms can be agreed. There is a further protection for the professional in that he has to be paid all fees owing before he entered into a collateral warranty (although this is provided for in the CIC form of collateral warranty). If there were no such provision, it is doubtful whether a professional could refuse to comply with a contractual obligation to give a collateral warranty, even where substantial fees were owing, and even where there was little prospect of recovering them. If the client were in repudiatory breach – see Section 8.3 – then the professional would be relieved of this obligation.

Some clients include in appointments an express provision that collateral warranties still have to be given, despite non-payment of fees, or that it is a condition precedent to the professional being paid that he enters into the agreed collateral warranties. This illustrates the importance to the client of having collateral warranties in order to be able to sell/let their development and hence their value to the client.

RIBA 2010 provides:

> 7.7 Where it is specified in the Project Data:
>
> 7.7.1 that the Architect will be required to enter into a collateral warranty or warranties in favour of funders, purchasers or first tenants and the terms of the warranty together with the names or categories of other parties who will sign such agreements are appended to this Agreement, the Architect shall enter into such agreement or agreements within a reasonable period of being requested to do so by the Client, providing that such warranties give no greater benefit to the beneficiaries than is given to the Client under this Agreement and all fees and other amounts properly due to the Architect have been paid.

The project data, however, only refers to the form of collateral warranty and makes provision neither for any form to be attached, nor for any other parties who are to give collateral warranties to be identified. It is not clear from this clause whether the professional is to be relieved from the obligation to sign collateral warranties if the other parties do not enter into such agreements – it would appear not because this is not expressed as a proviso or condition.

More seriously, the professional is required to give these collateral warranties to 'funders, purchasers and first tenants'. This would therefore cover all purchasers and all first tenants, not just the first purchaser and some of the tenants as is often agreed. Such a description could also cover residential owners/tenants who again would not normally be given collateral warranties. The project data have no provision dealing with the parties who are to receive the collateral warranties, only with the form of collateral warranty they are to get.

The CIC form is referred to in the project data but the parties can agree any other form. It is expressly stated, however, that such warranties can give no greater benefit to the beneficiaries than is given to the client and that all fees and other allowances properly due to the professional have been paid.

Although the clause refers to any additional collateral warranties giving no greater benefit than that given to the client, this is not much consolation for professionals as it is usual (and may be essential if the professional is to stay within his professional indemnity insurance) that the collateral warranties give lesser rights than those given to the client.

Clause 7.7.2 provides in similar terms for rights to be given by way of a third party rights schedules, although the same protection is not given in relation to either the rights giving no greater benefit or to the fees having been paid.

RICS 2011 gives a choice between collateral warranties and third party rights – see Section 21.13 – and in respect of collateral warranties says:

6.1(b) Where the Appendix states the Consultant is required to provide Collateral Warranties in favour of third parties, Clause 6.2 applies.

6.2 Within 14 days of the receipt by the Consultant of each written request from the Client, the Consultant executes a Collateral Warranty in favour of the person named in such request, provided such person is identified in the Appendix either by name or class as being a person entitled to a Collateral Warranty.

'Collateral Warranties' are defined as those 'in the relevant form annexed to the Appointment and signed by the parties by way of identification or in substantially the relevant forms as agreed between the parties'. The appendix provides for any purchaser, any tenant and/or any funder to be given a collateral warranty (not necessarily the first purchaser or tenant) or for the parties to stipulate the maximum number of collateral warranties that can be given to a purchaser or tenant. There is also space for other parties to be given collateral warranties.

21.7. The CIC forms of collateral warranty generally

Several standard forms of collateral warranty exist. Those produced by the CIC are collateral warranties for purchasers and tenants and for funders and, as has been said, are generally recognised in the industry as being reasonable and as has been seen are recommended by the ACE and the RIBA. Copies of both are included as Appendices 4 and 5, respectively. The CIC forms do not, of course, cover all of the matters that can be included in collateral warranties and advice should be taken on non-standard forms. The following sections concerning the CIC forms comment on their provisions and also illustrate some of the issues to which collateral warranties give rise.

21.8. The CIC purchaser/tenant collateral warranties

The CIC form provides for the consideration to be set out. This is normally a nominal amount but does not preclude a substantial sum being paid by the purchaser or tenant if the circumstances require. Consideration is needed if the collateral warranty is signed under hand in order for the collateral warranty to be an effective contract. If, however, the collateral warranty is signed as a deed, sufficient consideration is deemed to have been given.

Clause 1: The duty of care clause. This is the heart of the collateral warranty – the warranty or statement of the professional as to the work he has done. This states that the professional has exercised reasonable skill, care and diligence in the performance of his services to the client under the appointment. This is the common law duty of reasonable skill and care. Anything higher needs to be considered carefully. Just because the original appointment may impose a higher duty of care on the professional, there is no reason why the professional should warrant that a higher duty of care has been exercised in the collateral warranty.

If a collateral warranty is being signed after the services have been completed, the professional should not state that he 'will continue to exercise reasonable skill and care'. This could be construed as a continuing obligation on the professional to monitor his design or to check materials even after completion of the project.

Some forms of collateral warranties seek to extend the warranty so that it covers, for example, all the obligations under the appointment. This is undesirable – as it further extends the potential liability of the professional – and also unnecessary. The purchaser/tenant will need a remedy if the design was negligent but not, for example, if the professional had failed to comply with an obligation to comply with a certain programme for the design work.

It will be remembered that the warranty is also a statement and therefore needs to be accurate – see Section 14.3.

Clauses 2(a)–2(c): Limitation of liability. Clause 2(a) limits the damages recoverable from a professional by the purchaser or tenant to the reasonable cost of repair, renewal and/or reinstatement of any part or parts of the development. This is qualified by 'to the extent that the Purchaser/Tenant reasonably incurs such costs and/or the Purchaser/Tenant is or becomes liable either directly or by way of financial contribution for such costs'.

The objective of this limitation is to provide compensation to occupiers when the premises suffer physical damage as a result of the professional's failure to perform his services with the necessary degree of skill and care, but not for any other sorts of losses such as loss of rent, profits, re-accommodation, etc. This has come to be accepted by substantial parts of the industry as reflecting a fair balance between the risks to the professional of giving a collateral warranty, without which there might not be any liability at all to purchasers and tenants who subsequently occupy premises designed by the professional, and the risks to purchasers/tenants who may find themselves without a remedy if defects occur in the premises during their occupation, for which they are responsible.

It also recognises the fact that multiple claims can be made against a professional under collateral warranties which arise out of a single act of negligence and which could take the professional outside his professional indemnity insurance – see Section 21.3.

The reason why the limit refers to liability 'for financial contribution to such cost' is because the responsibility for carrying out any repairs could, as between the purchasers and tenants, be either to get the work done themselves or to pay in whole or in part for such works to be carried out.

Clause 2(a) is subject to clauses 2(b) and 2(c). Clause 2(b) is a standard net contribution clause (comparable to that in ACE Agreement 1), the purpose of which is to limit the professional's liability in circumstances where he is jointly liable with others to the amount of his net contribution if this is applicable – see Sections 16.9 and 16.10. The important part to note is the description of the other persons with whom the professional could become liable, that is 'all other consultants and advisers, contractors and sub-contractors involved in the Development'. The net contribution clause in bespoke collateral warranties can have a much more limited list, for example including only other consultants, which could leave the professional at risk of paying 100% if the person with whom he is jointly liable is a bankrupt contractor or sub-contractor.

Clause 2(c) says the professional is entitled in any proceedings to rely on any limitation or exclusion in the original appointment. This is essential in any collateral warranty, whether or not there is a limit in the collateral warranty itself. The professional should never be more liable to the purchaser/tenant than he would have been to his original client. The professional is also entitled to raise equivalent rights in defence of liability as it would have against the client under the original appointment. This too is essential in any collateral warranty. It is possible that under the appointment the client agreed with the professional that certain obligations should not be fulfilled or he could waive a requirement or he could agree that something was done in a different or lesser way. If that is the case, the professional ought to be able to rely on that in any claim by the purchaser/tenant – his position should be the same as it would have been had the client brought the claim. Equally, the client may have not in fact relied on the professional and/or may have been contributorily negligent. These defences too should be available to a professional in any claim by a purchaser/tenant.

Clause 2(d) says that the professional's obligations are not 'released or diminished by the appointment of any person by the Purchaser/Tenant to carry out any independent enquiry into any matter'. Normally a purchaser/tenant would be expected to carry out a survey before agreeing to purchase or take a leasehold interest. If that were done and the survey were negligent, the purchaser/tenant would have a remedy against the surveyor and there is an argument that the chain of causation as far as any claim against the professional is concerned has been broken. This provision could allow the purchaser/tenant to sue the professional even where there was a survey or indeed a report into any particular aspect of the premises. That would more certainly be the case if the word was 'liabilities' not 'obligations'. Mostly professionals are prepared to accept this but if in fact the purchaser/tenant has had a considerable input into the way the development was carried out or fitted out, and had his own consultants on which it relied, this clause ought to be deleted so that the courts can sort out the respective responsibilities on the usual basis. In the recent Scottish case of *Scottish Windows Services* v. *Hammer/CRM Facades (2010)* the court held that the architects were liable on the warranty even though the tenants had entered into leases actually knowing of the defects.

Clause 3: Deleterious/prohibited materials. This clause is a typical prohibited materials clause and is included because funders in particular and some purchasers/tenants require a specific confirmation about the specification of materials. It ought to be amended to reflect the obligations in the original appointment: if any, unless, of course, the latter is more onerous. If there is no such obligation in the appointment, this clause should be deleted as the collateral warranty may well be signed after the professional's duties in this respect are completed and he would be unable retrospectively to comply with it. If the professional is a quantity surveyor or planning supervisor (now CDM coordinator) it is unlikely he will have any duties in relation to the specification of materials and this is why the note says that in such cases the clause should be deleted.

Clause 4: Fees and expenses up to date. It may be of some value to the professional to have included in the collateral warranty an acknowledgement that all fees and expenses due up to date have been paid. The professional could raise with the client that executing the collateral warranty anticipates he has been paid up to date and that he cannot make a statement to that effect if it is not the case. Also, if he signs the collateral warranty with this clause but has not been paid up to date, the client might be able to argue that the professional has accepted that everything owing has been paid and the professional has waived his right to recover the outstanding fees and expenses. The professional could also consider whether the failure to pay gives rise to any entitlement to terminate the appointment and thus possibly bring to an end the obligation to give the collateral warranty – see Chapter 8. The failure to pay also has implications for the licence provisions under clause 6 – see below.

Clause 5: Authority of purchaser/tenant to issue directions or instructions. This is a useful reminder of the status of a purchaser/tenant in the construction process. The purchaser/tenant is not the client and is therefore not empowered to give instructions. If any are needed, they should be given by or through the client and any question of the extent of the instruction and/or additional payment agreed with the client. Without this, if the professional acted on the instructions of the purchaser/tenant, he could find that this is outside the professional's original appointment or even in breach of obligations under that appointment and that there is no entitlement to payment.

Clause 6: Licence to use documents, etc. This licence is in a standard form but it may be necessary to consider some restrictions on it – see Section 10.3. The point to note, however, is that it is 'subject to the Consultant having received payment of any fees properly due and owing'. This is a valuable condition enabling the professional to refuse to give a licence until it has been paid all the fees that are due to it – see comments on the funder's collateral warranty form in Section 21.9. However, if the professional has not been paid and clause 4 of the CIC form is not amended when the collateral warranty is signed, the professional may not be able to exercise his right to withhold the licence. The exception for documents for the health and safety file is because, regardless of payment, designers have a duty to give the CDM coordinator the relevant documents for the file under the CDM Regulations.

Clause 6: Maintenance of insurance. This needs to be checked to see that it correctly reflects the basis of the professional's insurance cover, for example if there is a limitation on some sorts of cover which differ from those described in this clause this should be stated. As to the amount, there is no reason for this to be the amount of the professional's insurance or the amount that the professional has undertaken to maintain under the original appointment. It should be an appropriate amount for the risks undertaken under the collateral warranty and no more, bearing in mind the limitations in clause 2.

The period for which insurance will be maintained will depend on the period of liability agreed under clause 10.

Clause 8: Number of assignments. The number of assignments should be as limited as possible – see Section 19.5. The CIC has taken the view that two is standard, but there is no reason why this should not be restricted further if the circumstances or professional indemnity insurers require it.

Clause 10: Time within which claims can be brought. The wording of the clause should be expanded as follows:

> No action or proceedings under or in connection with this Agreement, whether in contract or in tort, in negligence or for breach of statutory duty or otherwise shall be commenced against the Consultant after the expiry of _____ years from the date of practical completion of the relevant part of the Premises under the Building Contract or, in the event that practical completion is not achieved, the date that the Consultant finishes its services under the Appointment or such earlier date as may be prescribed by law . . .

The reason for this is that the wording as it currently stands in the CIC collateral warranty only limits the time within which claims for breach of the obligations under the collateral warranty can be brought. Although other claims may be more difficult to bring, it is possible that the purchaser/tenant could bring a claim in tort, in negligence, etc. and it is therefore desirable that all possible claims are subject to the same time limit.

Clause 11: Third party rights. This clause excludes third party rights under the Contracts (Rights of Third Parties) Act 1999 as is usual in contracts in the construction industry – see Section 2.3.

21.9. The CIC funder's collateral warranty

The CIC funder's collateral warranty is similar to the purchaser/tenant collateral warranty but includes 'step in' provisions. The following deals only with the differences between the funder's collateral warranty and the purchaser/tenant collateral warranty and also the 'step in' provisions. The other provisions are the same as in the purchaser/tenant collateral warranty and the same comments apply.

Clauses 1(a) and 1(b): Limitation of liability. There is no limit to the cost of repairs in the funder's warranty. Liability is limited to the professional's net contribution only, if this is applicable. There are two reasons for this. First, the funder is unlikely to be liable for the cost of repairs. Its losses will be those connected with financing the project, for example, financing charges, interest payments, etc. If the professional has been negligent and caused the project to be delayed, for example, the funder may wish to recover its extended financing charges. Such claims have to date rarely been brought. Most funder collateral warranties will contain no limitation except for a net contribution clause, although the latter is not invariably included.

The funder's main concern, however, is to be able to realise its security in the event of the client for the project becoming insolvent before it is completed. The funder will then have to decide whether to cut its losses and sell the development part-completed or complete the development itself. This can either be done by employing a whole new team (as the original contracts will probably have come to an end on the client's insolvency) or by taking over some or all of the existing team by 'stepping in' to the existing contracts.

The relative costs of both options will be an important consideration as well as the difficulty of persuading a new team to take over the responsibility, and therefore the liability, for all that has been designed and constructed to date.

A funder therefore wishes to keep its options open as to whether it takes over the professional's appointment in these circumstances and it is at this that the 'step in' rights are directed.

Clauses 5–9 contain the rights to step in. These clauses allow the funder to 'step in' and take over the original appointment in circumstances where there is a termination of the finance agreement by the funder and the professional would otherwise be able to exercise his right to terminate the appointment. The act of 'stepping in' is, in fact, a novation of the original appointment – see Section 19.7.

Clause 5 contains the acknowledgement that the client has paid all fees and expenses to the date of signing the collateral warranty and says expressly that the funder has no liability for fees and expenses until a notice has been given under clauses 5 and 6.

Clauses 5 and 6 set out the two circumstances where the funder may wish to exercise the right to step in. Clause 5 allows for this when the funder has terminated the finance agreement with the client before the professional has completed his services. The funder then gives notice to the professional and becomes the professional's client from then on. The clause also provides for an acknowledgement from the client that the professional can rely on the notice from the funder that the finance agreement has been terminated – thus relieving the professional from having to satisfy himself that the notice has been properly given. This is because if the notice had not been properly given and the professional started acting for the funder instead of his original client, he would be in breach of his obligations to his original client. The clause then goes on to add expressly that the acceptance of future instructions from the funder will not 'constitute any breach of the Consultant's obligations to the Client under the Appointment'.

Clause 6 deals with the situation where the professional may be entitled to terminate the appointment or treat is as having been repudiated – see Section 8.3 – or to discontinue (or suspend) the performance of the services. This latter situation could arise where there is a contractual right to suspend for non-payment or in other circumstances and/or as a result of the Construction Act (and amended Construction Act) right to suspend where payment in full has not been made and an effective notice of withholding/notice of intention to pay less has not been given – see Section 8.8. Before the professional exercises any of those rights he has to give 21 days' notice of his intention to do so to the funder and then if the funder gives the requisite notice again the funder steps in and becomes the professional's client under the appointment.

It is important that this notice is not of long duration. The professional will have to continue to work in such circumstances and if the funder decides not to step in these further fees will be at risk. Some funders ask for very extended periods but this is not reasonable, unless the funder agrees to pay the fees incurred.

Under the Construction Act the time limit for exercising the right to suspend is seven days (this was not changed by the amended Construction Act) and that time limit is also often included in bespoke appointments. These provisions in collateral warranties should therefore have a period of seven days for the professional to give notice if he is entitled to suspend, or else his rights under the Construction Act will be curtailed.

Clause 8 deals with the condition on which the funder is to take over the appointment. The funder's notice, whether under clause 6 or clause 7, has to confirm that the funder accepts liability for the payment of all fees and expenses under the appointment and for the performance of all the other client obligations (such as giving the necessary information, making the necessary decisions) including, expressly, payment of any fees and expenses outstanding at the date of the notice.

This provision is important. The circumstances in which this will become relevant, that is when the funder is thinking of stepping in and taking over the professional's appointment, will usually be, as has been said, when the client is insolvent or otherwise failing to complete the project. It is therefore likely that the client has not paid all that is due to his professional. If the professional is to agree that his appointment be taken over by the funder who becomes his client, this should only be on terms that any outstanding fees and expenses will be paid.

Some bespoke funder collateral warranties omit this provision or expressly exclude the obligation to pay outstanding fees. This is not reasonable as it would leave the professional with a claim against the original client which is unlikely to be met.

The clause then goes on to set out the effect of the funder stepping in – the appointment continues but with the funder becoming the client.

The reference in these clauses to the appointee of the funder is because funders can appoint an agent to act for them. This makes no difference to the professional as the funder will become liable to the professional if he steps in. The only protection needed is if, as is stated in clause 6, the appointee is entitled to give instructions: then the funder is to be liable as guarantor for the payment of all sums due to the professional from the appointee. The professional will thus be able to look to the funder itself for payment.

It is to be noted that the licence to the funder to use the professional's documents in clause 9 is also subject to the payment of fees. This is important particularly where a funder has a right to take over the development but decides not to take over the appointment of the professional but to appoint another professional. If the collateral warranty gave the funder a licence which was not subject to this proviso the funder would be able to use the professional's drawings and other documents to complete the development, even if the original professional had not been paid all outstanding fees unless the professional was protected against this by the relevant terms in his appointment.

21.10. Powers of attorney to execute collateral warranties and completing them in escrow

Any clause in an appointment whereby a professional agrees to appoint his client as his attorney for the purposes of executing the collateral warranty should the professional fail to do so is not acceptable. The problem about such a provision is that the professional's only recourse is against the client, should the client have signed a warranty in a form which the professional has not agreed and/or to a person to whom the professional has not agreed to give it. The professional will become liable to the third party in the terms of the collateral warranty because such a clause gives the client ostensible authority to sign the warranty in that form. Thus, the client could make the professional liable without limit to somebody to whom the professional has not agreed to give a warranty for an amount that the professional has also not agreed. The question then arises as to whether the client will be around and will have sufficient assets to pay to the professional all the damages, etc. that the professional may incur to that third party as a result of the client exceeding the authority given to him by the professional in respect of entering into collateral warranties. This might not become known until over 12 years after the professional has completed his services. This could well be the case where the client involved is a shell company or a company set up for one particular project only and is therefore almost inevitably not going to have sufficient funds.

As a compromise, a client might ask that the collateral warranties are executed in escrow; that is, that the professional executes them and then hands them to the client to complete the details of the person to whom they are to be given and dates them. He will then normally give a notice or a copy to the professional so that the professional knows to whom the liability under the collateral warranty has been incurred.

This deals with some but not all of the objections in relation to giving a power of attorney – the professional will at least know how many he has signed and that the form of collateral warranty is correct because he can check this. What he cannot know is whether the person to whom it is given is someone entitled to a collateral warranty under the terms of the appointment, for example, a first purchaser or tenant. Again, the only redress will be against the client if this is wrong. If the professional has to be told at the time by the client to whom it is about to be given, he will be able to raise objections immediately.

It will be seen from this that the dangers to the professional of either requirement from the client are very substantial and outweigh any arguments the client may try to advance about the prejudice to his development if he is unable to provide the collateral warranties he needs to the relevant purchasers and/or tenants.

It also again illustrates the value to the client of having such collateral warranties from his professionals.

21.11 Non-standard collateral warranties

All clauses in non-standard collateral warranties should be considered as carefully as the clauses in non-standard conditions of engagement. This section considers some specific issues that commonly arise in non-standard collateral warranties, but it is not an exhaustive commentary on all the matters that can be included in such collateral warranties.

The duty of care: as has been said, the professional should warrant no more than that he has exercised reasonable skill and care in the performance of his duties or services owed to his original client under his appointment. Even if the professional has accepted a higher duty of care to his client under the appointment, there is no reason why that higher level should apply and thus that the professional will be more likely to be negligent in respect of claims brought by third parties. If the professional had owed the third party a duty of care in tort, then that duty would have been to use reasonable skill and care.

There can be provisions which require the professional to state or acknowledge that he owes a direct duty of care to the beneficiary of the collateral warranty. This is unacceptable. The professional is acting for his client under the terms of his appointment with him and owes him a duty of care. He does not owe a duty of care to a potential purchaser/tenant of whose existence or identity he may well not be aware until after the project is completed. If the purchaser/tenant were involved in the project, perhaps as a result of fit-out works, that purchaser/tenant ought to be separately advised. If the professional were to accept an obligation to look after that purchaser/tenant's interests and thus accept a duty of care to the purchaser/tenant, he would be placed in a position of a conflict of interest.

A professional should only be liable to a purchaser/tenant or other third party if he has been in breach of his duty of care to his original client, judged in the context of what he was asked to do under the terms of his appointment and during the project itself.

Some forms of collateral warranties go further than this and require the professional to warrant that he has complied with all the obligations contained in the appointment. Again, for the reason set out above, this is not necessary and in fact after the project has been completed there can be little that is of relevance to a purchaser/tenant. Whether or not the professional has complied with obligations, for example, in relation to the delivery of information or client procedures, will not impact on the purchaser/tenant. Those obligations with which the purchaser/tenant could be concerned – the obligation to maintain professional indemnity insurance, those in relation to prohibited materials and the need for a licence to use the professional's documents – are generally included as separate direct obligations in the collateral warranty itself.

Prohibited materials clauses: these should be in the same terms as in the appointment or the professional will have to comply with the provisions of both, which could conflict. There would be no objection, however, to having a lesser obligation in the collateral warranty. For example, there is no reason why a professional should undertake to a purchaser/tenant that he will notify them if he becomes aware that prohibited materials are being used. Indeed, this may be impossible if, for example, the collateral warranty is not entered into and thus the identity of the purchaser/tenant known until after the project has been completed.

Copyright licence: again the licence should be in the same terms or lesser terms than that in the appointment. If the professional has had to agree to give copyright to the client, he cannot grant a licence to a purchaser/tenant. Likewise, if copyright has remained vested in the professional and the client has a licence to use his documents, etc. the professional cannot then agree in a collateral warranty to give copyright to the purchaser/tenant as this would prevent him from being able to give a licence to his client.

A more contentious point, however, may be whether or not the licence to the purchaser/tenant is to be subject to the payment of fees – see the comments under Section 21.9. The inclusion of such a provision may be of assistance to the professional if there are outstanding fees, particularly as bespoke collateral warranties rarely include an express acknowledgement that all fees due to date have been paid.

Licence provisions in bespoke collateral warranties can also include an obligation to provide the purchaser/tenant with copies of the professional's documents with or without reimbursement of the professional's reasonable costs. Care should be taken to ensure that this right only extends to documents which the purchaser/tenant needs for the purpose of its ownership or occupation and does not extend to other documents, including documents which would otherwise be privileged which would assist the purchaser/tenant in the event of a claim.

Insurance provisions: the obligation to maintain professional indemnity insurance should be in no more onerous terms than that in the appointment and again there is no reason why it should not be in lesser terms. Also, the amount of professional indemnity insurance the professional agrees to maintain should reflect his possible liability under the collateral warranty and should not necessarily be the amount he has agreed to maintain under the appointment. The length of time such insurance is to be maintained should also be appropriate for the collateral warranty – that could be six years rather than the 12 years that is usually requested.

Reliance on the professional: some bespoke collateral warranties have express provisions that the purchaser/tenant have relied on the professional in respect of certain matters or in respect of all matters within the scope of the appointment. This should be resisted. The aim of such a provision in to establish a duty of care and reliance under the Hedley Byrne principle – see Section 14.3. As has been said there, a professional owes no direct duty to a purchaser/tenant: his duty is owed to his original client, but it is possible that a professional can assume a responsibility to a third party in certain circumstances, although this cannot arise in a situation where the purchaser/tenant is not identified or known until after the project has been completed.

Many non-standard conditions of engagement accept, as is essential, that the professional should be entitled to rely on any exclusions or limitations in the appointment and raise the equivalent rights of defence of liability that he would have had if the client had brought the claim under the appointment – thus preserving the professional's position as it would have been under the appointment.

However, some collateral warranties, particularly funders' warranties, can seek to exclude from these reserved rights the professional's rights of set-off and counterclaim. Thus, for example, if the client still owed the professional fees of say £200 000 and the beneficiary of the warranty brought a claim, the professional would be able, without these words, to set off the £200 000 against the beneficiary's claim. With these exclusions the professional would be left only with his remedies against the client, who may no longer be around to meet the claim. Such exclusions therefore make the beneficiary's position better than the original client's and ought in the author's view to be deleted for this reason.

Such a provision can be coupled with a statement that the liability of the professional is not to be diminished or affected by any independent enquiry undertaken by the purchaser/tenant. The possible objections to this have been explained in Section 21.8, particularly where a purchaser/tenant who is thinking of making a significant investment in a property, whether through a purchase or under a lease, may well have had independent advice from a surveyor, valuer or others before doing so and in fact have relied on them rather than the professionals employed by the seller/lessor. Why should not the professional be able to plead this in any claim later brought against him under a collateral warranty?

Variations to the appointment: some collateral warranties seek to restrict the right of the professional and the client to vary the terms of the appointment without the purchaser/tenant's consent. This can be imposed by a purchaser/tenant who wishes, for example, to prevent the client and the professional from altering a particular design or standard from that originally required. This can be quite a burden administratively – if the client has to seek the purchaser/tenant's consent to every alteration. It is, however, mainly a problem for the client unless the professional agrees under the collateral warranty to inform the purchaser/tenant and/or not to act on any alteration unless it has been agreed by the purchaser/tenant. This could put him in any event in conflict with the duties to his own client. If, however, the collateral warranty is not signed until after the project is completed, this will in fact be of no effect.

With regard to liability outside the collateral warranty: as has been said, the purchaser/tenant rights against the professional should be restricted to those under the collateral warranty and any other rights should be expressly excluded. Some forms of collateral warranties provide that all the other rights of a purchaser/tenant are expressly preserved or are in addition to those under the collateral warranty.

21.12. Adjudication in collateral warranties

It now appears generally to be accepted that collateral warranties are not 'construction contracts' within the Construction Act or amended Construction Act – see Section 25.1 – except where (probably exceptionally) the professional has to give advice as a result of the terms of the warranty to the warranty-holder during the construction process or do anything else that would make it a construction contract in its own right. If the funder steps in and takes over the appointment, the new contract between the funder and the professional will be a construction contract if the original appointment was one, but the funder's warranty itself is not a construction contract.

Sometimes, however, bespoke forms of collateral warranty include provisions for adjudication. They will apply automatically if it is a 'construction contract' but professionals should not agree to include a contractual provision for adjudication.

The Construction Act intended to provide a rough and ready way of settling disputes during construction operations and was mainly directed at the contracting side of the industry. Usually, once a tenant, purchaser or others have taken possession, 'construction operations' will be over. Further, adjudication, which is aimed at providing a quick provisional decision, is not suitable for resolving the highly technical legal disputes that might arise under collateral warranties. These will in particular concern whether or not the professional has been in breach of his duty of care or of other obligations under the original appointment and the extent of the damages. Adjudication for disputes under collateral warranties which are not 'construction contracts' should be resisted.

Adjudication is dealt with in more detail in Chapter 25.

21.13. Third party rights schedules

Although the Contracts (Rights of Third Parties) Act 1999 has been in force for some considerable time, clients are not generally using the Act to confer rights on third parties equivalent to those under collateral warranties. It is thought that this is because the industry generally, and funders, purchasers and tenants in particular, are familiar with collateral warranties and their effect.

The principles of the Contracts (Rights of Third Parties) Act are set out in Section 2.3.

In the context of the third parties who might be given collateral warranties, these parties may be able to bring a claim against the professional under the Act to enforce a term in the professional's appointment, for example, a term that the professional perform its services with reasonable skill and care. Enforcement, in this context, could include bringing a claim for damages for breach of that term.

However, either the contract must expressly provide that the third party in question may do this or (more commonly) the term that the third party seeks to enforce must 'purport to confer a benefit on him'. Further, the third party must be expressly identified in the contract 'by name, as a member of a class or as answering a particular description'.

In relation to many appointments, there may well be people who could potentially qualify as third parties under the Act, including prospective purchasers, tenants or other users of the building or works being designed by the professional – but it would depend on the terms of the contract as to whether any such person has been sufficiently identified as someone on whom the contract purports to confer a benefit.

If any such person can bring a claim under the Act, the claim would be subject to the terms of the contract, just as if the claim was being brought by a party to the contract.

It ought to be a simple matter for an appointment to identify the parties who, it is agreed, should have rights under the Act, and to list the terms in the contract that those parties would be entitled to enforce, However, in practice, drafters of contracts have tended to set out, in long schedules to the appointment, further rights for the third parties and further terms for the exercise of those rights. The problem with this approach is that it strays beyond the policy of the Act, and is little different in practice to having collateral warranties. Indeed, a schedule of third party rights almost invariably resembles a collateral warranty and ought to be checked in the same way.

The parties who are to receive collateral warranties or third party rights can sometimes be entitled to either. They ought not, however, be entitled to both and this should be made clear.

The professional ought also to consider the total number of third parties who could be given third party rights and the extent of the liability to be incurred to them, again as if they were collateral warranties.

The ancillary matters dealing with third party rights in the appointment itself, in addition to specifying to whom they can be given, include when those rights are to vest – this is the equivalent of entering into a collateral warranty. This is normally done by the client agreeing this should be on the giving of a written notice to the professional identifying the person in whom the rights are to be vested and the nature of its interest in the project. This is desirable in any event, so that the professional is aware of the fact that third party rights have been given and can check that they have been given to someone who is entitled to them under the terms of the appointment.

The appointment is also likely to deal with amendments and variations to the appointment which could alter the entitlement to the rights and the vesting of those rights after those rights have in fact been vested and state this can be done without the consent of the recipient of those rights.

There is also usually an express agreement not to vary or depart from the terms and conditions of the appointment in such a way as to reduce materially or adversely affect the rights and remedies of the recipient without their consent.

Both these provisions could cause administrative difficulties for the client, as has been explained earlier in this chapter. They are there to protect the value of the rights to be given to the third party. If, however, the rights are not in fact vested until after the project has been completed, these provisions will not be relevant.

It is then usually stated that until a third party 'steps in' under any right to do so under the third party rights granted to them, the professional and the client can amend or vary or waive any term of the appointment and settle any dispute without the consent of the third party.

These provisions are necessary because of the requirements in the Contracts (Rights of Third Parties) Act which limit the parties to the contract's rights to rescind or vary the contract and make this subject to the third party's consent unless expressly provided otherwise.

21.14. The BPF warranties for funder (4th edition) and purchaser/tenant (3rd edition)

These are similar in some ways to the CIC standard forms of collateral warranty, but differ in a number of important respects and cannot be recommended without significant amendment.

Cost of repairs: the purchaser/tenant warranty has a provision for cost of repairs, as before, but now provides for an additional sum to be stipulated for other losses. This in itself means that the limitation of liability is less attractive to a professional than the equivalent in the CIC warranty and in some circumstances could take the professional outside his professional indemnity insurance cover, as has been explained. Also, this additional liability is not made expressly subject to clauses (b) and (c) (the net contribution clause and the equivalent rights of defence clause). It could well be argued that it is not intended to be. This could represent a serious extension of liability.

Net contribution: the net contribution clause has been revised. Those whom it is assumed will have provided contractual undertakings to the beneficiary are described as the 'Project Team' – and this will depend upon the definition of 'the Project Team' in the underlying appointment, if any. It is therefore possible that it will not be wide enough to cover all those who could become jointly liable with the professional. Unlike the CIC equivalent, the net contribution clause is not stated to be without prejudice to other limitations (e.g., limitations in the appointment incorporated into the warranty by clause 2(c)).

No greater liability: the equivalent rights clause is very similar to that in the CIC warranty, except that it refers only to limitation, which it might be argued does not include any outright exclusions of liability.

Assignment: assignment is limited (as in the CIC warranties) to twice only without consent (in earlier editions the parties could agree the number of assignments).

Limitation: the time for bringing claims is to be 12 years from practical completion or completion or abandonment of the services if earlier. Unlike the CIC warranty there is no provision for the parties to agree a shorter period.

Deleterious/prohibited materials: the warranty in relation to the professional's specification by reference to the Ove Arup publication (Sheehan and Ove Arup and Partners, 1997) may not replicate his duty concerning such materials in his appointment. There is a new statement concerning prohibited materials which requires the professional to see that such materials as are used in the project also accord with the Ove Arup guidelines. This does not reflect what a professional can do in practice. In addition, there may be no comparable duty in his appointment.

Insurance: unlike in earlier editions, the professional is not free to agree in relation to each warranty the amount of insurance and the length of time he maintains it, but is bound to the funder, purchaser or tenant in relation to these matters in the same terms as in his appointment. There is also a reference to Clause 8 of the appointment, which will only apply if the BPF Consultancy Agreement is used. The test is 'reasonable premium rates' instead of 'commercially reasonable rates'.

Licence: the licence is subject to the fees being paid (or tendered), as in the CIC warranties, but the BPF warranty does not include the exception for documents for the health and safety file. There are references to 'Intellectual Property' and 'Design Documents', when these definitions may not appear in the underlying appointment, and if they did not appear, would be undefined.

Third party rights: the Contracts (Rights of Third Parties) Act is not excluded.

21.15. What happens to collateral warranties if the client goes into liquidation?

There are two situations to be considered: the first is where the collateral warranties have already been entered into by the professional.

The professional would still be liable under the collateral warranty. His liability to the holder of the collateral warranty does not end simply because his client has gone into liquidation. The rights under the collateral warranty are separate stand-alone rights from those under the appointment.

Whether or not the professional has been paid by the client would make no difference to his liability under the collateral warranty unless that was a precondition to the professional becoming liable under the collateral warranty, which is extremely unlikely.

It has been asked whether the collateral warranty would be void in this situation because of the wording frequently found in collateral warranties – that the professional shall have no greater liability to the collateral warranty-holder than he would have had to his original client. The argument was that this would mean the professional's liability under the collateral warranty was now nil because the professional would no longer have any liability under the appointment.

That would not in fact be the case.

The professional's liability to his client under the appointment does not end simply because the client has gone into liquidation. A claim could still be brought through the liquidator, receiver or administrator (or trustee in bankruptcy if he is an individual) of that client until the time for bringing that claim is barred. If the claim was brought and the professional were liable he would have to pay whatever was due as a result. Even if the client is a company which is struck from the register, there are ways of getting companies restored to the register (although a registration application is itself subject to a limitation period). Liquidation, etc. may mean that a claim is slightly less likely to be brought, as the liquidator, etc. may find it difficult to fund such a claim.

The 'no greater liability' wording is directed at the position where the losses of the collateral warranty-holder may be different from and/or greater than those of the client because the professional's liability to the client was restricted in some way or ways. In that case the holder of the collateral warranty can collect no more than the client could.

The second situation is where the collateral warranties have not yet been entered into when the client goes into liquidation.

The question is whether the professional still has to enter into a collateral warranty in such circumstances, in particular where the client has failed to pay monies due to the professional.

It is first necessary to check the terms of the appointment. Some make it clear that the collateral warranties have to be given whether or not the professional is owed any money, in which case the professional would have to give them, except possibly where there has been a repudiatory breach – see Section 8.3. Some make it clear that collateral warranties are only to be given if the professional has been paid up to date, but this is rare.

If the professional exercises a right he has been given by his appointment to terminate that appointment because the client has not paid, or because of the client's insolvency, this may well bring to an end the professional's obligation to give collateral warranties in the future unless the appointment provides that all accrued rights and outstanding obligations still remain, as is usually the case. Otherwise, a valid termination generally relieves both parties of further performance of obligations.

There is often a right to the professional to terminate for 'material breach' under the appointment. However, 'material breach' is not necessarily sufficient to constitute repudiatory breach and so advice should also be taken before a professional uses this as an argument for not entering into collateral warranties.

If there are substantial outstanding fees at the point where the professional is being asked for the collateral warranty, the professional may be able to make use of a usual provision in the collateral warranty under which the professional has to acknowledge that all fees due to date have been paid (as in the CIC standard form) in order to persuade the client or the funder to pay outstanding fees.

Professional Services Agreements
ISBN 978-0-7277-4158-5

http://dx.doi.org/10.1680/psa.41585.345

Chapter 22
Guarantees and bonds

22.1. General

Historically, bonds were introduced to cover contractors' insolvency and the cost to a client of finding another contractor to complete the works (usually at an increased cost) and the cost of the delay to the project. Some clients such as government bodies or local authorities ask for performance bonds or performance guarantees from their professionals as well as from contractors. They are not appropriate for professionals and should be resisted where possible. Such bonds are moreover almost never available to professionals unless they are operating as the design and build contractor. There are sound reasons for this.

- Professionals do not deliver a product but a service. The nature of that service is normally only defined in general terms. Determining whether that service has been performed or not can be a complicated issue.
- A professional is subject to his duty of care in performing the required services; there is no exact standard. Without an exact standard, a surety is unlikely to offer a bond.

The risk of a professional's failure to perform is best managed through requiring him to maintain adequate professional indemnity insurance. This is usually much higher than the value of a bond. Therefore, a bond is normally of little value to a client.

The cost of bonds and guarantees has to be funded and this puts up the cost of providing the services, particularly of those who do not regularly need to provide bonds.

22.2. Basic characteristics

A bond is an undertaking under seal or signed as a deed to make a payment. There are two categories of bond.

(*a*) Single bond. This is an undertaking to pay a certain sum of money on a fixed date or on the happening of a certain event. Single bonds are very rare.
(*b*) Conditional bond (also known as a double bond). This falls into two categories:
 (*i*) on-demand bonds, under which payment must be made if the demand satisfies the conditions stated in the bond (e.g., presentation of an engineer's certificate or statement by the employer that the contractor is in default under the construction contract and that specified losses have been suffered);

(*ii*) default bonds, which are in the nature of guarantees, under which payment needs only be made when the claimant, usually the employer, establishes first a breach and second a loss. Frequently, the loss has to be established by means of court or arbitration proceedings.

A guarantee is a promise by the guarantor (usually) to pay money to a specified party (the beneficiary) if a third party defaults on a promise to pay money or perform obligations under a contract, for example a construction contract. The obligation under that construction contract is described as the primary obligation, and the obligation under the guarantee is described as the secondary obligation. A guarantee does not have to be executed under seal or signed as a deed, provided it is in writing and signed by the guarantor. Guarantees can include indemnities – see Chapter 17.

Bonds and guarantees suffer from imprecise and confusing terminology. For example, a document entitled 'performance bond' or 'default bond' is, in reality, normally a guarantee. The term 'conditional bond' (which may be either a conditional on-demand bond or a default bond) is frequently used to describe a document which is, in reality, a default bond (sometimes called an 'on default' bond). Further, one standard 'bond' in the industry (which in reality is a guarantee) is described as a 'guarantee bond'.

For those reasons, it is necessary to distinguish between a guarantee and a bond.

22.3. Guarantees

As explained, under a guarantee a guarantor (usually a bank, surety or insurance company) undertakes to see that a party (usually a contractor) fulfils its contract with another party (the client), or undertakes to perform the first party's obligations if that party fails to. If the contractor fails to fulfil its obligations, the contractor is liable for breach of the contract with the client. A guarantor's primary obligation is to see that the contractor performs the stated obligations (hence, these are often called 'performance guarantees' or, less accurately, 'performance bonds'), or the guarantor pays a sum by way of compensation.

Under a guarantee, the guarantor normally has all the defences open to the party whose performance is being guaranteed. This could include rights of set-off and cross-claims.

Further, the beneficiary claiming under the guarantee must prove the loss suffered as a result of the default of the primary obligor (usually the contractor). This means that it is not until the beneficiary has ascertained the extra costs of completing the works, the delay and any other loss or damage suffered, and, if necessary, proved this through court or arbitration proceedings (or in some cases adjudication), that payment has to be made by the guarantor. If, at the time of default, the obligor/contractor has outstanding claims (e.g., for variations, loss and expense and extensions of time) those claims will have to be ascertained prior to payment under the guarantee in order to determine the beneficiary's net loss.

22.4. On-demand bonds

This type of bond usually requires a simple demand made of the bondsman in good faith by the client, usually in the case of a contractor's bonds accompanied by an architect's or engineer's certificate. The demand can only be raised where there has been a breach of the condition stated in the bond.

The bondsman who receives a valid demand is not required to enquire into the actual loss suffered by the client – which may turn out to be less than the amount of the bond – although if the loss is demonstrably less than the sum demanded, it may be possible to challenge the amount payable in equity. Nothing is allowed for any cross-claims or set-off to which the professional may be entitled. On receipt of a valid demand the bondsman is then under an obligation to pay the full amount of the bond. On due performance of the condition stated, the bond is normally void.

The advantages to the client of such a bond are that the client has only to assert the breach/non-performance specified in the bond, make the notice in the prescribed manner and collect the money.

The disadvantage to the professional where he gives an on-demand bond is that the money paid by the bondsman to the client may be immediately recoverable from the professional under a counter-indemnity required by the bondsman as a condition of providing the bond. It is common for the counter-indemnity to be drafted in such a way that the professional has no choice but to pay immediately. If that is not the case, the professional may be able to avoid paying under the counter-indemnity if the professional can prove that the amount claimed by the beneficiary was not due, did not take into account any cross-claims, or had been claimed improperly.

If, for example, the client owed professional fees including additional fees of £20 000 and then as a result of an alleged default collected, say, £50 000 due under the bond, the professional is then £70 000 out of pocket. The professional has to initiate proceedings or arbitration (and incur the costs of doing that) in order to try to recover the money due from the client. There may also be additional interest charges to the bank because of the effect of calling the bond on the professional's overdraft.

If there was no bond, the professional would be able to bring a claim against the client for the outstanding fees. The client would then claim and have to prove both that there had been a breach of contract and that it had suffered damages. If the client's counterclaim was less than the amount of fees due or was totally unjustified, the court could order that something be paid on account by the client to the professional and during the process, or until the matter was settled, both sides would be out of pocket, and both sides would have to bear the costs of pursuing their claims. This sort of procedure provides an impetus to settle, a factor which is missing if the client has a bond.

Bonds also have implications for the professional's indemnity insurance. Some insurers may not be prepared to indemnify a professional against amounts payable under bonds unless both of the following apply.

(*a*) The amount payable is clearly in respect of some liability covered by the policy.
(*b*) The amount payable by the insurers equals or exceeds the amount paid by the surety.

22.5. Matters to be considered by professionals

Bonds for professionals are rarely standard. The following lists some of the points to be considered:

(*a*) The form of the bond must be agreed at the time of the appointment. Is the proposed bond attached to the appointment? Some proposed appointments oblige the professional to provide a bond, but no sample form is attached. Sometimes a form of bond is attached but the client is allowed to make 'reasonable revisions' to it. This is unacceptable as it may be difficult to reconcile what the client may think is reasonable with what a professional may think is reasonable.
(*b*) Is the surety required to pay once it has received a demand in a certain form, or are other conditions to be complied with, for example, must there be some assertion of non-performance or breach, or a condition requiring the surety to pay only on receipt of notice from the beneficiary of a failure by the professional to pay a sum ordered by a court? The latter is, of course, preferable.
(*c*) Can the surety question whether the sum is due in whole or in part? Generally this is precluded as the surety does not want to demand proof or require further evidence because of administrative inconvenience. However, it would be helpful if the surety were at least required to give notice to the professional before paying out to give him an opportunity to challenge what is being claimed.
(*d*) Do any matters discharge the liability of the surety, other than the completion of the services, for example, suspension of services, variations to the contract, termination of the contract, any waiver by the client, any other bond taken by the client, or any breach by the client? Sometimes it is essential for a professional to have a provision allowing the discharge of the bond in some of these, or similar, circumstances.
(*e*) How long does the bond last? Some bonds last for the length of the contract plus a further specified period, thus enabling a client to call on the bond after the professional's services are completed without having to formulate a detailed claim. In other forms of bond, once the amount of the bond is paid, or the last completion certificate for the whole project issued, the bond is discharged. Ideally, the bond should not extend beyond the completion of the services.
(*f*) Is the amount of the bond clear? The amount should not exceed 10% of the contract sum, or a reasonable percentage of the fee, and/or be reasonable for that particular project. If there are provisions which allow the sum to be called upon in tranches, the total aggregate sum must also be clear.
(*g*) What is the bond for? If it is for insolvency, a sum equal to 10% of the fee should be acceptable (as for contractors). If it is for liability for negligence, it may be for a much larger sum. However, it is not considered that professionals should ever give a bond in respect of negligence.

When the professional receives the draft bond from the surety, the bond details should be compared by the professional with the client's requirements before submission to the client. Copies of all bonds provided to clients by a professional organisation should be kept in a central register so that the professional can monitor them and ensure that they are discharged.

22.6. Parent company guarantees

Generally, guarantees are provided by parent companies to guarantee the performance of a subsidiary under a contract. Sometimes they require the guarantor to take over and complete the work. They should generally only be required if the subsidiary's assets are disproportionately small when compared to the value of the commission or if it is a shell company and not an established soundly financed company. A professional should not be required to give both a bond and a parent company guarantee as this would be covering the same risks twice.

Parent company guarantees should never extend the parent company's liability beyond that of the subsidiary company. They are sometimes coupled with an indemnity from the parent company to the client in respect of claims against the subsidiary. If so, the terms of the indemnity will need to be considered – see Chapter 17.

A form of wording such as the following might assist a professional in negotiating to remove a requirement for a parent company guarantee.

> [The Professional] is one of the UK's leading [consultancies, architectural practices, quantity surveying firms] . . . with a [xxx] million turnover and around [xxx] staff. It is thus not merely an operating company with negligible assets and there is no anticipation that [the professional] will become insolvent. We therefore see no reason why an undertaking from our holding company [xxx Holdings Ltd] to complete our services will assist you, not least because [xxx Holdings Ltd] does not itself provide [consultancy] services.

22.7. Advance payment guarantees

A client may under the appointment make advance payments to the professional to provide him with funds in respect of services ahead of the appointment being formalised. The client may wish to ensure that if the professional fails to deliver the services in accordance with the appointment, at the very least, he can recover the payments made to the professional. The advance payment guarantee would provide that if the professional fails to meet his contractual obligations the surety will refund the advance payments made by the client. The form of the guarantee will determine whether the client is entitled to repayment on demand.

In this context, it is important to avoid agreeing to give a performance guarantee as described in Section 22.3 in addition to signing an appointment. Where an advance payment guarantee is used in conjunction with a performance guarantee, it may result in the surety unwittingly guaranteeing an advance payment, if the performance guarantee remains extant notwithstanding the advance payment. Therefore, if an advance payment is to be made under the appointment, and an advance payment guarantee is required, an advance payment guarantee should be drafted so as to ensure that the parties' legal positions are properly set out.

Professional Services Agreements
ISBN 978-0-7277-4158-5

http://dx.doi.org/10.1680/psa.41585.351

Chapter 23
The Construction (Design and Management) Regulations 2007 (the CDM Regulations 2007)

23.1. Relevance to professionals

A professional may be affected by the Construction (Design and Management) Regulations. The original CDM Regulations 1994 have been replaced by the CDM Regulations 2007 which came into force on 6 April 2007. They also replace the Construction (Health, Safety and Welfare) Regulations 1996 which are in effect incorporated in the CDM Regulations.

The CDM Regulations 2007 apply to all construction work in Great Britain and its territorial sea. Some of the duties apply to all projects, whether notifiable or not. Additional duties apply where the project is notifiable. A project is notifiable if the construction work will last more than 30 days or involve more than 500 person days of construction work.

This chapter considers the effect of these Regulations on professionals as 'designers' and also sets out a summary of the duties imposed on the 'Client', the 'CDM coordinator' and other dutyholders under the CDM Regulations. As these are statutory requirements, it is unnecessary in conditions of appointment to set out the duties of the designer or CDM coordinator under the CDM Regulations.

However, many non-standard conditions of engagement do contain specific CDM obligations and often paraphrase them or include duties not imposed by the CDM Regulations at all. Before provisions in non-standard conditions of engagement can be considered, it is necessary to understand the background to the CDM Regulations, the criminal and civil liabilities imposed by the CDM Regulations, and the duties imposed on Clients, designers and CDM coordinators.

23.2. The statutory framework

The CDM Regulations 1994 were issued under the Health and Safety at Work, etc. Act 1974 to implement the 1992 Temporary or Mobile Construction Sites EC Directive. This imposed requirements and prohibitions on the health and safety aspects of 'construction work', which is very widely defined, and in any particular case the definition itself should be referred to. In summary, 'construction work' is the carrying out of any building, civil engineering or engineering construction works. It includes the construction, alteration, conversion, fitting out, commissioning, renovation, repair, upkeep or maintenance and decommissioning, demolition or dismantling of 'a structure'. It also covers the preparation for an intended structure, including site clearance or investigation; laying and installing foundations and de-assembly of prefabricated elements to form a structure; commissioning, maintenance and repair, and the installation of services which are normally found within a structure.

'Structure' is also defined very widely in the CDM Regulations and includes not only all traditional forms of structure but almost all other forms of construction, such as railway lines, docks, harbours, tunnels, bridges, reservoirs, pipes, cables, roads, airfield cables, sewers, sea defence works, earthworks, dams, underground tanks, as well as formwork and falsework and other temporary works and any fixed plant, the installation, commissioning, decommissioning or dismantling of which involves a risk of a person falling more than two metres. The CDM Regulations 2007 have not altered these wide definitions.

The CDM Regulations do not apply to construction work where the local authority is the enforcing authority.

The Health and Safety Commission has also issued an Approved Code of Practice 2007 (ACoP) which gives practical guidance on complying with the duties set out in the CDM Regulations 2007. It replaces the earlier Approved Code of Practice issued in 1994. Failure to comply with ACoP is not a criminal offence, but it can be used as proof that a person has contravened the CDM Regulations 2007 or a section of the Health and Safety at Work, etc. Act 1974.

23.3. Civil liability – breach of statutory duty

Civil liability may exist in the law of tort for breach of statutory duty. In order to claim, the potential claimant must show both that the statute or regulations are actionable and also that he has suffered damage as a result of the breach.

As to the first point, the general position is set out in the Health and Safety at Work Act 1974: breach of health and safety regulations issued under the Act may give rise to civil liability unless the regulations in question state otherwise. It is not possible to contract out of this statutory liability.

The CDM Regulations 1994 excluded civil liability except in two specific instances: a failure by the client to ensure that the construction phase of a project did not start in the absence of a health and safety plan and a failure by the principal contractor to take reasonable steps to prevent unauthorised access to the site. Neither of those regulations directly concerned designers. Further, neither would have been very likely to result in civil claims. In the case of the former, it would have been difficult in practice to show that any particular loss or injury had been caused by commencement of work without a health and safety plan. In case of the latter, it would generally have been easier for an injured trespasser to bring a claim under the Occupiers' Liability Act.

The position in respect of civil liability was the opposite under the Construction (Health, Safety and Welfare) Regulations 1996: any breach of those regulations could give rise to civil liability.

The position under the CDM Regulations 2007 is set out in Regulation 45.

Regulation 45 reads as follows:

> Breach of a duty imposed by the preceding provisions of these Regulations, other than those imposed by regulations 9(1)(b), 13(6) and (7), 16, 22(1)(c) and (l), 25(1), (2) and (4), 26 to 44 and Schedule 2, shall not confer a right of action in any civil proceedings insofar as that duty applies for the protection of a person who is not an employee of the person on whom the duty is placed.

Firstly, it would appear that the default position under the Health and Safety at Work Act applies to breach of a duty imposed under the various regulations that are listed. That is that there may be civil liability for breach of Regulations 9(1)(b), etc. The Regulations listed impose duties on clients, principal contractors and other contractors in four main areas.

(*a*) Controlling access to the site (mirroring the position under the CDM Regulations 1994).
(*b*) Not starting the construction phase until the principal contractor has prepared the construction phase health and safety plan (again mirroring the position under the CDM Regulations 1994).
(*c*) Provision of welfare facilities such as provision of sanitary conveniences, washing facilities, drinking water and so on (mirroring the position under the old Construction (Health, Safety and Welfare) Regulations). It seems perhaps a little surprising that a right of action should be given for failure to provide adequate welfare facilities.
(*d*) Compliance with detailed regulations such as stability of structures, demolition, use of explosives, excavations and so on (again largely mirroring the position under the old Safety Regulations).

These Regulations do not apply to designers.

Second, breach of any of the remaining Regulations by any dutyholder may give rise to civil liability subject to the proviso set out in the second part of Regulation 45.

The proviso in Regulation 45 appears to mean in effect that a dutyholder will not be liable for personal injury suffered by a person who is not an employee of his. So, for example, a designer might be liable to one of his own employees who had been injured as a result of a breach of the designer's duty but not (at least under the CDM Regulations 2007) to an employee of a contractor.

In order to complete the cause of action, the potential claimant, as in a claim for negligence, must also establish that he has suffered loss or damage and that the breach of duty has caused the loss in question. Issues of causation will remain central to claims against designers, whether brought under the Regulations or otherwise, and the extent to which an accident was caused by errors in design or workmanship or both.

Under general legal principles, a person can sue for breach of statutory duty if he belongs to a class of persons for whose benefit the statutory duty has been imposed. The overall public policy purpose of the CDM Regulations 2007, and indeed all health and safety legislation, is to protect workers from risk of injury. In practice, only an injured worker is likely to have a right of action, arising out of his personal injury.

That said, it would not be entirely surprising if enterprising lawyers do not argue that clients (and others) have wider rights of action under the CDM Regulations.

It is worth noting that a professional should not voluntarily assume contractual obligations to comply with CDM in his appointment. If he does, the client may then have a right to sue him for breach of the Regulations, although no such right is conferred by the Regulations themselves.

Although the extent of civil liability under the new Regulations is greater than under the old Regulations, it is not immediately obvious that claims are likely to be made which would not have been made previously. It seems unlikely that duties owed to employees under the Regulations will add materially to duties owed to employees under any other statutory regulations and at common law.

23.4. Criminal liability – prosecution and other sanctions

As to criminal liability, health and safety law including the CDM Regulations 2007 is part of the ordinary criminal law. It is sometimes regarded as 'regulatory' in nature, but that is to mischaracterise it. Breach of any of the Regulations is a criminal offence which is prosecuted in the ordinary criminal courts.

Enforcement is by the Health and Safety Executive, although if there has been a fatal accident which may result in manslaughter charges investigations are carried out by the police. Unlike the police, the HSE acts as both investigator and prosecutor: it decides whether prosecutions will be brought without referral to the Crown Prosecution Service. In practice, this means that less serious offences are prosecuted by HSE inspectors rather than specialist prosecutors. If the decision to prosecute is in the hands of an inspector who has also had charge of the investigation, the inspector may be predisposed to prosecute.

Maximum penalties depend on the sentencing court. In the Magistrates' Court the maximum is currently a fine of £20 000. In the Crown Court, it is an unlimited fine. Most health and safety prosecutions are in the Magistrates' Courts although even if a case is decided by magistrates it can be remitted to the Crown Court for sentence if the magistrates conclude that their sentencing powers are inadequate.

There are also sanctions short of prosecution. The HSE may issue improvement or prohibition notices. Improvement notices require the person to whom they are directed to take certain specified action, for example, improved training. Prohibition notices require works or a specified portion of works to be suspended. Both types of notices are used extensively on construction projects, although they tend to be addressed largely to contractors.

There were relatively few prosecutions under the old Regulations, especially of designers. In one prosecution of a designer under the CDM Regulations 1994 the case involved an accident at Speke Airport, Liverpool, where the designer was the designer for a new hangar. One of the hangar doors fell during the course of erection, causing fatal injuries to a workman, and the case against the designer was that his design did not include adequate information about the structure which might affect health and safety. The designer was acquitted.

The HSE lists successful prosecutions on its website (it does not list its failures). It appears that there were only four recorded convictions for breach of Regulation 13 of the CDM Regulations 1994 (i.e. the Regulation concerned with designers) over the five years up to 2007.

- The collapse of a first-floor roof during conversion of retail premises. The designer was fined £2500 for the Regulation 13 offence with a total fine of £12 500 plus total prosecution costs of £12 000.
- A 16-year-old labourer was struck and killed by a steel column which fell when holding-down bolts failed. The design of the bolts was inadequate to resist wind loads and inadequate information had been provided on allowable wind loads. There was also poor supervision of both design and erection. The designer was fined £40 000 for breach of Regulation 13, with a total fine of £100 000 plus total prosecution costs of £59 375.
- The defendant was the planning supervisor, designer and principal contractor for the extension of a warehouse building where an employee fell seven metres through a fragile roof-light. The allegations included failure to inform the client of his duties and failure to have pre-tender or construction phase health and safety plans. There was a fine of £2500 for the Regulation 13 offence, a total fine of £12 500 plus prosecution costs of £1378.
- An unsupported vertical excavation face: the allegations were inadequate ground investigation prior to excavation and lack of supporting material. There was a fine of £4000 for a breach of Regulation 13 with a total fine of £5000 and prosecution costs of £7000.

It will be noted from these examples that although the fines for Regulation 13 offences were modest, in each case additional offences were charged which resulted in higher total fines and also in each case the HSE was awarded costs.

Civil litigation concerning construction projects is often subject to the procedures and workings of the specialist Technology and Construction Court, where the judges are familiar with construction projects and issues. The criminal justice system is very different. The finders of fact are normally laymen, either magistrates or jurymen, and even if there is a professional judge (a district judge in the Magistrates' Court) he will not be a construction specialist. Reasoned judgements are not given so the basis of the decision to convict or acquit is not known. That factor makes it difficult to predict the effect of any new piece of criminal legislation.

The CDM Regulations 2007 will remain only one weapon in the armoury of the HSE. It is common for prosecutions to be brought alleging breach both of specific duties under particular health and safety regulations and of general duties under the Health and Safety at Work Act. The latter tend to be regarded as more serious and there is every reason to suppose that the HSE will continue to prosecute using both the Act and the CDM Regulations. In fact, prosecutors may well find it easier to present a case to magistrates or a jury which is based on breach of general duties under the Act rather than the far more technical duties imposed by specific regulations.

There are other consequences if a designer is found to be in breach of the Regulations.

In the first place, an HSE investigation, even if it does not result in prosecution, will almost inevitably require a significant input of management resources. The HSE will routinely use its statutory powers to require production of documents and to require individuals to be interviewed. Investigations frequently last for a period of many months, if not years, during which time both individuals and the company may have the threat of prosecution hanging over their heads.

Second, a conviction may be used in evidence in any subsequent civil proceedings. The fact of a conviction may make it far more likely that there will be civil proceedings and it is for that reason that some professional indemnity insurance policies include cover for costs of defending prosecutions.

Third, a prosecution and trial will be matters of public record. A trial, even one that results in an acquittal, will inevitably attract publicity which is likely to be unwelcome. The reputational damage arising out of a high profile trial can be enormous.

Finally, professionals in particular should be aware that invitations to tender will often require disclosure of health and safety convictions which could potentially lead to disqualification from tendering. A related and rather circular consequence is that a conviction under the CDM Regulations may itself tend to suggest that the offender is not competent for the purpose of accepting future appointments to which CDM applies.

23.5. The dutyholders

The CDM Regulations impose duties on 'dutyholders'. These include

- the Client
- designers
- contractors

- the CDM coordinator
- the principal contractor.

23.6. The 'Client'

In this chapter, references to 'the Client' are to the client under the CDM Regulations, who is not necessarily the designer's (or professional's) client.

It is essential to be able to identify 'the Client' for the purposes of the CDM Regulations 2007. Clients have extensive obligations under the CDM Regulations 2007.

The Client is defined in Regulation 2(1) as 'a person who in the course or furtherance of a business (a) seeks or accepts the services of another which may be used in the carrying out of a project for him or (b) carries out a project himself'.

There are no special provisions in the CDM Regulations 2007 for small projects or for domestic clients. A client not acting in the course or furtherance of a business will not be 'a Client' for the purpose of the CDM Regulations and therefore is not subject to CDM duties. A designer or contractor working for him (in the course of a business) will be subject to CDM duties. There will, however, be no CDM coordinator or principal contractor where the client is a domestic client, nor where the project is not notifiable.

Where there is more than one 'Client', one or more may elect to take on the duties of 'Client'. The role of Client's Agent, established under the CDM Regulations 1994, has been abolished.

23.7. The Client's duties

Clients now have a duty concerning arrangements for managing the project. This is new. There was no express duty like this on Clients under the CDM Regulations 1994.

Regulation 9 says:

> (1) Every client shall take reasonable steps to ensure that the arrangements made for managing the project (including the allocation of sufficient time and other resources) by persons with a duty under these Regulations (including the client himself) are suitable to ensure that:
> (*a*) the construction work can be carried out so far as is reasonably practicable without risk to the health and safety of any person;
> (*b*) the requirements of Schedule 2 are complied with in respect of any person carrying out the construction work; and
> (*c*) any structure designed for use as a workplace has been designed taking account of the provisions of the Workplace (Health, Safety and Welfare) Regulations 1992 which relate to the design of, and materials used in, the structure.
>
> (2) The client shall take reasonable steps to ensure that the arrangements referred to in paragraph (1) are maintained and reviewed throughout the project.

The requirements of Schedule 2 are welfare facilities for those carrying out construction work, incorporated from the Health, Safety and Welfare Regulations 1996.

It is worth noting the specific reference to the allocation of sufficient time and resources. A Client who seeks to reduce costs may be in breach of its CDM duty if the result is a curtailment of time or resources for carrying out any of the work safely.

Clients also have a duty to ensure that every designer and that every contractor appointed by the Client is promptly provided with pre-construction information.

There is a long definition of pre-construction information in Regulation 10, but broadly speaking it is all the information in the Client's possession (or which is reasonably obtainable) which is relevant to health and safety – more specifically, the purposes of providing this information is set out in Regulation 10(3), that is

(*a*) to ensure so far as is reasonably practicable the health and safety of persons:
- (*i*) engaged in the construction work,
- (*ii*) liable to be affected by the way in which it is carried out, and
- (*iii*) who will use the structure as a workplace; and

(*b*) without prejudice to sub-paragraph (a), to assist the persons to whom information is provided under this regulation:
- (*i*) to perform their duties under these Regulations, and
- (*ii*) to determine the resources referred to in Regulation 9(1) which they are to allocate for managing the project.

The information specifically includes the minimum amount of time that will be allowed to the contractors appointed by the Client for planning and preparation, as well as information generally about the site or the construction work and about the proposed use of the structure as a workplace.

The client will be entitled to considerable assistance in discharging this duty from the CDM coordinator, where one is appointed.

There are additional duties on the Client where the project is notifiable.

If the project is notifiable, the Client must appoint a CDM coordinator as soon as practicable after initial design work or other preparation work has begun. Pre-construction information must also be provided to the CDM coordinator.

After appointing a CDM coordinator, the Client must appoint a principal contractor. This must be done as soon as is practicable after the Client knows enough about the project to be able to select a suitable person.

The Client will be deemed to have the duties of a CDM Coordinator or principal contractor for any period for which no such person has been appointed. This is presumably intended to encourage the Client to appoint such persons promptly.

The Client must ensure that the construction phase does not start unless the principal contractor has prepared a construction phase plan that complies with the Regulations – that is the same as before, save that it is now ‘the construction phase plan’ instead of ‘the health and safety plan’ – and he is satisfied that the requirements in the Regulations relating to the provision of welfare facilities will be complied with during construction – that is new in the CDM Regulations 2007.

The Client must ensure that the CDM coordinator is provided with health and safety information for the health and safety file and he has various other duties in relation to the health and safety file which apply after construction, such as keeping it available for others who may need it and keeping it updated with relevant new information.

23.8. Designers

The CDM Regulations 2007 apply to any person who, in the course or furtherance of a business, prepares or modifies a design or arranges for or instructs any person under his control to do so relating to a structure or to a product or mechanical or electrical system intended for a structure.

Although it was the case under the previous Regulations that a Client and a contractor could also be a designer, the definition of a designer in the CDM Regulations 2007 makes this clear. The reference to ‘modification’ of a design is new, as is the reference to ‘instructing any person under his control to do so’ although an amendment to the old Regulations sought to cover this point. This could bring in responsibility for the work of a sub-consultant or self-employed person, depending on the level of control that the professional has of that person.

Designers’ responsibilities extend beyond considering the safe construction of their design. As before they need to consider the health and safety of those who will maintain, clean, repair, refurbish and eventually remove or demolish all or part of a structure (ACoP paragraph 111). There is also the new duty to consider the health and safety of users of workplaces.

The definition of designer has been extended so that ‘design’ not only relates to a structure with its very wide definition but to ‘a product or mechanical or electrical system intended for a particular structure’.

So people can also be designers if they specify the use of a particular method of work or material (which could bring in the client if he stipulates a particular layout for a building); or if they are people designing plant which forms part of the permanent structure, for example, a specialist provider of permanent fire extinguishing installations; or are purchasing materials where the choice has been left open.

Most construction professionals are likely to be ‘designers’ for the purposes of the CDM Regulations in relation to some, if not all, of their duties.

23.9. Design

There never was a definition of 'design', just a list of what design includes. This is basically unchanged. Design includes 'drawings, design details, specification and bill of quantities (including specification of articles or substances) relating to a structure' and 'calculations prepared for the purposes of a design'. There is still the reference to bills of quantities but it will only be those elements of the bills which could constitute a specification of articles or substances that could constitute 'design' or indeed any other matters in the bills that could determine the design of a structure.

ACoP paragraph 121 says that the designer's duties apply whenever designs are prepared and this includes concept designs, competitions, bids for grants, modification of existing designs and relevant work carried out as part of feasibility studies. It does not matter whether planning permission or funds have been secured, whether the project is notifiable or high risk or whether the client is a domestic client.

23.10. Designers' duties

All save two of the designers' duties apply to all projects. Two additional duties apply if the project is notifiable. If the client is a domestic client, however, the designer does not have to make the client aware of his duties and nor will the client have to check the competence of the designer. It should also be noted that there is no obligation on the client to appoint a CDM coordinator or principal contractor unless the project is notifiable.

23.10.1 Competence

There are two aspects relating to 'competence' that need to be considered. Under the CDM Regulations 1994, Regulations 8 and 9, no-one could 'arrange for a designer to prepare the design' unless they were reasonably satisfied as to his competence as to health and safety matters and as to the resources he proposed to devote to health and safety matters.

New Regulation 4(1) contains a similar duty. No-one – not necessarily the Client – can *appoint or engage* (not 'arrange for a designer to prepare a design') a designer unless he has taken reasonable steps to ensure that that designer is competent. This is a slightly different test and perhaps a little more onerous one for clients.

As to 'appoint', it is not thought this will mean there has to be a formal appointment duly signed, for example. It is more likely that a designer will be 'appointed or engaged' from the time he is asked to carry out something that constitutes 'design'.

'Competence' is competence in relation to performing health and safety requirements and avoiding contravening any health and safety prohibitions. (Regulation 4(1) is not therefore referring to competence as a designer.)

As far as a designer is concerned, two new strict duties concerning competence apply: Regulation 4(1)(b) says a designer cannot 'accept such an appointment . . . unless he is competent' or 'arrange for or instruct a worker to carry out or manage design . . . unless the worker is competent or under the supervision of a competent person'.

If there is an incident and it can be proved that a designer did not take health and safety properly into account in his design, it will probably follow that either the director/partner in the firm concerned or the person who prepared the relevant piece of design was not competent. There may therefore be two counts in the prosecution rather than one. It is also conceivable that there could be a prosecution if a design was found to be unsafe to construct even if it had not yet been built or was revised before it was built.

These provisions appear to be aimed at employers, to make them look more closely at the training of their employees and those who supervise them, and also to allow the HSE to prosecute those who manage and control the firm's work rather than those who actually prepare the design.

ACoP in Appendix 4 has a great deal about competence and how that is assessed – both by those assessing the firm's competence and how that of an individual within the firm should be assessed. A two-stage assessment of both firms and individuals is expected.

There is reference in the stage 1 assessment to an individual's qualifications, his or her continuing professional development and understanding of the risks arising from construction work.

Stage 2 assessments concentrate on that person's past experience of the type of work which the firm is asking him or her to do, to determine, for example, whether he or she needs to work under the supervision of someone who is competent and familiar with the work.

In Appendix 4 to ACoP there is also a set of core criteria and it is said that the stage 1 and stage 2 assessments should be made against these core criteria. It is strongly recommended therefore that designers take Appendix 4, which is a schedule divided into three columns entitled 'Criteria'; 'Standard to be achieved' and 'Examples of the evidence you could use to demonstrate you meet the required standard', and see how closely the firm or any person or organisation whose competence the designer is to assess meet these criteria. It also separates out from the general requirements the particular evidence that might be required from contractors, design organisations and CDM coordinators.

Other paragraphs in ACoP give guidance as to the packages designers have to put together to demonstrate competence, whether the designer comprises a large company or a company employing fewer than five people.

In particular, as a result of the specific requirement concerning the competence of employees or those supervising them, designers may need to consider whether they have enough evidence through their procedures or otherwise to be able to demonstrate if necessary and in the event of a prosecution that their employees, their supervisors and possibly sub-consultants and self-employed workers are competent to carry out or manage that particular design work or have an appropriate supervisor – who will need to be identified.

23.10.2 Making the Client aware of its duties

Under the CDM Regulations 1994, Regulation 13(1) stated 'no designer shall prepare a design unless he has taken reasonable steps to ensure that the client for the project is aware of his duties under the Regulations and of official guidance', that is the old ACoP.

Regulation 11(1) under the CDM Regulations 2007 says 'no designer shall commence work (not prepare a design as previously) in relation to a project unless any client for the project is aware of his duties under the Regulations'. There is no longer a reference to making the Client aware of the guidance provided by ACoP.

It is important to note that the reference is to 'the client for the project' who may not of course be the designer's client. This duty does not apply where the client is a domestic client.

In order to satisfy the old requirement, it was thought best to send the Client for the project a letter reminding him of his duties as one of the first things the designer did – even though this could mean the Client received dozens of such letters.

Now the designer has to consider whether, and be reasonably sure that, the Client is aware of his duties before he starts work. This requires the designer to exercise his judgement. This will be self-evident if the Client is checking the designer's competence or is about to appoint a CDM coordinator.

ACoP paragraph 123 says the duty to inform is aimed at the designer who has the initial or main contact with the Client and that other designers need take no action unless they have reason to suspect that Clients are not aware of their duties.

If in doubt, however, it would be best to send the Client a letter summarising his duties before the designer starts work as there could then be no doubt that the designer has complied with this Regulation. There is attached in Appendix 6 a summary of the Client's duties under the Regulations.

23.10.3 Design considerations

Under the CDM Regulations 1997, Regulation 13(2)(a) stated that designers had to

> ensure any design . . . included among the design considerations adequate regard to the need to:
> (*i*) avoid foreseeable risks to those constructing or cleaning or affected by the work of such persons;
> (*ii*) combat at source risks to those constructing or cleaning or affected by the work of such persons;
> (*iii*) give priority to measures which protect all such persons over those protecting each such person.

Under the CDM Regulations 2007, Regulation 11(3) separates out the two elements – the persons whose health and safety the designer is to have in mind and then what the designer has to do about it.

Regulation 11(3) says the designer shall 'in preparing or modifying a design which may be used in construction works in Great Britain' avoid foreseeable risks to the health and safety of any person

(*a*) carrying out construction work – this includes maintenance, demolition, etc.
(*b*) liable to be affected by such work
(*c*) cleaning 'any window or any transparent or translucent wall, ceiling or roof'
(*d*) maintaining the permanent fixtures and fittings of the structure
(*e*) using a structure designed as a workplace.

What the designer has to do is set out in Regulation 13(4). He has to

(*a*) eliminate hazards which may give rise to risks; and
(*b*) reduce risks from any remaining hazards

and in so doing shall give collective measures priority over individual measures.

This is comparable to the duties under the earlier Regulations – 'combating risks at source and giving priority to measures that protect all over those that protect the individual'.

The extent to which the designer was required to take these considerations into account under the earlier Regulations was 'to the extent that it is reasonable to expect the designer to address them at the time the design is prepared and to the extent that it is otherwise reasonably practicable to do so' (Regulation 13(3)).

Regulation 11(2) under the CDM Regulations 2007 says these duties shall be performed 'so far as is reasonably practicable taking due account of other relevant design considerations'.

The reference to so far as 'it is reasonable to expect the designer to address them at the time the design is prepared' in the earlier Regulations has been deleted. The obligation is now to keep these health and safety principles under review the whole time any design is being prepared or modified. They have to be 'foreseeable risks' (so if they were not foreseeable at the beginning but become foreseeable during the project the designer will need to address them) and it has to be 'reasonably practicable' to do so, which allows some leeway.

ACoP has some guidance in paragraphs 124 to 130 about eliminating hazards, and if that is not reasonably practicable, as to how the designer can reduce risks to an acceptable level, for example, by reducing the likelihood of harm or potential severity of the harm. But it also says that the amount of effort put in to eliminating hazards and reducing risks should depend on the degree of risk. 'There is little point in spending a lot of money, time and trouble on low risk issues. . . .' The focus should be on issues that are known to have the potential to cause significant harm and where there are known solutions that reduce the risks to everyone exposed. ACoP states 'Designers must not produce designs that cannot be constructed, maintained, used or demolished in reasonable safety'.

This is the essence of a 'reasonably practicable' test. As to 'taking due account of other relevant design considerations' it is assumed that this is a reference to the provisions made by other designers so that their designs can be constructed safely. It would not be sensible for one designer to insist on one sort of design if this negates or reduces the effectiveness of the arrangements being made by another designer. There is also, however, a new duty placed on all dutyholders under Regulation 7 to 'take account of the general principles of prevention in the performance of those duties during all stages of the project'. This duty is phrased as a strict obligation – designers 'shall take account of' those principles. The qualification contained in Regulation 13 – 'so far as is reasonably practicable taking due account of other relevant design considerations' – no longer applies. Schedule 1 to the Management of Health and Safety at Work Regulations has been updated since the 1994 CDM Regulations. They are now dated 1999 and the 'general principles of prevention' are in Appendix 7 to ACoP. ACoP says (paragraph 22) that designers should look at these 'when considering what precautions are necessary to control risks associated with a project'.

This duty relates to the performance of the designer's duties as a designer but it applies during all stages of a project.

Appendix 7 to ACoP sets out the 'general principles'. These are

(*a*) avoiding risks
(*b*) evaluating risks that cannot be avoided
(*c*) combating risks at source
(*f*) replacing the dangerous by the non-dangerous or less dangerous
(*h*) giving collective protective measures priority over individual protective measures.

All things the designer is already required to do under Regulation 11.

There are, however, four additional principles which more specifically apply to employers. It is more difficult to know how designers are to employ these in the context of preparing their designs. These are

(*d*) adapting work to the individual especially as regards the design of workplaces, choice of work equipment, working and production methods
(*e*) adapting to technical progress
(*f*) developing a coherent overall prevention policy covering technology, organisation of work, working conditions, social relationships, etc.
(*i*) giving appropriate instructions to employees.

23.10.4 The workplace

There is, as has been said, under Regulation 11(3) of the CDM Regulations 2007, the obligation on designers when preparing or modifying their designs to avoid foreseeable risks to the health and safety of any person using a structure designed as a workplace.

This is a further obligation in Regulation 11(5):

> In designing any structure for use as a workplace the designer shall take account of the provisions of the Workplace (Health, Safety and Welfare) Regulations 1992 which relate to the design of, and materials used in, the structure.

Again there are two elements the designer has to consider – the health and safety of people using the workplace and then the particular provisions that apply to the design of something that could be used as a workplace. How will the designer know it is to be used as a workplace? The Client has an obligation to tell designers of any proposed use as a workplace under Regulation 10(2). If he does not, the designer should ask the Client, even if he is not the designer's client. Nearly all places of work except ships are 'workplaces'.

The Workplace Regulations place duties on employers to ensure that suitable and sufficient arrangements are made for their employees. This covers everything from maintenance, ventilation, temperature, lighting, cleaning, room dimensions and space, work stations, conditions of floors and traffic routes, falling objects, windows, doors, gates, walls, skylights, ventilators, to organisation of traffic routes, escalators, sanitary conveniences, washing facilities, drinking water, accommodation for clothing, facilities for changing clothing and facilities for rest and meals.

Any designer producing a design for a Client for the project should have these things in mind because these are his Client's duties as an employer and the designer could be negligent if he did not make adequate provision subject to the usual caveats about what he was instructed to do or knew at the time. This would not necessarily be the case, however, if the designer was employed by someone else, for example, a contractor. It appears, however, that regardless of who his client is, a designer could be liable to prosecution if he does not take proper account of these workplace requirements in his design, both insofar as the permanent structure may be non-compliant with the Workplace Regulations and also if a worker's health and safety could be affected.

23.10.5 Information to be provided

Under the CDM Regulations 1994, Regulation 13 designers had to ensure their designs included 'adequate information about any aspect of the project or structure or materials (including articles and substances)' which might affect those constructing the project or those affected by their work. It did not say to whom this information was to be provided, but it did include information for the health and safety plan and file.

Under the CDM Regulations 2007 Regulation 11(6) designers have to take 'all reasonable steps to provide . . . sufficient information about aspects of the design of the structure or its construction or maintenance as will adequately assist (a) clients (b) other designers and (c) contractors to comply with their duties under these Regulations'. (If there is a residential client, this duty will not apply to them.)

The CDM coordinator does not feature in this list because Regulation 11 applies whatever the project. He is added to the list by Regulation 18(2) where the project is notifiable and therefore a CDM coordinator has to be appointed. This uses the same words as in Regulation 11 save that there is added particularly 'including his duties in relation to the health and safety file'.

Is 'sufficient information to adequately assist' the same as 'adequate information'? It is probably about the same but it is to help the other dutyholders with their 'duties', which goes beyond how the design is to be constructed and used safely, and would include information for the health and safety file and possibly the construction phase plan, and could also possibly include matters which will help them cooperate, coordinate, etc. in relation to design.

However, ACoP seems to concentrate on the way in which a project is to be constructed.

ACoP says (paragraphs 131 to 134) that the information to be provided should be project-specific and concentrate on significant risks which may not be obvious to those who use the design. Providing generic risk information about the prevention of falls is pointless because competent contractors will already know what needs to be done, but if the design gives rise to a specific and unusual fall-risk which may not be obvious to contractors, designers should provide information about this.

Significant risks, ACoP says, are not necessarily those that involve the greatest risks but those, including health risks that are not likely to be obvious to a competent contractor or other designers: unusual, or likely to be difficult to manage effectively.

ACoP does give some guidance about how designers are to give the information which has to be provided with the design. It says it should be brief, clear, precise and in a form suitable for the users, achieved through the following.

(*a*) 'Notes on drawings.' This is to be preferred because the notes are immediately available to those carrying out the works. These notes can refer to other documents if more detail is needed, and should be annotated to keep them up to date.

(*b*) 'Written information provided with the design.' This should be 'project-specific and should only contain information which will be useful to those constructing or maintaining the structure'. It would be useful if a note on the drawing referred to this information as it could become separated.

(*c*) 'Suggested construction sequences showing how the design could be erected safely where this is not obvious, for example, suggested sequences for putting up pre-cast panel concrete structures.' ACoP then says 'contractors may then adopt this method or adopt their own approach'.

If it is not possible to provide all the information at the same time, particularly where design work is continuing, the information should be released as the design develops but construction work should not be allowed to proceed unless all the information necessary for the work to be carried out safely has been provided.

If the information is to assist all dutyholders, it may be enough just to include it on drawings, etc. issued to the contractor and/or other designers unless there is something of specific concern to Clients, for example, which might affect the way they manage projects safely in terms of time and resources, in which case the information should also be issued to them, even if they are not the designer's client.

23.10.6 The duty to cooperate and coordinate

Under the CDM Regulations 1994, in Regulation 13 there was a duty on designers to cooperate with the planning supervisor and with any other designer 'so far as is necessary to enable each of them to comply with their health and safety requirements'. This obligation applied whatever the project and whether notifiable or not, although in the latter case there would have been no planning supervisor.

Under the CDM Regulations 2007 Regulation 5(1) designers have to seek the cooperation of and cooperate with 'any other person concerned in any project involving construction work at the same or an adjoining site so far as is necessary' to enable the designer and them to perform their health and safety requirements.

This goes further than the old Regulation 13 by bringing in all others whether or not dutyholders and whether or not on the site.

There is also a new duty to coordinate the designer's activities with 'all persons on whom a duty is placed' in a manner which 'ensures so far as is reasonably practicable the health and safety of persons (a) carrying out construction work and (b) affected by construction work' – Regulation 6.

ACoP says the object is to ensure that incompatibilities between designs are identified and resolved as early as possible and that the right information is provided in the pre-construction information.

It gives some examples of what could encourage cooperation (paragraphs 135 to 138).

(*a*) Setting up an integrated team of designers and contractors.
(*b*) The appointment of a lead designer where many designers are involved.
(*c*) Agreeing a common approach to risk reduction during design.
(*d*) Regular meetings of all the design team including the CDM coordinator with contractors and others.
(*e*) Regular reviews of developing designs.
(*f*) Site visits through which the designers can gain a direct insight into how the risks are managed in practice.

It is not thought the CDM Regulations 2007 place a duty on professionals visiting sites in their professional role to review and report on all aspects of health and safety. However, as under the earlier Regulations, where unsafe practices are noticed, these need to be drawn to the attention of the responsible person, the principal contractor and sub-contractor as appropriate.

Paragraph 21 suggests that complying with these duties to cooperate and coordinate may require something approaching the construction phase plan for higher risk activities such as

- structural alterations
- deep excavations and those in unstable or contaminated ground
- unusual working methods or safeguards
- ionising radiation or other significant health hazards
- nearby high voltage power lines
- a risk of falling into water which is or may become fast-flowing
- diving
- explosives
- heavy or complex lifting operations.

23.10.7 Additional duties where the project is notifiable

The first additional duty, as has already been mentioned, is to include the CDM coordinator as one of the dutyholders to whom 'sufficient information regarding the design of the structure or its construction and maintenance' has to be given, particularly for the health and safety file (Regulation 18(2)).

The other additional duty is that 'no designer shall commence work (other than initial design work) in relation to the project unless a CDM coordinator has been appointed for the project' (Regulation 18(1)).

There has to have been an actual appointment – the deemed appointment of the Client of himself under Regulation 14(4) does not apply to Regulation 18(1) – so the designer will need to know who the Client for the project is. He may not be the designer's client. As this appointment has to be in writing (Regulation 14(5)) the designer is entitled to ask for this.

The Client can appoint himself or any other competent person, for example, a designer or contractor, but the appointment needs to be as CDM coordinator.

It follows therefore that the designer has to include in his questions of his client, or the Client for the project if different, right at the beginning of a notifiable project a query as to whether he has appointed his CDM coordinator and if not to tell him when he has. It should be fairly easy to know when a project is notifiable because the criteria are really quite low, that is 30 days or 500 person days.

How far can a designer proceed with design work before he has to stop work unless a CDM coordinator has been appointed? There is no definition of 'initial design'.

Paragraph 66 of ACoP says the reason why the appointment of a CDM coordinator has to be made as soon as practicable after initial design work is to allow the Client to appraise their project needs and objectives, including the business case and any possible constraints on development, to enable them to decide whether or not to proceed with the project before appointing the CDM coordinator. The CDM coordinator needs to be in a position to be able to coordinate the design work and advise on the suitability and compatibility of designs and therefore they should be appointed before significant detailed design work begins. This is described as including 'preparation of the initial concept design and implementation of any strategic brief. As a scheme moves in to the detailed design stage it becomes more difficult to make fundamental changes to eliminate hazards and reduce risks associated with early design decisions'.

The HSE has given the view that initial design work does not include ACE outline proposals stage C3. It refers to work within and beyond ACE C3 Outline Proposals Stage and also within and beyond RIBA stage C as being beyond 'preliminary ' (sic) 'design'.

The ACE Agreements follow the same stages as the RIBA. Stage C under the earlier RIBA Plan of Work is 'The commencement of development of Strategic Brief into a full Project Brief. Preparation of outline proposals. Review of procurement route'.

That covers quite a lot of development work. The new RIBA Outline Plan of Work 2007 calls this Work Stage C 'Concept Stage' but it covers substantially the same ground. The duties within ACE Stage C 'Outline Proposals Stage' include preparation of outline proposals as well as development of the Strategic Brief into a full Brief.

It seems therefore that anything much after the initial feasibility and alternative designs and strategic brief or design brief and beyond the outline proposals in ACE Stage C3 would not be initial design. The same would be true in relation to Stages B and C in RIBA 2010. The development into a full brief setting out the Client's requirements is probably too far in view of the fact that the objective is to appoint the CDM coordinator as soon as possible and before the design/client requirements become 'set'. This would also reduce the scope for having to revise designs because of the health and safety considerations in others' designs. The progress of the design will obviously vary from project to project but it would be prudent to take the view that design work after outline proposals and/or after the preparation of the strategic brief or design brief is not 'initial design'.

Designers need to remember that this could also apply to initial design work for a contractor if they are employed by a contractor.

A prohibition on commencing work other than initial design work means a designer must stop work if a CDM coordinator has not been appointed, regardless of whether the client is pressing the designer to start work.

23.10.8 What designers do not have to do

ACoP in paragraph 143 lists things that designers do not have to do.

(*a*) Take into account or provide information about unforeseeable hazards and risks.
(*b*) Design for possible future uses of structures that cannot be reasonably anticipated from their design brief.

(*c*) Specify construction methods except where the design assumes or requires a particular construction or erection sequence or where a competent contractor might need such information.
(*d*) Exercise any health and safety management function over others.
(*e*) Worry about trivial risks.

23.10.9 Records

There is no specific obligation to keep record under the CDM Regulations 2007.

ACoP paragraphs 144 and 145 say:

> Designers are not legally required to keep records of the process through which they achieve a safe design but it can be useful to record why certain key decisions were made. Brief records of the points considered, the conclusions reached and the basis for those conclusions can be very helpful when designs are passed from one designer to another. This will reduce the likelihood of important decisions being revised by those who may not fully understand the implications of doing so. Too much paperwork is as bad as too little.

This is also good practice in any event in case of later prosecutions or claims, but equally of course could be used against the designer.

23.11. Obligations in appointments to comply with the CDM Regulations

It is not necessary for there to be any specific contractual duties to comply with the designer's duties under the Regulations. There is a statutory obligation to comply with them in any event.

The penalty for breach of the Regulations is prosecution. Civil liability may also arise but, as has been said, the Regulations themselves impose no civil liability on designers.

ACE Agreement 1 contains no reference to the CDM Regulations, although there is in some of the services schedules a reference to discussing the role of and the client's relationship with and need for the CDM coordinator.

RIBA 2010 includes the CDM coordinator in the 'Project Appointments' and 'Role Specifications' with a paraphrase of some of their duties and also refers to him under other role specifications. The design services do not refer specifically to any CDM duties except in relation to information in the health and safety file.

There are also references to the CDM coordinator in the other consultants' 'Role Specifications'. In work stage D 'Design Development' and work stage E 'Technical Design' there is a reference to investigating the effect of construction safety on 'Concept Design' and preparing technical details for construction safety respectively. Under stage K 'Construction to Practical Completion' the professional has an obligation to provide information for the health and safety file.

RICS 2011 has a specific obligation on the professional in clause 4.1 to comply at all times with the CDM Regulations to the extent they apply to the services or the project, with a separate obligation under clause 4.2 for any professional who is stated in the appendix to be a 'designer' for the purposes of the Regulations, for that designer to acknowledge that he is a 'designer' as defined by the Regulations and that he is to comply with his obligations and duties as such. No CDM obligations are imposed on the client.

PSC3 does not mention the CDM Regulations.

If the client is insisting on the professional undertaking a contractual duty to comply with the CDM Regulations, the professional should ask for a reciprocal obligation to be undertaken by the client if he is the Client for the project. The Client now has several duties that affect designers and the way designers carry out their CDM obligations (for which the designer may wish to sue him) and so the obligation should be reciprocal.

Professionals should not agree to perform duties which are paraphrases of the Regulations. This can be dangerous. The professional will then be required to comply with the paraphrased Regulations as a matter of contractual obligation, as well as the Regulations themselves. And the paraphrases could be wrong, more extensive than the equivalent duty under the Regulations, or interpret the Regulations in the way the courts do not.

Compliance with the Regulations could be time-consuming, particularly, for example, if other parties are dilatory in cooperating or providing information. So professionals need to consider if they have sufficiently allowed for this in their fees, for example

- the additional information that may be required and may need to be given to all involved
- the additional coordination and cooperation on health and safety matters including additional time and resources
- the extra consideration in relation to the designs because of workplace use/ requirements even if the designer's client is not the Client for the project.

What if, as a result of the cooperation/coordination with others, the professional has to change things? If it is for health and safety reasons, is the client going to exclude any additional fees for this? It will be easier to assess fees if the decisions which could affect health and safety matters in the designs are made early in the project, as the Regulations intend.

What if there are late design changes? What about fresh health and safety considerations as a result? Does the professional's appointment include a right to additional fees if the works are varied or if there is extra work or expense for reasons beyond the professional's reasonable control?

Obligations to comply with instructions will need to be checked. Many non-standard conditions of engagement include an unqualified obligation to comply with the client's instructions. What if this instruction is to continue with the services despite the failure to appoint a CDM coordinator or if there is a failure to give the professional the necessary information? A 'lawful' instruction might allow the professional to refuse to comply, as such an instruction would be unlawful, as would an obligation to give a 'reasonable' instruction. Alternatively, the professional could include a specific exception to the obligation to comply with an instruction, for example, 'save for an instruction by the client which does not comply with his obligations under CDM or which requires the professional to depart from his obligations under CDM'.

Professionals will need to check the obligations on the client in appointments to give them all the information they need. Sometimes non-standard conditions of engagement fail to include any obligations to give the designer information and then make it an entire agreement. A general obligation to give the professional the information he needs when he needs it can be used to get the information the professional needs under CDM, if the professional's client is the Client for the project. If there is not such an obligation, a reminder to the Client of his obligations under the Regulations to give such information could help.

The professional may need to check what his contract says about the programme. Obligations relating to the programme in non-standard conditions of engagement can be onerous and are sometimes completely unqualified. These should not be accepted in any event. As to the programme itself, it should, if it does not already, allow sufficient time for all the health and safety aspects to be carried out. The Client now has a duty to allocate time for this.

Warranties in appointments regarding competence and resources are and always have been otiose. Designers now have to be 'competent' before they can commence work and so such a warranty could be resisted on that ground.

It might be necessary for a professional to terminate his appointment if the Client will not comply with his CDM duties or is preventing the designer from doing so. Although it is thought unlikely that a court would find a professional in breach of contract in these circumstances, as health and safety is involved, it would be better if the professional can use a contractual right to terminate in such circumstances. ACE Agreement 1 includes a right to the professional, where the professional considers it irresponsible for the professional to perform all or part of the services (F9.6).

RIBA 2010 provides that the professional can suspend if the client is in material or persistent breach or if he is prevented from or impeded in the performance of the services for reasons beyond his control (8.1.2). The professional can then terminate if the breach is not remedied but the provisions about this are not entirely clear – see Section 8.9. The professional also has a right to terminate on reasonable notice (8.2.1) and so this provision could be used.

Under RICS 2011, the professional can only terminate if the project is cancelled, for material breach or on insolvency. In the absence of express obligations on the client to comply with the CDM Regulations this right in respect of material breach may not be sufficient.

Under PSC3 the professional can only terminate for non-payment (or insolvency) – clause 90.1 – but it is possible the prevention provisions in clause 18 could be used.

The new CDM duties, such as those to cooperate and coordinate and to give sufficient information to others, may result in more civil claims for contribution from designers. This may make it more important to have a net contribution clause as well as a limit of liability in appointments.

23.12. A designer's checklist

The following may be helpful as a checklist for professionals seeking to comply with the CDM Regulations 2007.

(*a*) Review procedures for demonstrating competence as designer.
(*b*) Review procedures for checking and demonstrating the requisite competence of the designer's employees or those supervising them.
(*c*) Check whether the designer's procedures include considering whether the Client for the project is aware of his duties *before* any design work starts.
(*d*) Ensure the designer's letter to the Client for the project refers to the Client's new duties.
(*e*) Include in procedures a mechanism to check whether the CDM coordinator has been appointed before the designer undertakes anything other than 'initial design'.
(*f*) Review the 'design considerations' to see that:
  (*i*) the additional people whose health and safety the designer has to consider are included (including those who will use it as a workplace)
  (*ii*) the general principles of prevention are taken into account
  (*iii*) the provisions of the Workplace Regulations have been taken into account – if appropriate
  (*iv*) the procedures allow for these considerations to be reviewed throughout the project.
(*g*) Review the health and safety information the designer has to give concerning his designs, how it is given and to whom. This could go beyond other designers, contractors and the CDM coordinator and could include the Client for the project.
(*h*) Check that everyone in the organisation is aware of the obligation to cooperate and coordinate with others regarding health and safety aspects of the design.
(*i*) Review procedures concerning record-keeping, in particular in relation to competence checks and key health and safety design decisions.
(*j*) Revise the criteria for the matters to be considered in relation to appointments.

23.13. The CDM coordinator's duties

The role of the CDM coordinator in effect replaces that of the planning supervisor. That role is abolished. His duties are in summary as follows.

- He has to notify the project to the Health and Safety Executive, by providing it with the particulars specified in the Regulations.
- He must do this as soon as is practicable after his appointment, all as set out in Regulation 21. The particulars must be signed by the Client and there must be a declaration by the Client that he is aware of his duties.

- He must give suitable and sufficient advice to the Client on undertaking the measures he needs to take to comply with the Regulations.
- He must take all reasonable steps to identify and collect the pre-construction information and provide to every designer such of the information as is relevant to him, and do the equivalent for every contractor appointed by the Client.
- He must ensure that suitable arrangements are made and implemented for the coordination of health and safety measures during planning and preparation for the construction phase. (There was no express duty like that under the CDM Regulations 1994.)
- He must take all reasonable steps to ensure that designers comply with their duties.
- He must prepare, where none exists, and review and update the health and safety file.
- He must liaise with the principal contractor over design developments that may affect planning and management, and take all reasonable steps to ensure cooperation between designers and the principal contractor during construction in relation to any design or change to a design.

23.14. The principal contractor's duties

The duties of the principal contractor, again in summary, are to

- plan, manage and monitor the construction phase in a way which ensures that, so far as is reasonably practicable, it is carried out without risks to health or safety. There was no express duty like that under the CDM Regulations 1994
- liaise with the CDM coordinator during construction in relation to any design or change to a design
- prepare the construction phase plan, keep it updated, arrange for it to be implemented.

The construction phase plan replaces (in effect) what was called the health and safety plan under the CDM Regulations 1994. As the new title makes clear, the plan relates only to construction work, and its preparation is now the exclusive responsibility of the principal contractor. It does not have to contain all the information prescribed under the old Regulations for the health and safety plan.

The principal contractor has other specific duties, including duties in relation to welfare facilities, site rules, other contractors, information for the health and safety file and the health, safety and welfare of workers engaged in the construction work, most of which is substantially the same as in the 1994 Regulations.

23.15. Contractors' duties

Contractors, in brief, have a duty to plan, manage and monitor construction work carried out by them or under their control in a way which ensures that, so far as is reasonably practicable, it is carried out without risks to health and safety. This again is new – there was no express duty like this on contractors under the CDM Regulations 1994.

Contractors also have duties relating to information and training, welfare facilities and health and safety on construction sites.

Contractors must not carry out construction work unless the Client is aware of his duties under the Regulations. This is a new duty for contractors.

Where the project is notifiable, contractors must not commence construction work unless, among other things, they have been provided with the name of the CDM coordinator and principal contractor. This is a new duty.

Contractors also have to provide certain information to the principal contractor and to comply with the principal contractor's directions and follow the construction phase plan.

23.16. Duties of all dutyholders

There are a number of duties which apply to all dutyholders. Briefly, these are

- the duty to check one's own competence and not to accept an appointment unless one is competent
- not to arrange for or instruct a worker to carry out or manage design or construction work unless the worker is competent or under the supervision of a competent person
- the duties to cooperate and seek cooperation
- the duty to coordinate
- the duty to take the general principles of prevention into account.

23.17. Duties on others

There are a few duties in the Regulations imposed on other people, which arise mainly from the old Construction (Health, Safety and Welfare) Regulations 1996.

There is also a duty which applies to any person working under the control of another person to report to that person anything which he is aware is likely to endanger the health or safety of himself or others.

23.18. Collateral warranties and third party rights for a designer/CDM coordinator

The CDM Regulations are concerned with obligations before and during the construction of the works only. The designer's obligations are confined to the design period, whereas the CDM coordinator's duties are carried out during the design and construction phases and both relate to the health and safety implications of design. CDM coordinators are also concerned with the preparation of the health and safety file. CDM coordinators are not, however, responsible for information in the file – that is the responsibility of the designers, contractors, sub-contractors or others who provided it. The CDM coordinator has no duties after the file has been handed to the Client at the end of the construction work.

It is therefore not appropriate that collateral warranties in respect of CDM obligations should be given to funders, tenants or purchasers, as the performance of these duties has no implications for purchasers and tenants who do not take an interest until after construction is completed (except perhaps where funders might step in and take over the project and thus become 'the Client' for that project). Further, the CDM Regulations have already imposed the penalty of a criminal prosecution and the possibility of a fine if there is a breach of the CDM Regulations by a designer or CDM coordinator, and specifically excluded civil liability – see Section 23.3.

It was not the intention of the legislators that the CDM Regulations should create new civil liabilities or extend to the civil liabilities that are created by warranties. If collateral warranties are given, therefore, care should be taken to preserve the right to argue that no additional civil liability can be created.

The same points apply to any request to give third party rights similar to collateral warranties to third parties.

If a professional has to agree to an obligation to comply with CDM duties in his appointment, it is all the more important that third party rights are excluded, as they are in the ACE Agreements, RIBA 2010, RICS 2011 and one of the options (Y(UK)3) to PSC3.

Unless excluded, third parties such as contractors could possibly bring a claim, for example, that they have not been given sufficient information about the professional's designs, or workers other than the designers' employees using premises the professional has designed could sue the professional if he has affected their health and safety. This is not a right given under the Regulations and so should not be assumed voluntarily.

Professional Services Agreements
ISBN 978-0-7277-4158-5

http://dx.doi.org/10.1680/psa.41585.377

Chapter 24
Law and jurisdiction, and dispute resolution

24.1. Law and jurisdiction

The law and jurisdiction that governs a professional services agreement may be dealt with expressly or be left to be decided when a dispute arises. Clearly, the former is sensible where there is any overseas element in, or related to, an appointment. In the absence of an express or implied choice of law, it will, when the occasion arises, be for the court to decide which law governs the contract and which court should hear any dispute. Factors taken into account include an assessment of the 'connections' of a contract with any law or jurisdiction, for example, the place where the contract was negotiated, the place where the contract was entered into, the place of performance of the contract, the location of the subject matter of the contract and the domicile, residence, nationality, place of incorporation and place of business of the parties.

The choice of law and jurisdiction can be important because the damages that may be awarded outside the UK (e.g., in the USA) may be greater than those that would be awarded in England under English law. Thus, if a professional has an English law clause in an agreement with a sub-consultant, but a foreign law clause in the principal consultancy agreement, there is a risk that the same sort of damages will not be payable under both. It is therefore essential that the sub-consultancy agreement is drafted 'back to back' with the principal agreement on this issue.

Dealing with the proceedings and attending court will be more time-consuming and expensive in some countries. Further, it may be necessary to check if there are any restrictions or exclusions in the professional indemnity insurance that limit the insurance cover available in different regions. Many have restrictions concerning work carried out in the USA and Canada.

In contracts with an overseas element, it is fairly common to find a clause which includes a provision such as 'the parties hereby submit to the exclusive jurisdiction of the English courts' or 'the parties submit to the non-exclusive jurisdiction of the English courts'.

Where there is an 'exclusive jurisdiction' clause, the contract requirement takes effect to the exclusion of the jurisdiction that would otherwise be imposed by, for example, the Brussels Convention or by the 'connection' test mentioned in the first paragraph above. A 'non-exclusive jurisdiction' clause is a way of conferring jurisdiction on the court stated in the clause, in addition to the jurisdictions that would have existed under the Brussels Convention or under the 'connection' test. It is a very persuasive factor in the choice of the appropriate court for proceedings, but not always decisive.

To avoid doubts as to which country's laws apply and questions of jurisdiction, the following can be used, as in ACE Agreement 1 in Part 1: The Memorandum of Agreement:

> A3 This Agreement shall be governed and construed in all respects in accordance with the laws of England and each party hereby submits to the non-exclusive jurisdiction of the English courts.

The Scottish versions of the Agreement have a Scottish law and jurisdiction clause.

RIBA 2010 provides:

> 1.6 This Agreement is subject to the law of England and Wales or Northern Ireland or Scotland as specified in the Project Data and subject to clause 9.3 [limits on the extent of any arbitration agreement] the parties submit to the exclusive jurisdiction of the specified jurisdiction.

The default position in the project data is the law of England and Wales.

RICS 2011 provides:

> 19 This Appointment is governed by and construed in accordance with English law.

If litigation is chosen as the method of dispute resolution, clause 16.11 provides that the parties submit to the exclusive jurisdiction of the courts of England and Wales.

PSC3 provides for the law of the contract to be set out in the contract data.

It is perhaps worth commenting that there is no such thing as British or UK law – the jurisdictions of England, England and Wales, Scotland and Northern Ireland are different. The English and Welsh jurisdictions used to be identical, but with the advent of the Welsh Assembly some laws in Wales can now be different from those in England.

24.2. Dispute resolution processes

There are various ways, apart from negotiation, in which a dispute can be resolved. The following sections deal briefly with litigation, arbitration, mediation, conciliation, expert determination and adjudication.

Litigation and arbitration are considered to be the traditional methods of resolving disputes. They are adversarial procedures where the conduct of the dispute is more in the control of the parties than is the case with more inquisitorial procedures such as adjudication or expert determination or those in non-common law jurisdictions. A final and binding decision is reached, subject only to rights of appeal where these exist.

The other processes are those normally referred to as alternative dispute resolution (ADR). There is no definition of the term ADR, but it is a useful way of referring to various dispute resolution processes. With one exception, they are consensual processes. They do not, unless the parties agree otherwise, result in final and binding decisions that exclude the 'traditional' processes of litigation or arbitration should the dispute need to be taken further. The exception is expert determination. This process would normally be commenced as a result of an express contractual clause, although it could be arranged *ad hoc* as circumstances require.

24.3. Litigation

Litigation is available in relation to any dispute unless the parties have agreed to resolve the dispute by a different process. This may be by way of an express term in the appointment or a separate *ad hoc* agreement after a dispute has arisen. Even where parties agree to resolve their dispute by means other than litigation, the court may retain residual powers that cannot be excluded.

The advantages of litigation are that the court possesses greater powers than arbitrators and the judges and the rooms for interlocutory hearings and trials are provided free by the state, although there are court fees. In appropriate cases, legal aid is available for litigation, but not for arbitration.

Some of the traditional criticisms of litigation are now out of date. The waiting time for trial dates has been reduced dramatically and radical changes to the rules governing civil litigation have streamlined the litigation process and trimmed costs. Changes include the introduction of pre-action protocols that require claimants to provide much more information to a prospective defendant prior to the commencement of proceedings. There is now a greater opportunity to investigate and, if appropriate, negotiate and settle claims prior to proceedings being commenced.

Litigation provides a forum where all parties involved in a particular problem or dispute, for example, a client, architect, consulting engineer, main contractor and sub-contractor, can have all the claims decided by the same judge at the same time.

24.4. Arbitration

The advantages of arbitration are perceived to be that it is a more informal and flexible procedure, the dispute can be dealt with by appropriately technically qualified arbitrators, and it is a private process. This latter point can be important to the parties. Courts are of course public. On the other hand, some argue that: (i) major arbitrations all too often mirror litigation procedures; (ii) it has been too easy to hijack, resulting in delays and disruption to the proceedings; (iii) the quality of arbitrators can be variable, and (iv) the parties have to pay for the time of the arbitrator and for the rooms for all hearings. Also, unless the appointment provides otherwise, arbitration is between the two parties to the contract only, meaning other parties cannot be joined in and all claims, including those for contribution, heard at the same time.

Those criticisms have been addressed in part by the Arbitration Act 1996 which gives the parties more autonomy and the arbitrators greater powers. Also, the arbitrator-appointing institutions have been seeking to maintain and improve the quality and training of arbitrators. A dispute before a good-quality, experienced arbitrator, keen to get on with the matter and not afraid to use his powers, can be very effective.

24.5. Provisions in the standard conditions of engagement

ACE Agreement 1 does not provide for arbitration and ultimately any dispute must be resolved by litigation.

RIBA 2010 provides that the parties can choose whether arbitration or litigation should apply. If it is arbitration, that does not apply to any enforcement of a decision of an adjudicator. In England and Wales or Northern Ireland, however, disputes regarding sums under £5000 or such other sum as is provided under Section 91 of the Arbitration Act can be litigated. The Construction Industry Model Adjudication Rules (CIMAR) are to apply and the arbitrator is not to have the power referred to in Section 38(3) of the Arbitration Act 1996, that is the power to order the claimant to provide security for costs. Where the law of Scotland applies, the arbitration is to be conducted under the rules of the current edition of the Scottish Arbitration Code for Domestic and International Arbitration.

RICS 2011 provides for the parties to choose between litigation or arbitration. In the case of arbitration the CIMAR apply.

PSC3 provides for 'the tribunal' to be specified in the contract data – part one and the guidance say the 'tribunal' will either be arbitration or litigation in the appropriate court, for example, 'trial by judge sitting as such in the High Court of Justice in London'. The parties are also expected to choose either option W1 which provides for adjudication where the Construction Act or amended Construction Act does not apply or to choose option W2 which provides for adjudication where the Construction Act or amended Construction Act do apply and then for matters to go to the tribunal. The contract data provide for the parties to set out in the case of arbitration the arbitration procedure, the place where the arbitration is to be held and the method of choosing the arbitrator.

24.6. Mediation

Mediation is a process in which a neutral third party – the mediator – seeks to assist the parties in reaching a negotiated settlement of the dispute. It is a consensual process, with the result that no party can be forced to participate in a mediation and, if it agrees to take part, can still walk away from the process at any time.

One of the benefits of mediation is that the solutions do not have to be simply the payment of money. Where long-term relationships exist between the parties, it is often the maintenance of those long-term relationships that is more important than determination of the precise amount of money due from one to the other by means of an adversarial dispute. The latter could lead to the relationship between the parties being damaged irreparably with the result that no further business will be conducted between them.

A mediator may be chosen by the parties or be selected by one of the bodies which holds itself out as providing mediation services. Once a mediator is appointed, it is usual for the parties to be asked to prepare a brief summary of their case prior to the date fixed for the mediation. Mediation is normally commenced with an initial session conducted by the mediator with all parties and their representatives present when each party is asked to present its case.

The mediator will usually make it a prerequisite of the mediation that each party will be represented by a senior executive who has the power to make decisions in relation to the dispute and to reach a settlement.

The parties then normally retire to private rooms. The mediator will visit the parties in turn in what are usually termed 'caucus' sessions. In these sessions, the mediator discusses the issues, seeks to clarify the parties' cases and to ascertain the parties' real interests and what they want or need out of the settlement. The mediator must treat all information given to him with the utmost confidence unless expressly authorised to release it.

Mediation is a relatively short process. The majority of mediations last no more than one or two days. As a result, the whole mediation process, from the appointment of the mediator to completion, can take place in three or four weeks.

ACE Agreement 1, RIBA 2010 and RICS 2011 all encourage mediation.

ACE Agreement 1 says:

> F12.1 The parties shall attempt in good faith to settle any dispute by mediation.

RIBA 2010 says:

> 9.1 The Client and the Architect may attempt to settle any dispute or difference arising under this Agreement by negotiation or mediation ...

RICS 2011 says:

> 16.1 The parties use reasonable endeavours to resolve any issue or dispute between them without delay by way of negotiation or, if they so choose, by mediation. If mediation is chosen, the parties endeavour to agree the identity of the mediator and either party may propose a list of up to 3 mediators to the other party. In default of agreement within 14 days of

submission of such proposal, a mediator may be nominated by the President or Vice-President of the Royal Institution of Chartered Surveyors on the application of either party.

16.2 Any mediation is carried out in accordance with the edition of the CEDR Model Mediation Procedure published by the Centre for Effective Dispute Resolution which is current at the date of this Appointment.

24.7. Conciliation

The terms 'conciliation' and 'mediation' are often used interchangeably where the conciliator performs a role identical to that of a mediator. Generally, the term 'mediation' has become more popular in recent years. There is no mandatory format for either mediation or conciliation, with the result that the processes are difficult to define with any degree of accuracy. It is believed that there are no mediation schemes in common usage in which the mediator forces a solution on the parties or recommends to them a particular solution. Conciliation, on the other hand, can be more proactive and a conciliator is frequently empowered to make recommendations or suggestions to the parties as to how the dispute should be resolved.

24.8. Expert determination

Expert determination is a process which is usually agreed by the parties to be final and conclusive, and, in the past, was used most often in relation to valuations of property and shares and for rent reviews by surveyors. Recently, it has become more popular in other fields, such as construction contracts, but it is rarely seen in professional appointments.

The procedure to be adopted will be determined by the expert and will vary depending on the nature of the dispute. In straightforward disputes, the matter can be disposed of on the basis of written statement of cases. In more complex disputes, the expert may ask such questions and make such investigations as he thinks necessary in the circumstances. Usually, the expert is not restricted to the evidence submitted to him by the parties. It is often the case that the parties are relying on the expert's knowledge and expertise in a particular field. This, in itself, can reduce the length of submissions necessary by the parties. Expert determination thus gives the parties and the expert a high degree of flexibility in determining the procedure, with no necessity for pleadings, disclosure, witness evidence and a hearing.

24.9. Adjudication

Adjudication is dealt with in detail in Chapter 25.

Professional Services Agreements
ISBN 978-0-7277-4158-5

http://dx.doi.org/10.1680/psa.41585.383

Chapter 25
Adjudication

25.1. Introduction

Most professional appointments will be 'construction contracts' under the Housing Grants, Construction and Regeneration Act 1996 (the Construction Act) – see Section 25.3 – and most will have clauses dealing with the reference of any dispute to adjudication.

Adjudication is a process by which parties refer a dispute to an impartial third party – the adjudicator – to decide the issue in dispute. The parties will normally provide an adjudicator with a range of powers to enable him to reach a decision. An adjudicator's decision can be agreed by the parties to be both final and binding, although usually it is simply binding until such time as one of the parties takes the matter to litigation or arbitration. It is now fairly rare to come across adjudication procedures that are not those required by the Construction Act. This chapter therefore deals with Construction Act adjudication only.

There is no right of appeal as such against an adjudicator's decision. Enforcement of an adjudicator's decision can be challenged on jurisdictional grounds or because the adjudicator infringed the rules of natural justice – if there has been bias shown or procedural irregularities.

25.2. Insurance considerations

As the adjudicator's decision under the Construction Act is enforceable in the courts, a professional will have a legal liability to comply with that decision. Assuming it is a monetary decision, the insured under a civil liability policy will be entitled to an indemnity from insurers, subject to any terms and conditions of the policy.

If an adjudicator decides against a professional in relation to, say, defects in design and/or inspection, there is a presumption that the adjudicator has decided that the professional has been negligent. This is particularly so if the relevant adjudication procedures require the adjudicator to reach a decision in accordance with the applicable law relating to the contract. On that basis, the insured, under a negligence policy, will be entitled to an indemnity from insurers, subject to the terms and conditions of the policy.

Where an adjudicator is entitled to reach a decision on the basis of a fair and commercially reasonable view, as under the TeCSA scheme, or at his absolute discretion, and with no reasons in a decision to show otherwise, insurers may be able to argue that there has been no finding of negligence against the professional and that they are not obliged to provide an indemnity under a negligence policy.

Most professional indemnity insurers provide cover for adjudicators' decisions, although in some cases this has been on the basis of an 'exclusion and buy-back', that is, it has to be bought as an extension of cover. Usually additional conditions are required, such as: (i) very early/immediate notice of adjudication is to be provided directly to insurers (compliance being a condition precedent to cover); (ii) cover is only to be provided where an adjudicator is under an obligation to reach a decision on the basis of the legal entitlements of the parties, and (iii) the onus is on the insured to show that an adverse adjudicator decision was the direct result of negligence by the insured in the conduct of professional activities.

25.3. Adjudication under the Construction Act 1996, the statutory Scheme and 'construction contracts'

The Construction Act 1996 gives either party to a 'construction contract' a right (i.e., it is not obligatory) at any time to refer a dispute arising under that contract to adjudication under a procedure that complies with the Act. In other words, a referring party has a right to refer a dispute to an adjudication procedure complying with at least the minimum requirements of the Act.

Whether there is a 'construction contract' will depend on satisfying the criteria in the Construction Act.

The Construction Act applies to a professional services agreement to carry out architectural, design or surveying work, or to provide advice on a building, engineering, interior or exterior decoration or on the laying-out of landscape, in each case in relation to 'construction operations'. 'Construction operations' are defined in the Construction Act and cover most forms of construction work (although it is not the same definition as under the CDM Regulations). The definition has been further clarified by the courts. The 'construction operations' must be carried out in England and Wales. If the parties wish to provide for adjudication in England and Wales in relation to a project overseas, then English law and a contractual right to adjudicate must be included in the agreement with relevant procedures incorporated.

The 'construction contract' also had to be in writing but the definition was quite wide and included agreements evidenced in writing or where there is a reference to the terms in writing. The Local Democracy, Economic Development and Construction Act 2009 ('the amended Construction Act') included a provision whereby adjudication under the Construction Act will be possible in relation to both written and oral 'construction contracts' – see Section 25.10. This will considerably widen the ability of the parties to take disputes to adjudication in respect of contracts entered into after 1 October 2011.

Most, but not all, construction professional services agreements will fall within the Act. Further investigation and possibly legal advice may be necessary to confirm this in the circumstances pertaining to any particular agreement.

The Construction Act requires that all 'construction contracts' contain certain minimum adjudication provisions. These are:

Section 108(2) The contract shall –

(*a*) enable a party to give notice at any time of his intention to refer a dispute to adjudication;

(*b*) provide a timetable with the object of securing the appointment of the adjudicator and referral of the dispute to him within 7 days of such notice;

(*c*) require the adjudicator to reach a decision within 28 days of referral or such longer period as is agreed by the parties after the dispute has been referred;

(*d*) allow the adjudicator to extend the period of 28 days by up to 14 days, with the consent of the party by whom the dispute was referred;

(*e*) impose a duty on the adjudicator to act impartially; and

(*f*) enable the adjudicator to take the initiative in ascertaining the facts and the law.

(3) The contract shall provide that the decision of the adjudicator is binding until the dispute is finally determined by legal proceedings, by arbitration (if the contract provides for arbitration or the parties otherwise agree to arbitration) or by agreement.
The parties may agree to accept the decision of the adjudicator as finally determining the dispute.

(4) The contract shall also provide that the adjudicator is not liable for anything done or omitted in the discharge or purported discharge of his functions as adjudicator unless the act or omission is in bad faith, and that any employee or agent of the adjudicator is similarly protected from liability.

The amended Construction Act adds for construction contracts entered into after 1 October 2011 the requirement that these provisions be 'in writing' in sub-sections (2) and (3) and also that the contract includes a provision in writing permitting the adjudicator to correct his decision so as to remove a clerical or typographical error arising by accident or omission – Section 108(3A).

If construction contracts do not contain the requisite provisions, the adjudication provisions in the Scheme for Construction Contracts (England and Wales) Regulations 1998 (the Scheme) or the amended Scheme will apply. The Scheme describes adjudicator appointment procedures and timescales, and states how the adjudicator must conduct the adjudication.

Thus, if no adjudication procedures are incorporated into an agreement or if an adjudication procedure already incorporated within a 'construction agreement' does not comply with the provisions in the Construction Act, the adjudication procedure under the Scheme or the amended Scheme applies instead, depending on when the construction contract was entered into. If one express part of the procedure in a 'construction contract' does not comply with the Construction Act or amended

Construction Act provisions, the whole of the Scheme or the amended Scheme procedure applies as appropriate in substitution for the original terms. This is different from the payment provisions – the Construction Act (or amended Construction Act) and the Scheme (or amended Scheme) only apply to the extent that the 'construction contract' is non-compliant and so they could, for example, supply one missing element only of the payment provisions.

Where the Scheme or amended Scheme is applicable, neither party can amend it. However, if the words of the Scheme or the amended Scheme are used as the basis of a contractually agreed adjudication procedure, they can be amended, as long as the resulting procedures still comply with the minimum requirements of the Construction Act or amended Construction Act.

25.4. Non-statutory standard adjudication procedures

There are a number of standard adjudication procedures published by construction industry-related organisations, such as the CIC, TeCSA and RICS. Some large organisations, such as the Highways Agency, also produce their own procedures. While most comply with the minimum adjudication requirements of the Construction Act, or amended Construction Act, often those minimum requirements are enhanced. Where such procedures are to be incorporated into professional services agreements, they need to be considered from both the professional's and insurer's points of view for acceptability and whether the professional is more likely to be the referring or responding party.

Matters to be considered include the following.

(*a*) Sometimes an adjudicator may be allowed or obliged to give reasons for a decision, sometimes not, and sometimes the agreement of both parties is needed. It is desirable that the option to give reasons is included, as this could help with establishing the grounds on which enforcement of that decision could be challenged.

(*b*) Some procedures require an adjudicator's decision to be based on the facts and applicable law; some are silent and/or may permit a decision on, say, commercial or fair and reasonable grounds. For reasons given earlier, the latter may not be acceptable to insurers.

(*c*) Some procedures permit the set-off of claims against the matter that is the subject of the dispute referred to the adjudicator, others do not. Whether this is acceptable will depend on the nature of the construction contract.

(*d*) Sometimes the disputes that can be referred will only be those 'arising under the contract'. Sometimes the disputes to be referred will be stated to be those arising 'under or in connection with the contract' or 'in relation to or arising out of' the contract. The wording of the Construction Act – 'arising under the contract' – should be used.

(*e*) An adjudicator must have a duty to act impartially as stipulated by the Construction Act. Sometimes this is expressed as 'fairly and impartially'. Other variations could also be included, including 'in accordance with natural justice'. It is best if the Construction Act wording 'to act impartially' is used.

(*f*) The adjudicator should be someone qualified to deal with disputes arising under the appointment. To be certain of the quality of the adjudicator, it is sensible to name the adjudicator or a selection of them in the agreement.

(*g*) The adjudicator's powers may differ regarding the allocation of fees and costs between the parties. Each party should bear its own costs.

(*h*) The scope of an adjudicator's indemnity can vary.

(*i*) The parties may or may not be entitled to be legally represented at any meetings an adjudicator may wish to have.

25.5. Non-standard *ad hoc* adjudication procedures

Sometimes individual organisations, such as contractors, will produce their own, even *ad hoc* project-specific, adjudication procedures. These are often biased towards the position the drafter believes his own organisation is most likely to be in, for example, a contractor as the responding party and a sub-contractor (or sub-professional) as the referring party. Sometimes the bias is obvious, such as the referring party having to bear both parties' costs (in order to deter disputes being referred) (but see Section 25.10), or the use of payment of any adjudication amounts awarded into joint signature shareholder accounts until a further decision has been made by arbitration or the courts (thus delaying actual payment). The adjudication procedure must comply with the express requirements of the Construction Act or amended Construction Act for construction contracts entered into after 1 October 2011 or the referring party will be able to use the Scheme instead. Where the agreement is not a 'construction contract', the parties are free to agree any adjudication procedure and will be obliged to comply with what is signed. These adjudication procedures therefore need reviewing, negotiating (and possibly not accepting), like any other conditions of engagement.

25.6. Referring party in an adjudication

Where professional services agreement disputes arise during the performance of those professional services, they are most likely to be claims for late or non-payment of fees by a client, with the professional as the referring party. If each party bears its own costs, the professional can only gain, and not be any worse off, from the adjudication process and may have the use of any money awarded, unless and until the other party goes to arbitration or the courts. As with any other claim for fees, the professional needs to consider whether referring a dispute concerning fees to adjudication will result in the client referring a dispute alleging breach of the duty of care on the part of the professional to adjudication.

25.7. Responding party in an adjudication

In some instances, a professional is more likely to be the responding party in an adjudication. For example, disputes arising under appointments or collateral warranties are likely to concern an alleged breach of contract by the professional, with the professional as the responding party. A professional will generally be in the position of no gain. If the adjudicator makes a decision against the professional, that decision has to be complied with straightaway and might involve large amounts of money, even if the dispute is later referred to arbitration or the courts. The availability of insurance cover for amounts higher than any excess is important, and insurers usually require that they are advised and involved as soon as possible. Immediate notification to insurers of the receipt of a notice of adjudication is essential.

25.8. Adjudication in collateral warranties

As has been explained in Section 21.12, it is rare that a collateral warranty will be a 'construction contract', and adjudication is not a suitable way of deciding the sort of complex disputes that arise out of collateral warranties. There is therefore no need for adjudication clauses in a collateral warranty.

25.9. Adjudication provisions in standard conditions of engagement

ACE Agreement 1 provides:

> F12.1 Where this Agreement is a construction contract within the meaning of [the Housing Grants, Construction and Regeneration Act 1996 as amended] either party may refer any dispute arising under this Agreement to adjudication in accordance with the Construction Industry Council Model Adjudication Procedure current at the time of the referral of the dispute. The adjudicator shall be appointed at the request of either party by the Association for Consultancy and Engineering. The statement of case to be sent by the referring party to the adjudicator in accordance with that Procedure shall not exceed eight single-sided sheets of A4-sized paper excluding any attachments.

If the appointment is not a 'construction contract' there will be no referral of any dispute to adjudication under ACE Agreement 1. The restriction on the statement of case is to try to avoid the practice some clients adopt, which is of serving very lengthy statements of case which do not properly deal with the dispute that is being referred.

RIBA 2010 provides:

> 9.1 The Client and the Architect may attempt to settle any dispute or difference arising under the Agreement by negotiation or mediation, if suitable, or either party may refer the matter to adjudication, arbitration or legal proceedings as specified in the Project Data.
>
> 9.2 Either party may give notice at any time of the intention to refer a dispute or difference to an adjudicator.
>
> 9.2.1 Referral of the dispute to such adjudicator shall be made within 7 days of such notice.
>
> 9.2.2 The appointment of the adjudicator shall be made in accordance with the procedures identified in the Project Data.
>
> 9.2.3 The parties may agree who shall act as adjudicator or the adjudicator shall be a person nominated at the request of either party by the nominator specified in the Project Data.

> 9.2.4 The adjudicator may allocate between the parties the costs relating to the adjudication, including the fees and expenses of the adjudicator, in accordance with the provisions of clause 5.20.

Although the client and the professional can attempt to settle any dispute or difference through negotiation or mediation and the parties may refer the matter to adjudication, arbitration or legal proceedings as specified in the project data, if the Construction Act applies either party will have a right to adjudicate at any time in any event, but this provision brings in a contractual right to adjudicate. As to the procedure when an adjudication arises, this is either to be the CIC model or the Scheme (or the RIBA adjudication scheme for consumer contracts where the client is a consumer).

The parties can agree who is to act as adjudicator, or else he is to be nominated by the nominator specified in the project data. (It will be the RIBA if nobody is nominated.) There is a specific right given to the adjudicator to allocate between the parties the costs relating to the adjudication including fees and expenses of the adjudicator in accordance with the provisions of clause 5.20 – which simply refers to costs to be determined by the tribunal.

As to clause 9.2.4, this provision may not be correct. The amended Construction Act has added a requirement that any contractual provision which concerns the allocation as between the parties of costs relating to the adjudication on a dispute is ineffective unless it is made in writing, is contained in the construction contract and confers power on the adjudicator to allocate his fees and expenses as between the parties or it is made in writing after the giving of notice of intention to refer the dispute to adjudication. Insofar as clause 9.2.4 deals with the adjudicator's fees and expenses then it is correct but the rest of the sub-clause could be ineffective.

RICS 2011 provides:

> 16.3 Regardless of any negotiation or mediation, either party may at any time refer any dispute under this Appointment to adjudication under the Scheme for Construction Contracts (England and Wales) Regulations 1998, Part I (as amended).
>
> 16.4 The Adjudicator's decision is binding until the dispute or difference is finally determined by arbitration or litigation as provided in Clause 16.7 or by agreement. If the adjudicator is not named or if the person named is not available, the parties agree the identity of the Adjudicator. If the parties fail to agree the identity of the Adjudicator within 5 days after one party notifies the other that it wishes to do so, the Adjudicator is appointed by the body stated in the Appendix.
>
> 16.5 Any dispute or difference in connection with the enforcement of a decision of the Adjudicator is referred to the English Courts.

As with RIBA 2010, this provision refers to 'any dispute' and so adjudication under the Scheme can take place even if the relevant appointment is not a 'construction contract'.

The body to appoint an adjudicator if the parties cannot agree is to be one of the President or Vice President of the RICS, the RIBA or the Chartered Institute of Arbitrators. The parties are to indicate which is to apply.

PSC3 provides options for adjudication procedures for when the Act applies and when it does not. Option W2.3 has been amended by the Amendments of September 2011 to bring in the amended Construction Act provisions concerning the changing of a decision, when an amount is to be paid following an amended decision and the adjudicator's right to allocate his fees and expenses between the parties. Where the Act applies, a bespoke adjudication procedure is included. PSC3 provides for an adjudicator to be named in the agreement.

25.10. Amendments to the adjudication provisions of the Construction Act 1996 by the Local Democracy, Economic Development and Construction Act 2009

The Local Democracy, Economic Development and Construction Act 2009 has come into force – see Section 7.7.

The effects of these changes on the payment provisions in the Construction Act are discussed in Section 7.19. The change to the requirement that the 'construction contract' be in writing is described in Section 25.3 and the changes to the contractual provisions concerning adjudication procedures, the addition of the right of the adjudicator to correct his decision and the requirements relating to a contractual provision to allocate costs are described in Sections 25.3 and 25.9.

Professional Services Agreements
ISBN 978-0-7277-4158-5

http://dx.doi.org/10.1680/psa.41585.391

Chapter 26
Management of risks

26.1. Risk management generally

Having identified and discussed ways of handling the risks inherent in professional services agreements, this final chapter discusses procedures to manage those risks. Each organisation will have different concerns and management structures and the following is therefore a broad outline only.

In order to minimise risks, a professional organisation should adhere to the following.

(*a*) Assess clauses in each potential agreement singly, and then in conjunction with the rest of the agreement, to determine the risks.
(*b*) Negotiate to delete or reduce the impact of the risks that are of concern.
(*c*) Consider its own financial capacity to meet the potential liability including the availability of insurance cover and its own assets. It needs to be remembered that the professional indemnity insurance available to meet any claim is that available at the time the claim comes in and so could be on a very different and/or reduced basis at the time.
(*d*) Make a commercial decision as to whether or not the agreement must have a limit of liability and/or a net contribution clause. More professional firms are insisting on this and it is the author's experience that more and more clients are accepting limits of liability.
(*e*) Make a commercial decision as to whether or not to sign the final agreement.

It will be realised that some forms of activities carry inordinately high risks, often in terms of consequential costs, for the amount of fee income involved, for example site investigations of potentially contaminated land.

26.2. Internal procedures

Because professional services can carry high potential risks, formal internal procedures are necessary to deal with appointments.

An initial step must be to establish the effective lines of communication between those in the organisation responsible for deciding what professional services agreements can be accepted and those knowledgeable about insurance cover, total company exposures, the possible consequences of claims, etc. Small practices may handle all such matters at partner/member/director level. Alternatively, it may be necessary for specialist members of staff to be available to provide advice to project managers on agreements and to assist in negotiations with clients.

There should also be standard procedures for dealing with the risks arising under agreements. For example, risks could be categorised in order of importance. At one end of the scale could be those requirements that are unacceptable (e.g., unqualified, wide-ranging indemnities or 'fitness for purpose' clauses) or outside company policy, for example, parent company guarantees. At the other end could be those requirements that could be accepted in certain circumstances (e.g., those increasing the standard of care above 'reasonable skill and care') for a particular client or a particular project.

It should then be decided which risks need to be considered and/or which need advice from the organisation's legal advisers, and which risks need sanctioning at which level within the professional's organisation. These decisions will take into account commercial aspects, such as the size and type of job, how much the work is wanted, maintenance of relationships with the proposers of the agreement, etc.

Acceptable alternative forms of words can be prepared for the more common problem clauses to aid negotiations if those are handled in-house. Legal advisers and insurers or brokers will need to be involved in this.

However, for each agreement, the final commercial decision as to whether or not to accept risk must be taken by a knowledgeable professional at the appropriate level in the organisation who has taken account of any necessary independent advice. Appropriate training may be necessary.

26.3. Conclusions

It is only by establishing, implementing and auditing the organisation's risk management procedures that a professional can expect to control this important, but potentially high-risk area of activity. When sufficient records become available, the results of these procedures should, in the long term, show that more effective control leads to reduced loss and fewer claims. The professional can also use this record to negotiate better insurance terms through brokers. A direct comparison should become possible between the additional management salary costs of implementing the strategy and the resulting savings in average annual uninsured losses and premium, thus providing an ongoing assessment of the cost-effectiveness of the risk management process.

Professional Services Agreements
ISBN 978-0-7277-4158-5

http://dx.doi.org/10.1680/psa.41585.393

Appendix 1
Building Information Modelling

The following summary of Building Information Modelling (BIM) has been drawn from a paper prepared by Christopher Cullen, a consultant of Beale and Company Solicitors LLP, and the author is grateful to Mr Cullen for his permission to reproduce this part of his paper.

A building information model is, in essence, an information resource in relation to a particular facility to be designed and constructed, but it is also a digital simulation of the facility that provides the ability to 'rehearse' the construction and updates itself automatically as more information is added. Thus the model will adjust automatically to design changes.

In its purest form, a BIM project would use a single data model for all purposes. Each participant would access the model, adding content that could be accessed immediately by all others. Exploration, analysis and evaluation would take place within the model with information being exported as contract documents, fabrication drawings, bills of materials, estimates, costs or other information. The model would also contain information for managing the facility as well as constructing it.

At present, however, there is rarely a single model on a complex project. More commonly, each participant will have its own model, which interrelates with the others. Further, the model (or, more accurately, the series of interrelated models) does not usually contain all the construction details required for a project. Thus, the contract documents will include some 2D information to supplement the information in the model.

Indeed BIM can simply be used by a designer as an aid to the traditional design process.

The technological challenges for BIM seem formidable. In theory, different models will communicate seamlessly. This, though, is described as an aspiration, not a reality, under current technology. Differences in capability between BIM software mean that there are real risks of data being changed or lost.

However, there are also perceived problems and obstacles to use of BIM of a business or legal nature, due largely to liability concerns. BIM can, it is thought, increase a designer's potential liability, and because designers do not share (at least, to the same extent as others) in the economic benefits of BIM, it is thought they have little incentive to adopt the BIM process.

There does indeed seem to be some caution among professional consultants in the US about the use of electronic material. The AIA (American Institute of Architects) owner–architect agreement states that material provided in electronic form must be the subject of a separate written agreement (though licences that have been produced since pursuant to this requirement are said to embody a more open approach). Documents published by the Engineers Joint Contract Documents Committee disallow any reliance on electronic information, and place the risk of errors and discrepancies on the receiving party: files in electronic media are stated to be furnished only for the convenience of the receiving party. There are clearly concerns about design responsibility when BIM designs change in response to input from others or due to embedded software. In virtually all jurisdictions in the US, design professionals of record must be in 'responsible charge' of the design. There are concerns too about ownership of material that is fed into a model along with material from others, about confidentiality, preservation of electronic data, and the legal status of the model. Such problems are exacerbated by, but are also of course partly the cause of, the lack of standard contract documents to address these matters.

Underlying many of these concerns is a general concern – or perceived concern – about collaboration generally. BIM, if used to full advantage, leads naturally to intensive communication and interdependence. Collaboration is not seen as a construction industry hallmark. Collaboration and interdependence blur the distinctions between the parties, but the legal systems are essentially individualistic, focusing on individual rights and responsibilities. An essential component of collaboration is the readiness and ability of a design professional to rely upon information provided by other participants and, to some extent, by the software itself. Reluctance to do this is likely to inhibit the sort of collaboration needed to make BIM truly effective.

This situation is leading to calls for fresh thinking on the forms of business relationships that characterise the construction industry and for arrangements that acknowledge the blurring and blending of roles and responsibilities. The most radical suggestion is a form of alliancing, which involves no liability between the parties, who agree that they will not sue each other, and a collective sharing of project risks, and a portion of compensation being tied to a successful outcome (pain share/gain share). BIM is said to be tailor made for alliancing.

Should BIM and collaboration in general cause the same business and legal concerns in the UK? There are reasons to think not.

Designers owe a professional duty to exercise reasonable skill and care and should only be liable if they are in breach of that duty. If a design is defective, that may indicate that the designer has failed to exercise reasonable skill and care, but it does not determine the point. In the US, the position of designers may be more vulnerable, due to what is known as the Spearin doctrine, under which plans and specifications provided for a construction project are warranted by the owner to be free from defect. This doctrine, established in a case in 1918, is intended to protect the contractor, who can not, as a result, be liable if it constructs a building to a defective design. Although the warranty is not given by the designer, it makes it all the more likely that the owner will seek redress from the designer if there is a problem with the design.

One can understand, therefore, why designers in the US may be wary of participating in a collaborative process to produce designs which will be subject to a warranty. One can also understand why contractors would also be wary of participating, as their participation could negate the warranty.

There is no Spearin doctrine in English law. Contractors are generally considered to be protected if they execute the design in accordance with the terms of the contract. There is some law to suggest that a contractor may have a duty to warn of a dangerous design, but it is not an area of the law that has developed very far. Nor, for designers, is there the same issue of formal design responsibility that there is in much of the US.

Another matter that may well cause concern in the US about the sort of collaboration envisaged in connection with BIM is the law of tort, in particular the Restatement of Torts Second section 552, which sets out the requirements for a negligent misrepresentation claim. In brief, these make a person liable for pecuniary loss caused to others in the course of a business as a result of their relying upon false information supplied by that person, if he fails to exercise reasonable care or competence in obtaining or communicating the information. It is not hard to see how this rule of law could make designers liable to others involved in a collaborative project as a result of their input into the project; the greater the degree of collaboration, the greater the risk of this happening.

No such principle of law applies in English law. The incremental approach of the common law to the development of duties of care has deliberately eschewed the adoption of an all-embracing principle. A person who gives wrong information to another person knowing that the other person is likely to rely on it may be held, in the circumstances, to have owed a duty to the other person, but it all depends on those circumstances. Any such duty in respect of economic loss has seldom been held to arise as between participants in a building project who are not in contract with each other, either during the planning and design stages or during the works themselves. The policy of the law in this area is for economic loss to be recovered, if it can be recovered, through the contractual route. As collaboration intensifies as a result of BIM, it is possible that the law may move in the direction of imposing duties of care in tort among the participants, but there is no reason at this stage to think that this is likely to happen.

The standard of professional care that will apply in any given situation is inherently flexible and develops and changes with the way that the practice of the profession itself develops and changes. The growth of collaboration in connection with BIM will no doubt give rise to questions as to what a designer should or should not do in hitherto unfamiliar situations, and appropriate standards of conduct will evolve, aided by guidance from professional bodies. Whether, for example, it is or is not appropriate for a designer to rely on a particular piece of information in a given situation will fall to be judged in the light of the relevant standards of the designer's profession at the relevant time. This is surely to be preferred, at least in principle, to endeavouring to draft contract terms specifying levels of permitted reliance or other standards of conduct or attempting to define a general obligation to collaborate. Contract drafters should generally resist telling professional people how to do their work; that principle should apply just as much where the work involves participating in the development of a building information model.

Concern has been expressed about the ownership of the model and of documents derived from it, when there have been so many different inputs. However, there is no reason why multiple ownership should cause particular problems. A person who contributes to a model will normally be thereby agreeing implicitly, if he has not already done so expressly, to his input being used on the project for which it is given. Consent can still be subject to payment of fees. There is no reason why the contractor or the project owner should be in a worse position. There would be no justification for ownership of the model being transferred to the project owner. All he will need is a licence to use the model in relation to the facility which is constructed from it.

Where there is to be a central model, the issue of management of the model would have to be addressed, and the sensible solution would seem to be to appoint a model manager with the appropriate expertise with responsibility for coordinating development of the model, including all matters involving transfer of data to and from it. Responsibility, however, is not the same thing as liability; the model manager, like any other professional, should be liable for failure to exercise the appropriate degree of skill and care but should not be made strictly liable for data corruption or loss. Reference is sometimes made to a liability gap arising from the contractual limitations that are usually found in software suppliers' standard terms and conditions. However, there is nothing inherently controversial about a loss for which no one is liable, and risk allocation, to cover situations for which there is no legal liability, is not the appropriate basis for a professional appointment. Professional consultants do not take risks in this sense and professional indemnity insurance does not insure against such risks.

At the end of the day, the project owner stands to gain most from BIM, is the only party in a position to insist on its use, and must be the party who bears the risk of any technological problems arising from BIM for which no one can be held legally liable. As the technology improves, this risk should grow less and project owners should be more inclined to require the use of BIM.

Professional Services Agreements
ISBN 978-0-7277-4158-5

http://dx.doi.org/10.1680/psa.41585.397

Appendix 2
CIC novation agreement

Construction Industry Council

NOVATION AGREEMENT

CIC/NovAgr
first edition 2004

Standard form of agreement for use where the appointment of a consultant is to be novated from an employer to a design and build contractor

ISBN 1 898671 36 2

Construction Industry Council

NOVATION AGREEMENT

CIC/NovAgr
first edition 2004

Standard form of agreement for use where the appointment of a consultant is to be novated from an employer to a design and build contractor

NOVATION AGREEMENT

CIC/NovAgr
first edition

For use where the appointment of a consultant is to be novated from an employer to a design and build contractor.

THIS AGREEMENT

is made the .... day of .... 20.......................

Between

1 Insert name of Employer

1. The Employer ....[1]

2 Insert company number if applicable and place of registration if incorporated outside England and Wales

company number .... and place of registration ....[2]

whose address/registered office is at

....

....

and

3 Insert name of Consultant

2. The Consultant ....[3]

company number .... and place of registration ....[2]

whose address/registered office is at

....

....

and

4 Insert name of Contractor

3. The Contractor ....[4]

company number .... and place of registration ....[2]

whose address/registered office is at

....

....

CIC/NovAgr
first edition

NOVATION AGREEMENT

Whereas

A. On or about ....[5] the Employer appointed the Consultant (the 'Appointment') to provide professional services in connection with

....[6]

at ....[7]

(the 'Project') including the design of certain elements of the Project.

B. The Employer has entered or proposes to enter into a contract with the Contractor (the 'Main Contract') for all or part of the design and the construction of the Project in accordance with (among other things) designs prepared by the Consultant pursuant to the Appointment.

5 *Insert date Consultant appointed*

6 *Insert description of works*

7 *Insert address of Project*

Now it is hereby agreed

1. As from the date of this Agreement the Employer releases and discharges the Consultant from the further performance of the Consultant's services and other obligations under the Appointment save for:

   (a) any obligations in the Appointment to provide warranties in favour of third parties at the request of the Employer; and

   (b) any obligations in the Appointment as to confidentiality to the extent that these do not conflict with the duties owed by the Consultant to the Contractor arising under this Agreement

   and, subject to all fees properly due and owing under the Appointment as at the date of this Agreement having been paid to the Consultant, the Consultant releases and discharges the Employer from the further performance of the Employer's obligations under the Appointment.

2. In respect of those services and any other obligations to be performed under the Appointment that have yet to be performed by the Consultant at the date of this Agreement, the Consultant agrees to perform those services and obligations (subject to the variations if any set out in Schedule 1 to this Agreement) for the Contractor as if the Contractor were a party to the Appointment in place of the Employer and in accordance with and subject to the terms of the Appointment (including without restriction any limitation or exclusion of liability therein). The Consultant shall have no liability to the Employer arising out of the performance or non-performance of such services and obligations.

3. In respect of those obligations to be performed under the Appointment that have yet to be performed by the Employer at the date of this Agreement (with the exception of any outstanding obligation to pay any fees properly due and owing under the Appointment as at the date of this Agreement), the Contractor agrees to perform those obligations (subject to the variations if any set out in Schedule 2 to this

NOVATION AGREEMENT

CIC/NovAgr
first edition

Agreement) as if the Contractor were a party to the Appointment in place of the Employer and in accordance with and subject to the terms of the Appointment (including without restriction any limitation or exclusion of liability therein). The Employer shall have no liability to the Consultant arising out of the performance or non-performance of such obligations, subject to all fees properly due and owing under the Appointment as at the date of this Agreement having been paid to the Consultant.

4. (a) Insofar as the Contractor is responsible under the Main Contract for services that have been performed by the Consultant for the Employer prior to the date of this Agreement, the Consultant warrants to the Contractor that all such services have been performed for the Employer in accordance with the terms of the Appointment.

   (b) In any claim for loss suffered by the Contractor that is alleged to have arisen as a result of a breach by the Consultant of the warranty in this clause, the Consultant shall be entitled to rely on any limitation in the Appointment (including without restriction any limitation or exclusion of liability therein) and to raise the equivalent rights in defence of liability for such loss as if the claim were being brought by the Employer rather than the Contractor, save that the Consultant shall not be absolved from liability to the Contractor for such loss merely by virtue of the fact that the loss has not been suffered by the Employer.

5. This Agreement shall be governed by and construed in all respects in accordance with the law that governs the Appointment. Where this Agreement is governed by the law of England and Wales or Northern Ireland, nothing in this Agreement confers or purports to confer on any third party any benefit or any right to enforce any term of this Agreement pursuant to the Contracts (Rights of Third Parties) Act 1999.

Where the Agreement is to be executed under hand and not as a deed

As witness the hands of the parties hereto

Signed by or on behalf of the Employer

8 *Name and signature of person signing on behalf of Employer*

.... 8

Signed by or on behalf of the Consultant

9 *Name and signature of person signing on behalf of Consultant*

.... 9

Signed by or on behalf of the Contractor

10 *Name and signature of person signing on behalf of Contractor*

.... 10

CIC/NovAgr
first edition

NOVATION AGREEMENT

Where the Agreement is to be executed as a deed

In witness whereof the parties have executed this Agreement as a deed the day and year first before written

[WHERE THE CONSULTANT IS A SOLE PRACTITIONER OR PARTNERSHIP]

Signed as a deed by the Consultant

....[11] — 11 *Name and signature of sole practitioner or first partner*

in the presence of ....[12] — 12 *Name and signature of witness*

of ....[13] — 13 *Address of witness*

and by ....[14] — 14 *Name and signature of additional partner*

in the presence of ....[12]

of ....[13]

and by ....[14]

in the presence of ....[12]

of ....[13]

[WHERE THE CONSULTANT IS A LIMITED LIABILITY PARTNERSHIP OR COMPANY]

Executed as a deed by the Consultant

by ....[15] — 15 *Name and signature of member or director*

and ....[16] — 16 *Name and signature of member or director or company secretary*

[WHERE THE EMPLOYER IS AN INDIVIDUAL OR PARTNERSHIP]

Signed as a deed by the Employer

....[17] — 17 *Name and signature of individual or first partner*

in the presence of ....[12]

of ....[13]

NOVATION AGREEMENT

CIC/NovAgr
first edition

14 *Name and signature of additional partner*

and by 14

12 *Name and signature of witness*

in the presence of 12

13 *Address of witness*

of 13

and by 14

in the presence of 12

of 13

[WHERE THE EMPLOYER IS A LIMITED LIABILITY PARTNERSHIP OR COMPANY]

Executed as a deed by the Employer

15 *Name and signature of member or director*

by 15

16 *Name and signature of member or director or company secretary*

and 16

[WHERE THE CONTRACTOR IS A PARTNERSHIP]

Signed as a deed by the Contractor

17 *Name and signature of individual or first partner*

.... 17

in the presence of 12

of 13

and by 14

in the presence of 12

of 13

[WHERE THE CONTRACTOR IS A LIMITED LIABILITY PARTNERSHIP OR COMPANY]

Executed as a deed by the Contractor

by 15

and 16

CIC/NovAgr
first edition

NOVATION AGREEMENT

SCHEDULE 1

Variations to the Consultant's services and other obligations

SCHEDULE 2

Variations to the Contractor's obligations

NOVATION AGREEMENT

CIC/NovAgr
first edition

GUIDANCE

1. Novation has been legally defined as taking place where there is a contract in existence and some new contract is substituted for it, with the old contract usually being discharged. The expression has more recently tended to be used for a situation where the acts to be performed under the old contract remain the same, but are to be performed by different parties. However, the effect of a novation is not to assign or transfer a right or liability, but rather to extinguish the original contract and replace it by another.

NOVATION OF PROFESSIONAL APPOINTMENTS

2. Thus, where, in the construction industry, a consultant has been engaged by a building owner or developer to perform professional services and is then engaged instead by a design and build contractor, one contract (between employer and consultant) is substituted by another (between contractor and consultant). The contractor takes on the obligations of the employer and (usually) the employer is discharged from performing these obligations. The consultant carries on performing his services but does so now for the contractor instead of the employer.

3. However, this novation usually takes place at a time when at least some of the obligations of both employer and consultant have already been performed. Thus the new contract is, in effect, a contract for the performance, by the consultant and the contractor, of those obligations that remain to be performed. Also, the employer retains its interest in the project, and the extent to which the contractor assumes responsibility for executing the design as well as the construction of the project will vary, according to the terms of the building contract.

4. Another important point about this novation is that the interests and concerns of the employer and the contractor are not the same. Under the new contract, the consultant may perform the same services for the contractor as, under the old contract, he would have performed for the employer. However, he may not do so in the same way, as he is now under a duty to take the specific interests and concerns of the contractor into account. Likewise, if he had been engaged by the contractor from the outset (or at any other time prior to novation), he might have performed his services differently from the way he actually did for the employer.

5. The CIC Novation Agreement aims to reflect and address the realities inherent in this situation.

THE PARTIES

6. The Agreement is between all three parties, the original client (the 'Employer'), the Consultant and the Contractor.

THE RECITALS

7. The recitals A and B recite the Consultant's original appointment by the Employer (the 'Appointment') and the design and build contract between the Employer and the Contractor (the 'Main Contract').

8. Recital A provides for the possibility that the exact date of the Consultant's appointment by the Employer may not be known. The date is likely to be known if there is a formal written appointment, as this would normally be dated, but appointments of consultants are sometimes relatively informal, though still giving rise to a binding contract between the parties. Details of the Project must also be entered in recital A.

9. Regarding recital B, the Agreement may be entered into before or after the Main Contract, although the identity of the Contractor will be known. It is also recited that the Consultant's services under the Appointment include the preparation of designs and that the Contractor has been engaged to design as well as construct the project.

CLAUSE 1

10. Under clause 1, save as mentioned in paragraph 11, the Consultant is released by the Employer from the further performance of his obligations under the Appointment. The Consultant is not released from outstanding liabilities to the Employer arising out of work already performed.

11. Two possible obligations on the part of the Consultant survive this release (clause 1(a) and (b)). If there is an obligation on the part of the Consultant in the Appointment, not yet performed, to give collateral warranties at the request of the Employer, the Employer may still enforce this obligation. This obligation could be enforced by the Contractor but employers may prefer to retain the right to do so themselves. Any obligation in the Appointment as to confidentiality is also retained, as long as it does not conflict with duties now owed to the Contractor.

12. Likewise, the Employer is released by the Consultant from further performance of its obligations under the Appointment – subject to all outstanding fees having been paid. The release will not therefore take effect until the fees have been paid. It would be better, though, for the Agreement not to be signed until the fees have been paid.

13. As with the Consultant, the Employer is not released from outstanding liabilities to the Consultant accruing prior to novation.

CLAUSE 2

14. Under clause 2, the Consultant undertakes to perform all outstanding services and any other outstanding obligations under the Appointment for the Contractor.

15. The Agreement provides for the possibility that, in the light of the novation, the parties may wish to vary the Consultant's services and the fees paid by the Contractor. Some of the services that are yet to be performed may not now be appropriate, since they are to be performed for the Contractor rather than the Employer. An example would be assisting the Employer in any dispute with the Contractor; clearly the Consultant could not now do this. Could he assist the Contractor in any dispute with the Employer? This too might not be thought feasible, since the Consultant was formerly engaged by the Employer. There may also be services that should be worded differently now they are to be performed for the Contractor, and additional services that the Contractor would like the Consultant to perform. ***Any such variations must be agreed, and recorded in Schedule 1. If there are no variations, this should be stated.***

16. The Consultant is to perform his outstanding services and other obligations (subject to any such agreed variations) in accordance with and subject to the terms of the Appointment, as if the Contractor were a party to the Appointment in place of the Employer. Thus any limitation of liability in the Appointment would apply to the new contract as it did to the old.

17. It is expressly stated that the Consultant shall have no liability to the Employer in respect of services performed following novation.

CLAUSE 3

18. Under clause 3, the Contractor agrees to perform all outstanding obligations under the Appointment, for example, to pay the Consultant's fees (except for fees that were already due at the date of novation, which the Employer is obliged to pay).

19. Again, the parties may wish to agree variations to these obligations, for example as to the fees payable, and these should be recorded in Schedule 2. ***If there are no variations, this should be stated.***

20. The Contractor is to perform the outstanding obligations in accordance with and subject to the terms of the Appointment; thus any limitation of liability in the Appointment would apply to the new contract as it did to the old.

21. It is expressly stated that the Employer will have no obligations to the Consultant to perform the obligations now to be performed by the Contractor (subject to the outstanding fees due at novation having been paid).

CLAUSE 4

22. Under the Main Contract, the Contractor will be responsible to the Employer for the design of the project, and this design responsibility may include design work undertaken by the Consultant prior to novation. The warranty in clause 4 may allow the Contractor to seek redress from the Consultant for any deficiency in the performance of his services prior to novation, insofar as the Contractor is responsible to the Employer for the services in question. For example, under the JCT Major Project Form the Contractor is not responsible for the adequacy of the

design contained within the Requirements. If the Main Contract is on that form, therefore, the warranty under clause 4 will not apply to any designs prepared by the Consultant contained within the Requirements.

23. Any such deficiency in the performance of the Consultant's services prior to novation for which the Contractor seeks redress under the warranty must also constitute a breach of duty owed by the Consultant to the Employer under the old contract in order for the Contractor to have a right to redress. Thus, in the event of a claim being brought by the Contractor against the Consultant for breach of the warranty, the Consultant will be able, with one exception, to rely on any limitations of liability in the Appointment and any other defences that would have been available if the claim were being brought by the Employer. The exception is any defence that the Employer has not suffered the loss being claimed. It is quite likely that the Employer will not, in fact, have suffered any loss as a result of a breach of duty by the Consultant in preparing designs prior to novation, as it may well have passed the risk of all such losses to the Contractor. The Agreement, therefore, prohibits the Consultant from raising this 'no loss' argument as a defence, thus allaying a concern that has sometimes been expressed about the possible effect of the Scottish case of *Blyth & Blyth Ltd v Carillion Construction Ltd.*

24. Before agreeing to novation, the Consultant may wish to check that the Contractor is also taking over the appointments of all other consultants and other parties engaged in connection with the project. If any such party is not engaged by the Contractor, it may not be possible for the Consultant to make a claim for contribution against that party for any loss or damage for which the Consultant is held liable. This problem could be avoided if there is a net contribution clause in the Consultant's appointment, which would then apply to any claim brought under the novation agreement, whether by virtue of clause 2 (for a claim in respect of services performed following novation) or clause 4(b) (for a claim under the warranty in clause 4(a) in respect of services performed prior to novation). The novation agreement does not itself, however, contain a net contribution clause.

CLAUSE 5

25. The Agreement is subject to the same law that governs the Appointment. Where English law or the law of Northern Ireland applies, the Contracts (Rights of Third Parties) Act 1999 is excluded.

SIGNING

26. The document should be executed as a deed if the Appointment is so executed. Otherwise it should be signed under hand.

WARRANTY FROM THE CONSULTANT TO THE EMPLOYER

27. As explained in paragraph 15, the Consultant remains liable to the Employer for services performed pre-novation. Further, in engaging the Contractor to design and build the Project, the Employer has recourse against the Contractor for all work done (including design work) post-novation. The Employer does not therefore need a warranty from the Consultant in respect of post-novation services. It will be noted, for example, that in the case of the JCT Major Project Form, no such warranty is contemplated. If however, the Consultant agrees to give the Employer a warranty, an appropriate form can be found on the CIC's website, at www.cic.org.uk. It takes the form of a supplemental agreement (CIC/NovAgr/EWar) which is ***only*** suitable for use with this Agreement.

ACKNOWLEDGEMENT

The Construction Industry Council acknowledges with thanks the assistance given by Beale & Company, solicitors, in drafting the Agreement.

DISCLAIMER

These notes are for general guidance only and are not intended to affect the interpretation of the agreement. Legal / professional advice should be sought to cover any particular situation.

Professional Services Agreements
ISBN 978-0-7277-4158-5

http://dx.doi.org/10.1680/psa.41585.409

Appendix 3
Note on switch and *ab initio* novations

There are in the construction industry at present two types of novation for the situation where a consultant is to be novated from a client to a design and build contractor. These are the 'switch' and the *ab initio* forms.

In order to explain the differences between them, it is necessary to start with some basic principles about novation.

Novation process – rights and obligations

Novation describes the process by which the two parties to a contract deal with the rights and obligations under that contract. In a consultant's contract, the rights of either party are basically the rights to sue the other. These rights can be assigned to third parties without consent unless the contract provides otherwise. This is referred to simply as 'assignment' and happens all the time.

'Obligations' have always been treated differently. In simple terms, the client's obligation is to pay the consultant and the consultant's obligation is to perform the services. These obligations cannot be assigned to a third party without consent unless the contract provides otherwise. The reason for this is clear. The client does not want to wake up one morning to find a totally different consultant (one he has not chosen and of whose experience he has no knowledge) performing the services for which he engaged his original consultant. Similarly, a consultant does not want to have to accept as a new client a person of whom he may have had bad experiences in the past, or a person who has no assets with which to pay him.

New contract

When rights and obligations are transferred or assigned, this is called 'novation'. What happens as a result is that the second client steps into the shoes of the first client. In law, the first contract is extinguished and a new contract replaces it – hence the name.

There are circumstances when consultants may be happy to accept a substitution of one client for another, for example when an affiliate company of the client takes over the project.

Past obligations

Novation cannot cover obligations already performed; that is the services carried out to date, because these are not going to be re-performed for the new client. The consultant is not going to start his design all over again. Novation will, however, cover the liabilities arising out of the past performance of those obligations. So if a consultant has been negligent before the second client took over, the second client will be able to sue the consultant in respect of that negligence and the first client will not.

It is in this area that the problems concerning novation to a design and build contractor primarily arise. What is the position to be concerning the services that have been performed for the client up to the date of the novation?

Differences between novation to another client and to a design and build contractor

It needs to be appreciated that there is a very real difference where the novation is to be to a second client and where the novation is to be to a design and build contractor.

In design and build novation the original client does not drop out of the picture – he continues to carry out the same project. The contractor is to construct that project for the original client and only takes over the consultant's appointment, not the project itself. The client and the design and build contractor have very different interests.

- The client wants the project carried out to high standards within his budget and timescale and then will sell or use the building.
- The contractor wants to construct the project as efficiently and economically as possible, make a profit and then leave.

In all this, the tender price of the contractor is crucial. The client is looking for a competitive price. The contractor wants to know as accurately as possible what it is going to cost him to build it.

When novation takes place

Generally, novation takes place when the building contract is entered into and therefore after the contractor has tendered for the work, but it could take place earlier than that. Also, since it is to be a design and build contract, the contractor at that stage will be taking over responsibility for some or all of the design.

What is being novated

What is being novated at that time? Only the obligations under the original appointment that have yet to be performed by the consultant. This could be, at one end of the scale, when nearly all the design has been completed, or at the other end when the design is still at a preliminary stage – for example, only that necessary to put together the Employer's Requirements.

As I have said, obligations that have already been performed cannot be novated unless the consultant is to do them again.

Nature of liability arising out of services performed prior to novation

As to the liabilities arising out of the design done prior to novation, that is in respect of obligations already performed by the consultant: these can only ever have been performed for the original client even if the design and build contractor may now take the benefit of them – that is, will now be able to sue the consultant in place of the original client.

This is because the law says that a consultant does not owe duties in a vacuum. He owes a duty to a particular person and his relationship to that person determines the scope of his duty and thus the kind of loss for which he is liable (and therefore for which he is liable to that person's assignee: the design and build contractor in a design and build novation).

So the relevant question in law relating to the liability of a consultant for loss arising from his performance prior to novation is 'Was it the kind of loss that the consultant owed a duty to the original client to use reasonable skill and care to prevent?'.

Blyth & Blyth

It will be apparent from this how the sort of situation highlighted by the Blyth & Blyth case (*Blyth & Blyth* v. *Carillion Construction Ltd (2001)* 79 Con LR 142) arose. There the engineers had done some design for the client. The contractor used the consultant's information in the Employer's Requirements prior to tender to estimate how much steel reinforcement would be needed, and costed on that basis. Thereafter, the building contract was let and the consultants novated. The contractor then discovered that much more steel reinforcement was needed and sued the consultants on the basis of alleged deficiencies in the Employer's Requirements, which resulted in their tender price being too low. As the design and build contractor had accepted responsibility for the design he had to bear this loss and could not recover it from the client.

However, the court said this was *not* the kind of loss the consultant was under any duty to his client to use reasonable skill and care to avoid. Consultants may advise clients on aspects concerning tenders but they do not become responsible to the client if the design and build contractor makes insufficient allowance for any particular item in his tender as this is not a loss the client will suffer.

The judge said in that case that to make Blyth & Blyth responsible to the design and build contractor in these circumstances would involve a retrospective recasting of their duty of care so that it was owed to the contractor and not to the client, and the novation agreement in that case had not achieved that.

Nub of the debate

This brings us to the central point of the whole debate about *ab initio* and switch novation agreements. It is really an argument about to whom and in respect of what should the consultant be liable in respect of services performed prior to the actual novation.

Novation *ab initio*

Design/build novation *ab initio*, the standard form for which is produced by the CLLS, is based is on the traditional form of novation, when the first client transfers his interest in the project to a second client and the second client steps into the shoes of the first client as far as the consultant's appointment is concerned and is treated as if he had always been a party to that contract, in place of the first client.

But the way in which these *ab initio* forms seek to make the consultant liable to the contractor for services carried out for the client prior to the novation is to create the fiction that the consultant's client has been the contractor from the beginning – the retrospective recasting of the duty of care the judge referred to in the Blyth & Blyth case.

Does this make sense?

This does not make any sense in reality. How can a consultant agree to the legal fiction that he has always worked for the contractor? Up until novation the consultant is working to the requirements of his client and he should have his interests only in mind.

He has not *in fact* been employed by the contractor at that time and was therefore under no duty to advise him as to the sufficiency of his tender price, nor as to what contingencies and allowances he should provide in his price. Indeed, had he done so at the time he would have been in conflict with the duties he owed to his client and acting in a manner which was adverse to that client's interests.

If, however, the consultant agrees to this retrospective duty of care the contractor will have a remedy if he can show that if he had been properly advised by the consultant, on the basis that the consultant was acting for him at that time, the contractor would not have got his tender price wrong.

Other wording

It is not necessary for the words *ab initio* to be used to create this sort of liability.

There are other ways in which the same result can be achieved. For example the novation agreement could

- make the consultant give the contractor a warranty that he owed *the contractor* a duty of care from the outset
- make the consultant acknowledge that the contractor has relied on all information and advice given to the client
- contain an acknowledgement that any losses suffered by the design and build contractor were within the contemplation of the consultant when he first entered into his appointment with the client.

These too are all based on *fictions*.

What is the consultant liable for in respect of services before the novation?

The consultant does not escape all liability for services performed prior to the novation. The design and build contractor will be able to sue the consultant in respect of errors committed prior to novation if it resulted in the kind of loss for which the consultant was under a duty of care to his original client to avoid, for example the cost of repairing defective work resulting from a defect in the consultant's design prior to the novation. But this is a relatively limited protection for a design and build contractor and certainly does not help him if he has got his tender price in relation to the consultant's design wrong.

Client's position under *ab initio* novation

Clients are not concerned about the liabilities their consultants incur to third parties – in this case the design and build contractor *before* he becomes the consultant's employer. They do, however, want their design and build contractor to take on full one-point responsibility for all design and construction, and if an *ab initio* form of novation agreement helps to achieve this the client will try to procure one from their consultants.

The client's position is not affected either way – having passed over all responsibility to the design and build contractor he has no need to reserve any rights to sue his consultant in respect of any errors that may have been committed prior to novation. (He may of course want a warranty back from the consultant in respect of the post-novation services on a 'have your cake and eat it' basis, but that's another story.)

When liability to the contractor can arise outside any novation

Should a consultant volunteer advice to the design and build contractor during the tender period and/or answer queries from the design and build contractor on which the contractor relies, then the consultant will be liable to the contractor if he does so negligently on the Hedley Byrne principle, regardless of any terms in his appointment or any novation agreement. On the absence of any agreed novation the *Galliford Try Infrastructure Ltd* v. *Mott MacDonald Ltd (2008)* CILL 2612 TCC case was put on this basis – that Motts had made representations to Galliford Try as to the prop forces that could be taken by the concrete slab part of the floor and/or that the steelwork would not be required to perform this function. The judge found there was no representation of this sort by Motts, no reliance by Galliford Try and that Motts had not been negligent.

Switch

This is not a novation in the traditional sense. The 'switch' recognises the reality of the design and build situation. The consultant works for his client at the outset and then from a certain stage works for the contractor. The same result could be achieved by bringing to an end the original appointment by the client comprising the services to be performed up to novation, and having a new appointment between the design and build contractor and the consultant comprising the services the consultant is to perform thereafter for the design and build contractor. Under a 'switch' form of novation the consultant switches from acting for one client – the original client – to acting for another client – the design and build contractor at the point of novation.

CIC novation agreement

This is a 'switch' type of novation. The way the CIC novation agreement achieves this is by the consultant remaining liable to the client in respect of the work done for him prior to the novation. Post-novation, the consultant agrees to work for the contractor and to be liable to him. The contractor agrees to step into the shoes of the original client post-novation and thenceforth act as the client. Thus, the 'switch' happens at the point of novation.

Under the construction contract the contractor will accept some or all responsibility for design, and this will generally include an element of design done pre-novation (i.e., at a time when the consultant was working for the client). The contractor will want a remedy against the consultant who did the design work to cover him in the event that he suffers loss due to an error in that design. The consultant therefore gives a warranty to the contractor in respect of services undertaken pre-novation. He warrants to the contractor that, insofar as the contractor is responsible under the main contract for pre-novation services undertaken by the consultant, such services have been performed *for the client* in accordance with the terms of the appointment. Then if he has been in breach of the duty owed to his client, and the contractor suffers loss as a result, the contractor will be able to sue him.

An important caveat is included in the CIC agreement: 'save that the Consultant shall not be absolved from liability to the Contractor for such loss merely by virtue of that fact that the loss has not been suffered by the Employer'. The consultant cannot therefore raise the 'no loss' argument, thus allaying a concern that has been expressed about the possible effect of the Blyth & Blyth case.

Services, fees, other changes

The CIC novation agreement allows for the necessary distinction to be made regarding the services, so that the services to be performed by the consultant for the client are properly recorded, and those to be performed for the contractor are similarly properly recorded. If the consultant is, say, inspecting workmanship on behalf of the contractor, his obligations will be different from those he would have if he was inspecting on behalf of the client – he will not issue certificates, for example. Similarly, the pre- and post-novation fee can be set out and any other necessary changes made to the obligations in the appointment to reflect the fact that the consultant is now acting for the design and build contractor. There is a schedule attached to the CIC novation agreement where the necessary changes in the services and any other obligations can be recorded.

If the consultant agrees to give the client a warranty for work done post-novation, an appropriate form (a supplemental contract) can be found on the CIC's website (at www.cic.org.uk (liability and contracts).

Conclusion

The CIC novation agreement should be the only one recommended to consultants for the design and build situation. It has been endorsed by all the relevant consultant bodies. This will also mean that the consultants will be engaged on a proper basis and not in a position of conflict of interest and/or where their duties of confidentiality are broken. Several projects have used it.

A final cautionary note: we are already seeing adjudication and claims arising out of the novation of consultant to design and build contractors and these will only get worse (i) as time goes on and more projects are carried out on this basis; and (ii) due to the effects of the recession people (particularly contractors) start suing to try and recover the monies they have lost.

Professional Services Agreements
ISBN 978-0-7277-4158-5

http://dx.doi.org/10.1680/psa.41585.415

Appendix 4
CIC purchaser/tenant collateral warranty

Construction Industry Council

COLLATERAL WARRANTY CONSULTANT - PURCHASER/TENANT

CIC/ConsWa/P&T
first edition 2003

Standard form of agreement for use where a collateral warranty is to be given by a consultant to a purchaser or tenant of the whole or part of a commercial or industrial development

Full members of the Construction Industry Council • Association for Project Management • Association of Building Engineers • Association of Consultant Architects • Association of Consultant Building Surveyors • Association of Consulting Engineers • Association of Cost Engineers • Association of Planning Supervisors • British Institute of Architectural Technologists • British Institute of Facilities Management • Building Research Establishment • Building Services Research and Information Association • Chartered Institute of Building • Chartered Institution of Building Services Engineers • Construction Industry Research and Information Association • Consultant Quantity Surveyors Association • District Surveyors Association • Ground Forum • Institute of Clerks of Works of Great Britain • Institute of Highways Incorporated Engineers • Institute of Maintenance and Building Management • Institute of Plumbing • Institution of Civil Engineering Surveyors • Institution of Civil Engineers • Institution of Highways and Transportation • Institution of Structural Engineers • Landscape Institute • National Federation of Demolition Contractors • National House-Building Council • Royal Institute of British Architects • Royal Institution of Chartered Surveyors • Royal Town Planning Institute • Steel Construction Institute • Timber Research And Development Association • **Associate members** • Association of Civil Engineering Departments • Association of Heads of Surveying • British Association of Construction Heads • Construction Health & Safety Group • Chartered Institute of Marketing Construction Industry Group • Construction Industry Computing Association • Council of Professors of Building Engineering and Management • Faculty of Building • Federation of Property Societies • Institute of Construction Management • Royal School of Military Engineering • Standing Conference of Heads of Schools of Architecture • Society of Construction Law • SPONGE • Local Government Technical Advisers Group • Technology and Construction Solicitors' Association

COLLATERAL WARRANTY

CIC/ConsWa/P&T
first edition

For use where a collateral warranty is to be given by a consultant to a purchaser or tenant of the whole or part of a building project in a commercial or industrial development.

THIS AGREEMENT

is made the .... day of.. 20........................

Between

1 *Insert name of Consultant*

1. ....[1]

whose address/registered office is at

....

....

(the 'Consultant') and

2 *Insert name of Purchaser/Tenant*

2. ....[2]

whose address/registered office is at ....

....

3 *Delete as appropriate throughout Agreement*

(the 'Purchaser'/'Tenant' [3]).

Whereas

4 *Insert name of Consultant's Client*

A. The Purchaser/Tenant and ....[4]

5 *Delete as appropriate*

(the 'Client') have entered into an agreement to purchase/an agreement to lease/ a lease of [5]

6 *Insert description of Premises*

....[6]

7 *Delete if Premises form the whole of Development*

(the 'Premises') [being a part of [7]]

8 *Insert address of Development*

....[8]

9 *Delete if Premises form part only of Development*

(the 'Development'). [The Premises are also referred to in this Agreement as the 'Development'. [9]]

COLLATERAL WARRANTY CIC/ConsWa/P&T first edition

10 *Insert date of Appointment*

B. By a contract dated .... [10] (the 'Appointment'), which shall include any variation thereto, the Client has appointed the Consultant

11 *Insert discipline*

as ....[11]

in connection with the Development.

12 *Delete as appropriate*

C. The Client has entered/may enter[12] into a contract

13 *If Building Contract signed, insert date; if not, delete word 'dated'*

[dated ....[13]] (the 'Building Contract')

14 *Insert name of contractor or 'a contractor to be selected'*

with ....[14]

(the 'Contractor') for the construction of the Development.

Now it is hereby agreed

15 *Insert figure*

In consideration of the payment of £..................[15] by the Purchaser/Tenant to the Consultant, receipt of which the Consultant hereby acknowledges:

16 *Delete if services completed*

1. The Consultant warrants to the Purchaser/Tenant that it has exercised [and will continue to exercise[16]] reasonable skill care and diligence in the performance of its services to the Client under the Appointment.

2. In the event of any breach of this Agreement:

   (a) Subject to sub-clauses (b) and (c), the Consultant shall be liable for the reasonable costs of repair, renewal and/or reinstatement of any part or parts of the Development to the extent that the Purchaser/Tenant reasonably incurs such costs and/or the Purchaser/Tenant is or becomes liable either directly or by way of financial contribution for such costs. The Consultant shall not be liable for other losses incurred by the Purchaser/Tenant.

   (b) Without prejudice to any other exclusion or limitation of liability, damages, loss, expense or costs the Consultant's liability for such costs of the repair, renewal and/or reinstatement in question shall be further limited to that proportion thereof as it would be just and equitable to require the Consultant to pay having regard to the extent of the Consultant's responsibility for the same and on the assumptions that:

      (i) all other consultants and advisers, contractors and subcontractors involved in the Development have provided contractual undertakings on terms no less onerous than those set out in clause 1 to the Purchaser/Tenant in respect of the carrying out of their obligations in connection with the Development; and

      (ii) there are no exclusions of or limitations of liability nor joint insurance or co-insurance provisions between the Purchaser/Tenant and any other party referred to in this clause 2 and any such other party who is responsible to any extent for such costs is contractually liable to the Purchaser/Tenant for the same; and

COLLATERAL WARRANTY CIC/ConsWa/P&T first edition

(iii) all the parties referred to in this clause 2 have paid to the Purchaser/Tenant such proportion of such costs which it would be just and equitable for them to pay having regard to the extent of their responsibility for the same.

(c) The Consultant shall be entitled in any action or proceedings by the Purchaser/Tenant to rely on any limitation or exclusion in the Appointment and to raise the equivalent rights in defence of liability as it would have against the Client under the Appointment.

(d) the obligations of the Consultant under or pursuant to this Agreement shall not be released or diminished by the appointment of any person by the Purchaser/Tenant to carry out any independent enquiry into any relevant matter.

17 Ensure clause is consistent with any deleterious clause in Appointment; delete if Consultant is a QS or planning supervisor

18 Delete if services completed

3. [17] The Consultant has exercised [and will continue to exercise[18]] reasonable skill and care to see that, unless authorised by the Client in writing or, where such authorisation is given orally, confirmed by the Consultant to the Client in writing, materials specified by it for use in the Development are in accordance with the guidelines contained in the edition of the publication *Good Practice in Selection of Construction Materials* (Ove Arup & Partners) current at the date of its specification.

19 Delete clause if not appropriate

4. [19] The Consultant acknowledges that the Client has paid all fees and expenses properly due and owing to the Consultant under the Appointment up to the date of this Agreement.

5. The Purchaser/Tenant shall have no authority to issue any direction or instruction to the Consultant in relation to the Appointment.

20 Delete clause if Consultant does not retain copyright/amend as appropriate according to the terms of Appointment

6. [20] The copyright in all drawings, reports, models, specifications, bills of quantities, calculations and such other documents and information prepared by or on behalf of the Consultant pursuant to the Appointment (the 'Documents') shall remain vested in the Consultant. Subject (save in the case of the documents for the health and safety file) to the Consultant having received payment of any fees properly due and owing as at the date of exercise of the licence, the Purchaser/Tenant and its appointee shall have a licence to copy and use the Documents and to reproduce the designs and content of them for any purpose related to the Premises including, but without limitation, the construction, completion, maintenance, letting, promotion, advertisement, reinstatement, refurbishment and repair of the Premises. Such licence shall enable the Purchaser/Tenant and its appointee to copy and use the Documents for the extension of the Premises but such use shall not include a licence to reproduce the designs contained in them for any extension of the Premises. The Consultant shall not be liable for any use by the Purchaser/Tenant or its appointee of any of the Documents for any purpose other than that for which the same were prepared by or on behalf of the Consultant.

21 Insert amount

22 Insert amount; if there is no such limit, delete words in brackets

23 Insert period

7. The Consultant shall maintain professional indemnity insurance in an amount each year of not less than £ ....[21] in respect of each and every occurrence or series of occurrences arising out of one event [but limited to £ ....[22] in the aggregate for claims arising out of or in connection with pollution or contamination and/or asbestos and/or date recognition] for a period of ....[23] years from the date of practical completion of the Premises under the Building Contract and provided that such insurance is available at commercially reasonable rates. The Consultant shall inform the Purchaser/Tenant if such insurance ceases to be available at commercially

reasonable rates in order that the Consultant and the Purchaser/Tenant can discuss the means of best protecting their respective positions. The Consultant shall, on reasonable request by the Purchaser/Tenant, provide documentary evidence that such insurance is being maintained.

24 *Delete clause if the law of Scotland applies*

8. [24] The Purchaser/Tenant may assign by way of absolute legal assignment only the benefit of this Agreement to a third party who also takes an assignment of the Purchaser's/Tenant's interest in the Premises (the 'First Assignee'). The First Assignee may assign by way of absolute legal assignment only the benefit of this Agreement to a third party who also takes an assignment of the First Assignee's interest in the Premises. Any such assignment shall only be effective if written notice thereof is given to the Consultant. No further or other assignment of this Agreement shall be permitted.

25 *Delete clause if the law of England and Wales applies*

8S. [25] The Purchaser/Tenant may assign or transfer the rights under this Agreement without the consent of the Consultant to another person (the 'First Assignee') who also takes an assignation of the Purchaser's/Tenant's whole interest in the Premises. The First Assignee may assign or transfer the rights under this Agreement without the consent of the Consultant to another person taking an assignation of the First Assignee's whole interest in the Premises. Any such assignation shall only be effective if written notice thereof is given to the Consultant in accordance with clause 9. No further or other assignation or transfer of this Agreement shall be permitted.

9. Any notice to be given by the Consultant shall be deemed to be duly given if it is delivered by hand or sent by recorded (signed for) or special delivery to the Purchaser/Tenant at the above mentioned address, its registered office or principal place of business for the time being; and any notice given by the Purchaser/Tenant shall be deemed to be duly given if it is delivered by hand or sent by recorded (signed for) or special delivery to the Consultant at the above mentioned address, its registered office or principal place of business for the time being. Any such notices, if sent by recorded (signed for) or special delivery, shall be deemed to have been received forty eight hours after being posted (subject to proof to the contrary).

26 *Insert period*

10. No action or proceedings for any breach of this Agreement shall be commenced against the Consultant after the expiry of ....[26] years from the date of practical completion of the relevant part of the Premises under the Building Contract or, in the event that practical completion is not achieved, the date that the Consultant finishes its services under the Appointment.

27 *Delete clause if the law of Scotland applies*

11. [27] Nothing in this Agreement confers or purports to confer on any third party any benefit or any right to enforce any term of this Agreement pursuant to the Contracts (Rights of Third Parties) Act 1999.

COLLATERAL WARRANTY

CIC/ConsWa/P&T
first edition

FOR USE WHERE THE APPLICABLE LAW IS THAT OF ENGLAND AND WALES

12. This Agreement is subject to the law of England and Wales and the parties hereto submit to the jurisdiction of the courts of England and Wales.

Where the Agreement is to be executed under hand and not as a deed

As witness the hands of the parties hereto

Signed by or on behalf of the Consultant

28 *Name and signature of person signing on behalf of Consultant* 28

Signed by or on behalf of the Purchaser/Tenant

29 *Name and signature of person signing on behalf of Purchaser/Tenant* 29

Where the Agreement is to be executed as a deed

In witness whereof the parties have executed this Agreement as a deed the day and year first before written

[WHERE THE CONSULTANT IS A SOLE PRACTITIONER OR PARTNERSHIP]

Signed as a deed by the Consultant

30 *Name and signature of sole practitioner or first partner* 30

31 *Name and signature of witness* in the presence of 31

32 *Address of witness* of 32

33 *Name and signature of additional partner* and by 33

in the presence of 31

of 32

and by 33

in the presence of 31

of 32

COLLATERAL WARRANTY CIC/ConsWa/P&T first edition

[WHERE THE CONSULTANT IS A LIMITED LIABILITY PARTNERSHIP OR COMPANY]

Executed as a deed by the Consultant

34 Name and signature of member or director

by 34

35 Name and signature of member or director or company secretary

and 35

[WHERE THE PURCHASER/TENANT IS AN INDIVIDUAL OR PARTNERSHIP]

Signed as a deed by the Purchaser/Tenant

36 Name and signature of individual or first partner

.... 36

31 Name and signature of witness

in the presence of 31

32 Address of witness

of 32

33 Name and signature of additional partner

and by 33

in the presence of 31

of 32

and by 33

in the presence of 31

of 32

and by 33

in the presence of 31

of 32

[WHERE THE PURCHASER/TENANT IS A LIMITED LIABILITY PARTNERSHIP OR COMPANY]

Executed as a deed by the Purchaser/Tenant

by 34

and 35

COLLATERAL WARRANTY

CIC/ConsWa/P&T
first edition

FOR USE WHERE THE APPLICABLE LAW IS THAT OF SCOTLAND

12. This Agreement is subject to the law of Scotland and the parties hereto submit to the jurisdiction of the courts of Scotland.

In witness whereof these presents consisting of this and the four preceding pages are (under declaration that the insertion of the names and addresses of the said parties and the alterations to the recitals and to clauses 7 and 10 and the deletion of clauses 8 and 11 and ....[28] which are initialled by the parties were made prior to the subscription of this Agreement by the parties) subscribed as follows:

28 *Insert details of any other alterations or deletions*

they are subscribed [for and on behalf of [29]] [by [30]] the said Consultant [by [29]]

29 *Delete when subscription by party signing on own behalf*
30 *Delete when subscription on behalf of party*
31 *Enter full name of person subscribing including middle names; where signing on behalf of party, capacity of person signing should be added after signature (where on behalf of a company 'director' or 'company secretary', a partnership 'partner', and an LLP 'member' 'of said party'); where any other form of execution is being considered, or where party is any other corporate body, legal advice should be sought*

....[31]

at ....[32]

on ....[33]

which subscription is witnessed by

34 *Enter full name of witness including middle names*

....[34]

35 *Enter address of witness*

of ....[35]

.... (Consultant)

.... (Consultant witness)

they are subscribed [for and on behalf of [29]] [by [30]] the said Purchaser/Tenant [by [29]]

....[31]

32 *Enter location of place of signing*

at ....[32]

33 *Enter date of signing*

on ....[33]

which subscription is witnessed by

....[34]

of ....[35]

.... (Purchaser/Tenant)

.... (Purchaser/Tenant Witness)

Professional Services Agreements
ISBN 978-0-7277-4158-5

http://dx.doi.org/10.1680/psa.41585.425

Appendix 5
CIC funder's collateral warranty

Construction Industry Council

COLLATERAL WARRANTY CONSULTANT - FUNDER

CIC/ConsWa/F
first edition 2003

Standard form of agreement for use where a collateral warranty is to be given by a consultant to a funder of a commercial or industrial development

Full members of the Construction Industry Council • Association for Project Management • Association of Building Engineers • Association of Consultant Architects • Association of Consultant Building Surveyors • Association of Consulting Engineers • Association of Cost Engineers • Association of Planning Supervisors • British Institute of Architectural Technologists • British Institute of Facilities Management • Building Research Establishment • Building Services Research and Information Association • Chartered Institute of Building • Chartered Institution of Building Services Engineers • Construction Industry Research and Information Association • Consultant Quantity Surveyors Association • District Surveyors Association • Ground Forum • Institute of Clerks of Works of Great Britain • Institute of Highways Incorporated Engineers • Institute of Maintenance and Building Management • Institute of Plumbing • Institution of Civil Engineering Surveyors • Institution of Civil Engineers • Institution of Highways and Transportation • Institution of Structural Engineers • Landscape Institute • National Federation of Demolition Contractors • National House-Building Council • Royal Institute of British Architects • Royal Institution of Chartered Surveyors • Royal Town Planning Institute • Steel Construction Institute • Timber Research And Development Association • **Associate members** • Association of Civil Engineering Departments • Association of Heads of Surveying • British Association of Construction Heads • Construction Health & Safety Group • Chartered Institute of Marketing Construction Industry Group • Construction Industry Computing Association • Council of Professors of Building Engineering and Management • Faculty of Building • Federation of Property Societies • Institute of Construction Management • Royal School of Military Engineering • Standing Conference of Heads of Schools of Architecture • Society of Construction Law • SPONGE • Local Government Technical Advisers Group • Technology and Construction Solicitors' Association

COLLATERAL WARRANTY

CIC/ConsWa/F
first edition

For use where a collateral warranty is to be given by a consultant to a funder of a commercial or industrial development.

THIS AGREEMENT

is made the .... day of .... 20......................

Between

1 *Insert name of Consultant*

1. ....[1]

whose address/registered office is at

....

....

(the 'Consultant') and

2 *Insert name of Consultant's Client*

2. ....[2]

whose address/registered office is at ....

....

(the 'Client') and

3 *Insert name of Funder*

3. ....[3]

whose address/registered office is at ....

....

(the 'Funder').

CIC/ConsWa/F
first edition

COLLATERAL WARRANTY

Whereas

A. The Funder and the Client have entered into an agreement ('the Finance Agreement') for the provision of finance in connection with the carrying out of

4 *Insert description of works* .... [4]

5 *Insert address of Development* at .... [5]

(the 'Development').

6 *Insert date of Appointment* B. By a contract dated .... [6] (the 'Appointment'),

which shall include any variation thereto, the Client has appointed the Consultant

7 *Insert discipline* as .... [7]

in connection with the Development.

8 *Delete as appropriate* C. The Client has entered/may enter[8] into a contract

9 *If Building Contract signed, insert date; if not, delete word 'dated'* [dated .... [9]] (the 'Building Contract')

10 *Insert name of contractor or 'a contractor to be selected'* with .... [10]

(the 'Contractor') for the construction of the Development.

Now it is hereby agreed

11 *Insert figure* In consideration of the payment of £ .... [11] by the Funder to the Consultant, receipt of which the Consultant hereby acknowledges:

1. The Consultant warrants to the Funder that it has exercised and will continue to exercise reasonable skill care and diligence in the performance of its services to the Client under the Appointment.

2. In the event of any breach of this Agreement giving rise to losses recoverable by the Funder as a result of such breach (the 'Funder's losses'):

   (a) Without prejudice to any other exclusion or limitation of liability, damages, loss, expense or costs the Consultant's liability for any claim or claims under this Agreement shall be limited to that proportion of the Funder's losses as it would be just and equitable to require the Consultant to pay having regard to the extent of the Consultant's responsibility for the same and on the assumptions that:

      (i) all other consultants and advisers, contractors and subcontractors involved in the Development have provided contractual undertakings on terms no less onerous than those set out in clause I to the Funder in respect of the carrying out of their obligations in connection with the Development; and

(ii) there are no exclusions of or limitations of liability nor joint insurance or co-insurance provisions between the Funder and any other party referred to in this clause 2 and any such other party who is responsible to any extent for the Funder's losses is contractually liable to the Funder for the same; and

(iii) all the parties referred to in this clause 2 have paid to the Funder such sum as it would be just and equitable for them to pay having regard to the extent of their responsibility for the Funder's losses.

(b) The Consultant shall be entitled in any action or proceedings by the Funder to rely on any limitation or exclusion in the Appointment and to raise the equivalent rights in defence of liability as it would have against the Client under the Appointment.

(c) the obligations of the Consultant under or pursuant to this Agreement shall not be released or diminished by the appointment of any person by the Funder to carry out any independent enquiry into any relevant matter.

12 Ensure clause is consistent with any deleterious clause in Appointment; delete if Consultant is a QS or planning supervisor

3. [12] The Consultant has exercised and will continue to exercise reasonable skill and care to see that, unless authorised by the Client in writing or, where such authorisation is given orally, confirmed by the Consultant to the Client in writing, materials specified by it for use in the Development are in accordance with the guidelines contained in the edition of the publication *Good Practice in Selection of Construction Materials* (Ove Arup & Partners) current at the date of its specification.

4. The Funder has no authority to issue any direction or instruction to the Consultant in relation to the performance of the Consultant's services under the Appointment unless and until the Funder has given notice under clause 6 or 7.

5. The Consultant acknowledges that the Client has paid all fees and expenses properly due and owing to the Consultant under the Appointment up to the date of this Agreement. The Funder has no liability to the Consultant in respect of fees and expenses under the Appointment unless and until the Funder has given notice under clause 6 or 7.

6. The Consultant agrees that, in the event of the termination of the Finance Agreement by the Funder before completion of the services under the Appointment, the Consultant will, if so required by notice in writing given by the Funder and subject to clause 8, accept the instructions of the Funder or its appointee to the exclusion of the Client in respect of the Development upon the terms and conditions of the Appointment. The Client acknowledges that the Consultant shall be entitled to rely on a notice given to the Consultant by the Funder under this clause as conclusive evidence for the purposes of this Agreement of the termination of the Finance Agreement by the Funder; and further acknowledges that such acceptance of the instructions of the Funder to the exclusion of the Client shall not constitute any breach of the Consultant's obligations to the Client under the Appointment.

7. The Consultant further agrees that it will not without first giving the Funder not less than twenty one days' notice in writing exercise any right it may have to terminate the Appointment or to treat the same as having been repudiated by the Client or to discontinue the performance of any services to be performed by the Consultant

pursuant thereto. Such right to terminate the Appointment with the Client or treat the same as having been repudiated or discontinue performance shall cease if, within such period of notice and subject to clause 8, the Funder shall give notice in writing to the Consultant requiring the Consultant to accept the instructions of the Funder or its appointee to the exclusion of the Client in respect of the Development upon the terms and conditions of the Appointment.

8. It shall be a condition of any notice given by the Funder under clause 6 or 7 that the Funder or its appointee accepts liability for payment of the fees and expenses payable to the Consultant under the Appointment and for performance of the Client's obligations including payment of any fees and expenses outstanding at the date of such notice. Upon the issue of any notice by the Funder under clause 6 or 7, the Appointment shall continue in full force and effect as if no right of termination on the part of the Consultant, nor any right of the Consultant to treat the Appointment as having been repudiated by the Client, nor to discontinue the performance of any services, had arisen and the Consultant shall be liable to the Funder and its appointee under the Appointment in place of its liability to the Client. If any notice given by the Funder under clause 6 or 7 requires the Consultant to accept the instructions of the Funder's appointee, the Funder shall be liable to the Consultant as guarantor for the payment of all sums from time to time due to the Consultant from the Funder's appointee.

13 *Delete clause if Consultant does not retain copyright/amend as appropriate according to the terms of Appointment*

9. [13] The copyright in all drawings, reports, models, specifications, bills of quantities, calculations and such other documents and information prepared by or on behalf of the Consultant pursuant to the Appointment (the 'Documents') shall remain vested in the Consultant. Subject (save in the case of the documents for the health and safety file) to the Consultant having received payment of any fees properly due and owing as at the date of exercise of the licence, the Funder and its appointee shall have a licence to copy and use the Documents and to reproduce the designs and content of them for any purpose related to the Development including, but without limitation, the construction, completion, maintenance, letting, promotion, advertisement, reinstatement, refurbishment and repair of the Development. Such licence shall enable the Funder and its appointee to copy and use the Documents for the extension of the Development but such use shall not include a licence to reproduce the designs contained in them for any extension of the Development. The Consultant shall not be liable for any use by the Funder or its appointee of any of the Documents for any purpose other than that for which the same were prepared by or on behalf of the Consultant.

14 *Insert amount*

15 *Insert amount; if there is no such limit, delete words in brackets*

16 *Insert period*

10. The Consultant shall maintain professional indemnity insurance in an amount each year of not less than £...............[14] in respect of each and every occurrence or series of occurrences arising out of one event [but limited to £..............[15] in the aggregate for claims arising out of or in connection with pollution or contamination or asbestos and/or date recognition] for a period of ....[16] years from the date of practical completion of the Development under the Building Contract and provided that such insurance is available at commercially reasonable rates. The Consultant shall inform the Funder if such insurance ceases to be available at commercially reasonable rates in order that the Consultant and the Funder can discuss the means of best protecting their respective positions. The Consultant shall, on reasonable request by the Funder, provide documentary evidence that such insurance is being maintained.

COLLATERAL WARRANTY

CIC/ConsWa/F
first edition

17 *Delete clause if the law of Scotland applies*

11. [17] The Funder may assign by way of absolute legal assignment only the benefit of this Agreement to a third party who is also providing finance or re-finance in connection with the carrying out of the Development. Any such assignment shall only be effective if written notice thereof is given to the Consultant. No further or other assignment of this Agreement shall be permitted.

18 *Delete clause if the law of England and Wales applies*

11S.[18] The Funder may assign or transfer the rights under this Agreement without the consent of the Consultant to another person who is also providing finance or re-finance in connection with the carrying out of the Development. Any such assignation shall only be effective if written notice thereof is given to the Consultant in accordance with clause 12. No further or other assignation or transfer of this Agreement shall be permitted.

12. Any notice to be given by the Consultant shall be deemed to be duly given if it is delivered by hand or sent by recorded (signed for) or special delivery to the Funder at the above mentioned address, its registered office or principal place of business for the time being; and any notice given by the Funder shall be deemed to be duly given if it is delivered by hand or sent by recorded (signed for) or special delivery to the Consultant at the above mentioned address, its registered office or principal place of business for the time being. Any such notices, if sent by recorded (signed for) or special delivery, shall be deemed to have been received forty eight hours after being posted (subject to proof to the contrary).

19 *Insert period*

13. No action or proceedings for any breach of this Agreement shall be commenced against the Consultant after the expiry of ....[19] years from the date of practical completion of the relevant part of the Development under the Building Contract or, in the event that practical completion is not achieved, the date that the Consultant finishes its services under the Appointment.

20 *Delete clause if the law of Scotland applies*

14. [20] Nothing in this Agreement confers or purports to confer on any third party any benefit or any right to enforce any term of this Agreement pursuant to the Contracts (Rights of Third Parties) Act 1999.

COLLATERAL WARRANTY

CIC/ConsWa/F
first edition

FOR USE WHERE THE APPLICABLE LAW IS THAT OF ENGLAND AND WALES

15. This Agreement is subject to the law of England and Wales and the parties hereto submit to the jurisdiction of the courts of England and Wales.

Where the Agreement is to be executed under hand and not as a deed

As witness the hands of the parties hereto

Signed by or on behalf of the Consultant

21 Name and signature of person signing on behalf of Consultant 21

Signed by or on behalf of the Client

22 Name and signature of person signing on behalf of Client 22

Signed by or on behalf of the Funder

23 Name and signature of person signing on behalf of Funder 23

Where the Agreement is to be executed as a deed

In witness whereof the parties have executed this Agreement as a deed the day and year first before written

[WHERE THE CONSULTANT IS A SOLE PRACTITIONER OR PARTNERSHIP]

Signed as a deed by the Consultant

24 Name and signature of sole practitioner or first partner 24

25 Name and signature of witness in the presence of 25

26 Address of witness of 26

27 Name and signature of additional partner and by 27

in the presence of 25

of 26

and by 27

in the presence of 25

of 26

 six

COLLATERAL WARRANTY

CIC/ConsWa/F
first edition

[WHERE THE CONSULTANT IS A LIMITED LIABILITY PARTNERSHIP OR COMPANY]

Executed as a deed by the Consultant

28 Name and signature of member or director

by 28

29 Name and signature of member or director or company secretary

and 29

[WHERE THE CLIENT IS AN INDIVIDUAL OR PARTNERSHIP]

Signed as a deed by the Client

30 Name and signature of individual or first partner

.... 30

25 Name and signature of witness

in the presence of 25

26 Address of witness

of 26

27 Name and signature of additional partner

and by 27

in the presence of 25

of 26

and by 27

in the presence of 25

of 26

[WHERE THE CLIENT IS A LIMITED LIABILITY PARTNERSHIP OR COMPANY]

Executed as a deed by the Client

by 28

and 29

[WHERE THE FUNDER IS A LIMITED LIABILITY PARTNERSHIP OR COMPANY]

Executed as a deed by the Funder

by 28

and 29

COLLATERAL WARRANTY

CIC/ConsWa/F
first edition

FOR USE WHERE THE APPLICABLE LAW IS THAT OF SCOTLAND

15. This Agreement is subject to the law of Scotland and the parties hereto submit to the jurisdiction of the courts of Scotland.

In witness whereof these presents consisting of this and the five preceding pages are (under declaration that the insertion of the names and addresses of the said parties and the alterations to the recitals and to clauses 10 and 13 and the deletion of clauses 11 and 14 and ....[21] which are initialled by the parties were made prior to the subscription of this Agreement by the parties) subscribed as follows:

they are subscribed [for and on behalf of [22]] [by [23]] the said Consultant [by [22]]

....[24]

at ....[25]

on ....[26]

which subscription is witnessed by

....[27]

of ....[28]

.... (Consultant)

.... (Consultant witness)

they are subscribed [for and on behalf of [22]] [by [23]] the said Funder [by [22]]

....[24]

at ....[25]

on ....[26]

which subscription is witnessed by

....[27]

of ....[28]

.... (Funder)

.... (Funder Witness)

21 *Insert details of any other alterations or deletions*

22 *Delete when subscription by party signing on own behalf*

23 *Delete when subscription on behalf of party*

24 *Enter full name of person subscribing including middle names; where signing on behalf of party, capacity of person signing should be added after signature (where on behalf of a company 'director' or 'company secretary', a partnership 'partner', and an LLP 'member' 'of said party'); where any other form of execution is being considered, or where party is any other corporate body, legal advice should be sought*

25 *Enter location of place of signing*

26 *Enter date of signing*

27 *Enter full name of witness including middle names*

28 *Enter address of witness*

COLLATERAL WARRANTY CIC/ConsWa/F first edition

22 *Delete when subscription by party signing on own behalf*

23 *Delete when subscription on behalf of party*

24 *Enter full name of person subscribing including middle names; capacity of person signing should be added after signature (where on behalf of a company 'director' or 'company secretary', a partnership 'partner', and an LLP 'member' 'of said party'); where any other form of execution is being considered, or where party is any other corporate body, legal advice should be sought*

25 *Enter location of place of signing*

26 *Enter date of signing*

27 *Enter full name of witness including middle names*

28 *Enter address of witness*

they are subscribed [for and on behalf of [22]] [by [23]] the said Client [by [22]]

....[24]

at ....[25]

on ....[26]

which subscription is witnessed by

....[27]

of ....[28]

.... (Client)

.... (Client Witness)

Professional Services Agreements
ISBN 978-0-7277-4158-5

http://dx.doi.org/10.1680/psa.41585.437

Appendix 6
Summary of the Client's duties under the CDM Regulations 2007

Specimen letter to 'the Client' from a designer

The CDM Regulations 2007 may apply to your proposed project. If so, these Regulations impose duties on you as 'the Client', for example, in relation to managing the health and safety aspects of the project, the information to be provided and the appointment of a CDM coordinator. We, as designers, are not able to commence works unless 'the Client' is aware of his duties. In summary, these duties are the following.

- Check the competence of all designers, contractors and other team members you propose to engage for the project and that they are adequately resourced and appointed early enough for the work they have to do.
- Allow sufficient time for all stages of the construction process.
- Cooperate with others so they can comply with their duties under the Regulations,
- Coordinate your work with others to ensure the safety of those carrying out the construction work or who may be affected by such work.
- Provide promptly pre-construction and other information to every designer, contractor, the CDM coordinator and others which they need to plan or manage the health and safety aspects of their work.
- Ensure that there are reasonable management arrangements in place throughout the project so that the construction work can be carried out safely, and that welfare facilities are in place and any structure designed as a workplace takes account of the Workplace (Health, Safety and Welfare) Regulations 1992.
- Appoint a competent CDM coordinator as soon as is practicable or fulfil the role yourself, if you are competent to do so. If you do not do this, the Regulations do not allow us to proceed beyond initial design work.
- Appoint a competent principal contractor or fulfil the role yourself, if you are competent to do so.
- Make sure that the construction phase does not start unless there are and will be suitable welfare facilities available on site and that there is a compliant construction phase plan in place.
- Provide information relating to the health and safety file to the CDM coordinator.
- Retain and provide access to the health and safety file and revise it as necessary.

Professional Services Agreements
ISBN 978-0-7277-4158-5

http://dx.doi.org/10.1680/psa.41585.439

Bibliography

Cullen C (No date) Summary of Building Information Modelling. (s.n.): (s.l.).

HSE (2007) Managing Health and Safety in Construction. Construction (Design and Management) Regulations 2007. (CDM) Approved Code of Practice (ACoP). HSE Books, London. Available online from http://www.hse.gov.uk/construction/cdm/acop.htm.

NHS (1995) Agreement for the Appointment of Architects and Engineers for Commissions in the National Health Service. TSO, London, UK.

RIBA (1999) *SFA/99: Standard Form of Appointment for an Architect.* Riba Publishing, London, UK.

RIBA (2007) Agreements. RIBA Publishing, London, UK.

Sheehan T and Ove Arup and Partners (1997) *Good Practice in the Selection of Construction Materials.* British Property Federation, London, UK.

Wallace IND (1986) *Construction Contract: Principles and Policy in Tort and Contract.* Sweet & Maxwell, London, UK.

Websites

Architects' Registration Board. See http://www.arb.org.uk for further details (accessed 10/08/2011).

Association of Consultant Architects publications. See http://www.acarchitects.co.uk/publications.htm for further details (accessed 10/08/2011).

Association for Consultancy and Engineering agreements. See http://agreements.acenet.co.uk/aceagreements/40 for further details (accessed 10/08/2011).

British and Irish Legal Information Institute (BAILII). See http://www.bailii.org for further details (accessed 10/08/2011).

British Property Federation. See http://www.bpf.org.uk for further details (accessed 10/08/2011).

British Standards. See http://shop.bsigroup.com for further details (accessed 10/08/2011).

British Standards Institution/Eurocodes. See http://www.bsigroup.com/templates/FourTabContent.aspx?id=147778&epslanguage=EN for further details (accessed 10/08/2011).

Brussels Convention. See http://curia.europa.eu/common/recdoc/convention/en/c-textes/brux-idx.htm for further details (accessed 10/08/2011).

Chartered Institute of Architectural Technologists. See http://www.ciat.org.uk for further details (accessed 10/08/2011).

Chartered Institute of Building Services Engineers. See http://www.cibse.org for further details (accessed 10/08/2011).

Construction Industry Council Consultants' Contract Conditions. See http://www.cicshop.co.uk/acatalog/CIC_Consultants__Contract.html#a23 for further details (accessed 10/08/2011).

City of London Law Society. See http://www.citysolicitors.org.uk/Default.aspx?sID=18 for further details (accessed 10/08/2011).

Council of Mortgage Lenders. See http://www.cml.org.uk/cml/home for further details (accessed 10/08/2011).

Companies House. See http://www.companieshouse.gov.uk for further details (accessed 10/08/2011).

Health and Safety Executive. See http://www.hse.gov.uk for further details (accessed 11/08/2011).

Institution of Civil Engineers Conditions of Contract. See http://www.ice.org.uk/topics/lawandcontracts/ICE-Conditions-of-Contract for further details (accessed 11/08/2011).

International Federation of Consulting Engineers. See http://www.fidic.org for further details (accessed 11/08/2011).

Joint Contracts Tribunal Ltd contracts. See http://www.jctcontracts.com/JCT/contracts/index.jsp for further details (accessed 11/08/2011).

Model Forms. See http://www.theiet.org/publishing/model-forms/index.cfm for further details (accessed 11/08/2011).

National House-Building Council. See http://www.nhbc.co.uk/ for further details (accessed 11/08/2011).

New Engineering Contract contracts. See http://www.neccontract.com/ for further details (accessed 11/08/2011).

Royal Institute of British Architects agreements. See http://www.ribabookshops.com/agreements/RIBAAgreementsSite/About.aspx for further details (accessed 11/08/2011).

RIBA Plan of Work. See http://www.pedr.co.uk/textpage.asp?menu=1a&sortorder=130&area=main for further details (accessed 11/08/2011).

Royal Institution of Chartered Surveyors Standard Form of Consultant's Appointment. See http://www.ricsbooks.com/productInfo.asp?product_id=17275 for further details (accessed 11/08/2011).

Scottish Arbitration Code for Domestic and International Arbitration 1999. See http://www.scia.co.uk/ScotArbC.htm for further details (accessed 18/08/2011).

Technology and Construction Solicitors' Association. See http://www.tecsa.org.uk/ for further details (accessed 11/08/2011).

Legislation/Standards/Regulations

Arbitration Act 1996. See http://www.legislation.gov.uk/ukpga/1996/23/contents for further details (accessed 11/08/2011).

Civil Liability (Contribution) Act 1978. See http://www.legislation.gov.uk/ukpga/1978/47/contents for further details (accessed 11/08/2011).

Contracts (Rights of Third Parties) Act 1999. See http://www.legislation.gov.uk/ukpga/1999/31/contents for further details (accessed 11/08/2011).

Copyright, Designs and Patents Act 1988. See http://www.legislation.gov.uk/ukpga/1988/48/contents for further details (accessed 11/08/2011).

County Courts Act 1984. See http://www.legislation.gov.uk/ukpga/1984/28/contents for further details (accessed 11/08/2011).

Enterprise Act 2002. See http://www.legislation.gov.uk/ukpga/2002/40/contents for further details (accessed 11/08/2011).

Enterprise Act 2002 (Insolvency) Order 2003. See http://www.legislation.gov.uk/uksi/2003/2096/contents/made for further details (accessed 11/08/2011).

Freedom of Information Act 2000. See http://www.legislation.gov.uk/ukpga/2000/36/contents for further details (accessed 11/08/2011).

Health and Safety at Work, etc. Act 1974. See http://www.legislation.gov.uk/ukpga/1974/37/contents for further details (accessed 11/08/2011).

Housing Grants, Construction and Regeneration Act 1996 (the Construction Act 1996). See http://www.legislation.gov.uk/ukpga/1996/53/contents for further details (accessed 11/08/2011).

Late Payment of Commercial Debts (Interest) Act 1998. See http://www.legislation.gov.uk/ukpga/1998/20/contents for further details (accessed 11/08/2011).

Law of Property Act 1925. See http://www.legislation.gov.uk/ukpga/Geo5/15-16/20/contents for further details (accessed 07/08/2011).

Limited Liability Partnership Act 2000. See http://www.legislation.gov.uk/ukpga/2000/12/contents for further details (accessed 11/08/2011).

Limitation Act 1980. See http://www.legislation.gov.uk/ukpga/1980/58/contents for further details (accessed 11/08/2011).

Local Democracy, Economic Development and Construction Act 2009. See http://www.legislation.gov.uk/ukpga/2009/20/contents for further details (accessed 11/08/2011).

Local Government Act 1972. See http://www.legislation.gov.uk/ukpga/1972/70/contents for further details (accessed 11/08/2011).

Misrepresentation Act 1967. See http://www.legislation.gov.uk/ukpga/1967/7/contents for further details (accessed 11/08/2011).

Occupiers' Liability Act 1984. See http://www.legislation.gov.uk/ukpga/1984/3 for further details (accessed 11/08/2011).

Partnership Act 1890. See http://www.legislation.gov.uk/ukpga/Vict/53-54/39/contents for further details (accessed 11/08/2011).

Provision of Services Regulations 2009. See http://www.legislation.gov.uk/ukdsi/2009/9780111486276/contents for further details (accessed 07/08/2011).

Sale and Supply of Goods Act 1994. See http://www.legislation.gov.uk/ukpga/1994/35/contents for further details (accessed 11/08/2011).

Senior Courts Act (formerly the Supreme Court Act 1981). See http://www.legislation.gov.uk/ukpga/1981/54/contents for further details (accessed 11/08/2011).

Supply of Goods and Services Act 1982. See http://www.legislation.gov.uk/ukpga/1982/29/contents for further details (accessed 11/08/2011).

Temporary or Mobile Construction Sites 1992 (EU Directive 92/57/EEC). See http://europa.eu/legislation_summaries/employment_and_social_policy/health_hygiene_safety_at_work/c11120_en.htm for further details (accessed 08/08/2011).

The Construction (Design and Management) Regulations 1994. See http://www.legislation.gov.uk/uksi/1994/3140/contents/made for further details (accessed 08/08/2011).

The Construction (Design and Management) Regulations 2007. See http://www.legislation.gov.uk/uksi/2007/320/contents/made for further details (accessed 08/08/2011).

The Construction (Health, Safety and Welfare) Regulations 1996. See http://www.legislation.gov.uk/uksi/1996/1592/contents/made for further details (accessed 08/08/2011).

The Provision of Services Regulations 2009. See http://www.legislation.gov.uk/ukdsi/2009/9780111486276/contents for further details (accessed 11/08/2011).
The Scheme for Construction Contracts (England and Wales) Regulations 1998. See http://www.legislation.gov.uk/uksi/1998/649/contents/made for further details (accessed 11/08/2011).
The Unfair Terms in Consumer Contracts Regulations 1994. See http://www.legislation.gov.uk/uksi/1994/3159/contents/made for further details (accessed 11/08/2011).
The Unfair Terms in Consumer Contracts Regulations 1999. See http://www.legislation.gov.uk/uksi/1999/2083/contents/made for further details (accessed 11/08/2011).
Third Parties (Rights Against Insurers) Act 1930. See http://www.legislation.gov.uk/ukpga/Geo5/20-21/25/contents for further details (accessed 11/08/2011).
Unfair Contract Terms Act 1977. See http://www.legislation.gov.uk/ukpga/1977/50/contents for further details (accessed 11/08/2011).
Workplace (Health, Safety and Welfare) Regulations (1992) See http://www.legislation.gov.uk/uksi/1992/3004/contents/made for further details (accessed 18/08/2011).

Table of cases

ADT Limited v. *BDO Binder Hamlyn (1995)* LTL 7/12/95 – unreported elsewhere
Alderslade v. *Hendon Laundry Limited (1945)* 1 All E 244
Alfred McAlpine Co v. *Panatown (2000)* LTL 27/7/2000
American Cyanamid Co v. *Ethicon (1975)* 1 All ER 504
Anglian Water Services v. *Laing O'Rourke (2010)* EWHC 1529, TCC
Aspen UK Limited v. *Pectel Limited* (2008) EWHC 2804 (Comm) v. *Blyth & Blyth* v. *Carillion Construction Ltd (2001)* 79 Con LR 142
Axa Sun Life Services plc v. *Campbell Martin Limited [2011]* EWCA Civ 133
Blyth & Blyth v. *Carillion Construction Ltd (2001)* 79 Con LR 142
Bolam v. *Friern Hospital Management Committee (1957)* 1 WLR 583
Bosky v. *Shearwater Property Holdings (1992)* 61 BLR 64
British Fermentation Products Limited v. *Compare Reavell Limited (1999)* 2 All ER (Comm) 389
British Sugar plc v. *NEI Power Projects Limited and Another (1997)* 87 BLR 42
Caparo Industries plc v. *Dickman (1990)* 2 AC 605
Chesham Properties Ltd v. *Bucknall Austin Project Management Services Ltd & Ors (1996)* 53 Con LR 1
Chown v. *Parrott (1863)* 14 CBNS 74
Co-operative Retail Service v. *Taylor Young Partnership (2002)* UKHL 17. See (2002) 1 All ER (Comm) 918
Coco v. *A N Clark (Engineers) Limited (1969)* RPC 41
Costain Ltd v. *Charles Haswell & Partners Ltd (2009)* EWHC B25 (TCC)
Croshaw v. *Pritchard (1899)* 16 TLR 45
Deepak Fertilizers & Petrochemical Corporation v. *Davy Kee* (1) ICI Chemicals (2) (1998) 2 Lloyd's Rep 139
Domsalla (t/a Domsalla Building Services) v. *Dyason (2007)* EWHC 1174 (TCC)
Eckersley v. *Binnie and Partners (1988)* 18 Con LR 78
EDS v. *BSkyB (2010)* EWHC 86 (TCC)
Farmers Build Ltd v. *Carier Ltd (1999)* RPC 461

Fitzroy Robinson Limited v. *Mentmore Towers Limited (2009)* EWHC 1552 (TCC)
Galliford Try Infrastructure Ltd v. *Mott MacDonald Ltd (2008)* CILL 2612 TCC
George Mitchell v. *Finney Lock Seeds (1983)* 3 WLR 163
Glasgow Airport v. *Kirkman & Bradford (2007)* CSOH 52 and (2007) CSIH 47
Hart Investments Ltd v. *Fidler (2006)* EWHC 2857 (TCC)
Hedley Byrne v. *Heller (1963)* 2 All ER 575
IBM UK Ltd v. *Rockware Glass Ltd (1980)* FSR 335
Jobson v. *Johnson (1989)* 1 WLR 1026, Court of Appeal
John Mowlem & Company v. *Eagle Star and Others (1995)* 44 Con LR 134
Kajima UK Engineering Ltd v. *The Underwriter Insurance Company Ltd (2008)* EWHC 83 (TCC)
Langstane Housing Association Limited v. *Riverside Construction (Aberdeen) Limited (2009)* CSOH 52
Linden Gardens Trust Ltd v. *Lenesta Sludge Disposals Ltd (1994)* 1 AC 85, HL
Mayhaven v. *DABI (2009)*
McCullagh v. *Lane Fox & Partners Limited (1996)* 1 EGLR 35
McGlinn v. *Waltham Contractors Ltd & Others (2007)* EWHC 149 (TCC)
McKenzie v. *Potts and Dobson Chapman (1995)* CA 50 Con LR 40
Midland Land Reclamation Limited and Leicestershire County Council v. *Warren Energy Limited (1997)* LTL 8-9-97 – unreported elsewhere
Merrett v. *Babb (2001)* All ER (D) 183
Moores v. *Yakeley Associates (TCC October 1998)* 1999 52 Con LR 76
Munckenbeck and Marshall v. *Harold (2005)* EWHC 356 (TCC)
Murphy v. *Brentwood District Council (1991)* AC 398
Omega Trust Company Limited and Banque Finindus v. *Wright Son & Pepper and Barker & Co (1998)* 3 PNLR 337
Platform Funding Limited v. *Bank of Scotland plc (formerly Halifax plc) (2008)* LTL 32 EG 83 (CS) Court of Appeal
Regus (UK) Limited v. *Epcot Solutions Limited (2008)* EWCA Civ 361 – A3/2007/1092 Court of Appeal
Rhodia International Holdings Limited and Another v. *Huntsman International LLP (2007)* EWHC 292
Robinson v. *P E Jones (Contractors) Limited (2011)* EWCA Civ 9
Rotherham MBC v. *Frank Haslam (1996)* 59 Con LR 33
RTS Flexible Systems v. *Molkerie Alois Müller Gmbh & Co KG (UK Production) (2010)* UKSC 14; (2010) WLR (D) 75
Salvage Association v. *Cap Financial Services (1995)* (unreported 9 July 1993)
Sanjay Lachhami & Another v. *Destination Canada (UK) Limited (1997)* 13 Const LJ 279
Scottish Windows Services v. *Hammer/CRM Facades (2010)* CSOH 42
St Albans City and District Council v. *International Computers Limited (1996)* 4 All ER 487
St Martins Corporation Limited v. *Sir Robert McAlpine & Sons Limited (1994)* AC 83 57 BLR 66 House of Lords
Sheffield District Railway Company v. *Great Central Railway Company (1911)* 27 TLR 251
Sir Lindsay Parkinson Ltd v. *Commissioner of Crown Works and Public Buildings (1949)* 2 KB 632

South Australia Asset Management Corporation v. *York Montague Limited* 80 BLR 1, HL (*also reported (1997)* AC 191)
St Martins Corporation Limited v. *Sir Robert McAlpine & Sons Limited (1994)* AC 83 57 BLR 66
Stephen Donald Architects Limited v. *King (2003)* EWHC 1867
Technotrade Limited v. *Larkstore Limited (2005)* EWHC 2742 (TCC); (2006) 1 WLR 2926 (CA)
The Canada Steamship Lines v. *R (1952) AC 192*
Thomas Marshall (Exports) v. *Guinle (1976)* 1979 CH 227/1978 3 All ER 193
Thorn v. *London Corpn (1876*) 1 App Cas 120
Watts and Watts v. *Morrow (1991)* 4 All ER 937, (1991) 1 WLR 1421
Yorkshire Water v. *Sun Alliance (1996)* 1997 2 Lloyd's Rep 21, Court of Appeal

Professional Services Agreements
ISBN 978-0-7277-4158-5

http://dx.doi.org/10.1680/psa.41585.445

Index